Reading Practice

Reading Practice

THE PURSUIT OF NATURAL KNOWLEDGE FROM MANUSCRIPT TO PRINT

Melissa Reynolds

The University of Chicago Press CHICAGO AND LONDON

The University of Chicago Press, Chicago 60637
The University of Chicago Press, Ltd., London

Published 2024
Printed in the United States of America

33 32 31 30 29 28 27 26 25 24   1 2 3 4 5

ISBN-13: 978-0-226-82362-1 (cloth)
ISBN-13: 978-0-226-83489-4 (paper)
ISBN-13: 978-0-226-82363-8 (e-book)
DOI: https://doi.org/10.7208/chicago/9780226823638.001.0001

Library of Congress Cataloging-in-Publication Data

Names: Reynolds, Melissa B., author.
Title: Reading practice : the pursuit of natural knowledge from manuscript to print / Melissa Reynolds.
Other titles: Pursuit of natural knowledge from manuscript to print
Description: Chicago ; London : The University of Chicago Press, 2024. | Includes bibliographical references and index.
Identifiers: LCCN 2023057804 | ISBN 9780226823621 (cloth) | ISBN 9780226834894 (paperback) | ISBN 9780226823638 (ebook)
Subjects: LCSH: Communication in science—England—History—To 1500. | Communication in science—England—History—16th century. | Books and reading—England—History—To 1500. | Books and reading—England—History—16th century. | Scientific literature—England—History—To 1500. | Scientific literature—England—History—16th century. | Manuscripts—Social aspects—England. | Science—Social aspects—England. | Science—England—History—To 1500. | Science—England—History—To 1600.
Classification: LCC Q225.2.G7 R43 2024 | DDC 508.42—dc23/eng/20240104
LC record available at https://lccn.loc.gov/2023057804

♾ This paper meets the requirements of ANSI/NISO Z39.48-1992 (Permanence of Paper).

In memory of my father, Frank W. Buckner (1949–2019), who taught me to find joy in the pursuit of every kind of knowledge

Contents

Figures and Tables

FIGURES

TABLES

Abbreviations

Beinecke ▸ Beinecke Library, Yale University, New Haven, CT

BL ▸ British Library, London, UK

Bodleian ▸ The Bodleian Libraries, Oxford, UK

CCP ▸ Jonathan Mackman and Matthew Stevens, *Court of Common Pleas: The National Archives, Cp40 1399–1500* (London: Centre for Metropolitan History, 2010), *British History Online*, https://www.british-history.ac.uk/no-series/common-pleas/1399-1500

CUL ▸ Cambridge University Library, Cambridge, UK

EEBO ▸ *Early English Books Online*, https://www.proquest.com/eebo

Glasgow ▸ University of Glasgow Library, Glasgow

Huntington ▸ The Huntington Library, San Marino, CA

KJV ▸ *The Bible: Authorized King James Version*, edited by Robert Carroll and Stephen Prickett (Oxford: Oxford University Press, 2008)

MLM ▸ The Morgan Library and Museum, New York

MWME ▸ George Keiser, ed., *A Manual of the Writings in Middle English, 1050–1500, Volume X: Works of Science and Information* (New Haven, CT: Connecticut Academy of Arts and Sciences, 1998)

NLM ▸ National Library of Medicine, Bethesda, MD

ODNB ▸ Sir David Cannadine, ed., *Oxford Dictionary of National Biography*, online ed. (Oxford: Oxford University Press, 2004), https://www.oxforddnb.com

SOTR ▸ *Statutes of the Realm, Vols. 1–4: 1235–1624* (Westminster: House of Commons, 1819), www.heinonline.org

STC ▸ Katherine Pantzer, Alfred W. Pollard, and G. R. Redgrave, eds., *A Short-Title Catalogue of Books Printed in England, Scotland, and Ireland, and of English Books Printed Abroad, 1475–1640*, revised and enlarged 2nd ed. (London: Bibliographic Society, 1976–1993)

TCC ▸ Trinity College Cambridge, Cambridge, UK

TNA ▸ The National Archives of the United Kingdom, Kew, UK

USTC ▸ Andrew Pettegree et al., eds., *Universal Short Title Catalogue* (University of Saint Andrews), https://ustc.ac.uk

Wellcome ▸ Wellcome Library, London, UK

A Note on Transcriptions and Titles

English words were spelled all kinds of ways in the fifteenth and sixteenth centuries, depending on a scribe's local dialect or (so it sometimes seems) a printer's mood. Scribes and printers also often abbreviated words using superscript lines or apostrophes to indicate missing consonants or syllables. Though these odd spellings are charming, they can make for difficult reading. With novice readers of Middle and early modern English in mind, I have modernized spelling and punctuation when selections from manuscripts and early printed books are quoted in the text: silently extending abbreviations, changing *u* to *v* and *i* to *j* where necessary, and changing the characters thorn and yogh to *th-* and *y* or *g* where needed. Diplomatic transcriptions of quotations from fifteenth-century manuscript sources can be found in the notes.

Early modern printed books often had very long titles—some can run to as much as a paragraph in length! Rather than reproduce these lengthy titles in the text, I use short titles. For example, in chapter 4 I use the shortened title *The treasure of pore men* to refer to the publication *Here begynneth a newe boke of medecynes intytulyd or callyd the treasure of pore men whiche sheweth many dyuerse good medecines for dyuerse certayn dysseases as in the table of this present boke more playnly shall appere. the boke of medecines.* I have also shortened and standardized early printers' colophons in the notes, citing only the city of publication, the printer's name, and the date of publication. Finally, in several chapters I discuss multiple editions of the same text, which were often published under different titles. In these cases, I use one short title to refer to all editions so readers can recognize that I am referring to the same text. However, in the first citation for every printed book, I include English Short Title Catalogue or Universal Short Title Catalogue numbers so readers can easily locate an edition.

Introduction

Think back to the last time you had a scratchy throat or an itchy rash. Did you Google your symptoms? Scan pages of search results for information to alleviate your concerns? Now consider what you might do if you needed to learn some mundane but necessary hands-on skill, like dressing a wound or growing a vegetable garden from seeds. Would you watch a YouTube video that broke down the task, step-by-step?[1] If you can imagine yourself doing either of these things, then you are not so different from the fifteenth- and sixteenth-century readers you will encounter in this book, all of whom were engaged in the pursuit of useful, practical, natural knowledge in what counted as YouTube or Google six hundred years ago. Now imagine once more that you're Googling your symptoms or searching YouTube. An advertisement pops up on the screen for a diet supplement or fitness regimen. This was not what you were looking for on the internet, and yet it, too, is supposedly useful knowledge about your health. Let's say curiosity gets the better of you, and you follow one of those ads into a corner of the internet replete with misinformation. What do you do then? How do you identify a bogus claim or false advertisement? With so much useful knowledge about your health, your body, and the environment at your fingertips, how do you determine what information is effective, useful, or true?

Six hundred years ago, English people seeking medical advice or instructions for planting herbs were faced with similar questions. Like us, they found themselves with newfound access to a wealth of useful, natural knowledge in a world that was being rapidly transformed by new media. Like us, they wondered how to make sense of it all. This book tells the story of how those ordinary fifteenth- and sixteenth-century English people learned to assess and evaluate information through everyday interactions with manuscripts and books like almanacs, medical recipe collections, astrological tracts, surgical treatises, herbals, manuals on gardening or animal husbandry, and prognostications. These were the books English people read for over two centuries when they needed to cure a fever, ease backache, plant a garden, tend to a lame horse, calculate the best time to administer medicine, make simple household goods, or forecast the weather. They were filled with recipes and recipe-like texts that promised readers that they could wrestle order from the disorder of the natural world by reading the patterns of the divinely ordered universe.

Before 1375 or so, most English people would not have dreamed of owning such a book. For most of the European Middle Ages, books of natural knowledge had circulated almost entirely among monks or university clerics, the only people who could read the Latin in which medieval medicine and science were written. Though ordinary medieval English folk certainly tended wounds, planted crops, and watched the stars just as often as did monks or university professors, they didn't record their knowledge in books, nor could they turn to books to guide their actions. They possessed what we would call experiential or observational knowledge—the sort of knowledge prized by historians of medicine and science but afforded very little status within medieval culture. In medieval Europe, churchmen were the most learned members of society. They were trained to read Latin manuscripts filled with the wisdom of the ancient philosophers or the church fathers, which made them certain that the most valuable and authoritative knowledge was that which had been written down and preserved in books. That reverence for books permeated medieval culture, so that when Middle English translations of authoritative Latin texts began to appear in the last quarter of the fourteenth century, even the English laity were thrilled at the prospect of collecting these texts in books of their own.

Over the course of the fifteenth century, non-elite English people commissioned or created hundreds if not thousands of manuscripts filled with texts that guided them through the *practices* of healing, tending crops, making medicines, or forecasting the weather. They could do so because manuscripts became less expensive and more accessible over the course of the fifteenth century, thanks to rising rates of literacy among the laity, the introduction of paper as a writing medium, and the adoption of faster, cursive scripts for writing. This book examines 182 of these "practical manuscripts," as I call them, to reconstruct how premodern English people made sense of their health, their bodies, and their relationship to the natural world. In their everyday interactions with Middle English recipes and instructional texts—many of which were translations of Latin writings that were hundreds of years old by the fifteenth century—lay English readers became participants in a scholarly conversation that stretched back to antiquity. To these readers, Hippocrates and Galen weren't distant authorities whose word was law. Rather, they were trusted guides, whose advice could be excerpted, rearranged, recombined, and even altered when it suited their needs. English readers and writers of practical manuscripts felt free to exercise their own faculties of discernment, assessing or critiquing authoritative knowledge, and occasionally even inserting their voices alongside those of the ancients in commentaries or notes jotted in the margins of their recipe collections.

And then, just as English readers had begun to master the manuscript as a medium for accessing and assessing natural knowledge, the printing press arrived on English shores. Within a decade following the establishment of William Caxton's print shop in Westminster in 1476, English printers began mining fifteenth-century manuscript sources for popular medical recipes or herbal remedies that they would publish, over and over again, in competing editions. In fact, much of the same natural knowledge that had circulated in manuscript in fifteenth-century England recirculated in vernacular printed books throughout the first half of the sixteenth century. More so than other information economies in continental Europe, the English print market was dominated by the vernacular. Nearly sixty percent of the books published in England before 1501 were published in English, double the rate of most comparable European vernaculars.[2] This trend continued throughout the sixteenth century, as English printers saw no point in competing with continental publishing houses that had already captured the international market for learned medicine and science.[3] By and large, books printed in England were meant for English readers. And what could be better suited to meet these readers' needs than a corpus of Middle English medical recipes, herbals, plague tracts, surgical treatises, and agricultural manuals that had circulated for decades in hundreds of fifteenth-century manuscripts?

English publishers' preference for the vernacular—to include vernacular medicine and science—kept England on the periphery of learned European medical and scientific culture for much of the fifteenth and sixteenth centuries. Yet, at the same time, it put non-elite readers at the center of the English information economy, perhaps inadvertently making them arbiters of information about the body, health, and the natural world. Just as these readers had critiqued and evaluated authoritative knowledge in their manuscript collections, so too did they assess and evaluate the dozens of printed almanacs, recipe books, herbals, or prognostications for sale in bookshops around St. Paul's Cathedral in London. Their critical faculties were put to the test, too, because English printers—like today's internet advertisers—developed marketing strategies to entice readers to purchase *their* edition of an herbal, recipe collection, almanac, surgery, or book of secrets, rather than one for sale from their rivals. By the later sixteenth century, printers had been so successful at convincing English readers of the novelty and originality of printed practical books that many of these readers began to turn away from printed books and back to the manuscript collections of their great-great-grandparents' generation. They found in fifteenth-century manuscripts just what they were looking for: knowledge that was both old *and* English, and thus perfectly suited to meet the needs of Elizabethan collectors hoping to uncover evidence of an exceptional English past. For these sixteenth-century readers, practical manuscripts became critical sources for the invention of a distinctly English tradition of natural knowledge.

After years of reading, assessing, and evaluating natural knowledge *in English*, these Elizabethan readers were certain of their critical faculties and convinced of the merits of textual criticism. They had come of age in a world replete with an "infinite multitude of books," as the famous Protestant polemicist John Foxe put it.[4] Moreover, by the close of the sixteenth century, England was moving from the periphery to the center of European intellectual culture, particularly when it came to engaging with the natural world. Historians of science have tended to explain this shift in the same language used by England's greatest sixteenth-century natural philosopher, Sir Francis Bacon.[5] In Bacon's *Advancement of Learning*, the story of English science is one of ingenuity and invention led by ordinary practitioners like printers or sailors. According to Bacon, these "ordinary" people could advance knowledge because they were undisciplined by the scholastic natural philosophy that had ruled European universities since the thirteenth century but had since degenerated into obsolescence.[6] Because these practitioners weren't conditioned to revere long-dead authorities, because they weren't weighted down by a musty textual tradition, they could trust the authority of their own experience.

The problem with this oppositional framing—textual tradition versus experiential knowledge—is revealed in the hundreds of practical manuscripts and printed practical books read in England between 1400 and 1600, which show innovation developing in response to textual tradition. Fifteenth-century readers of practical manuscripts revered ancient authorities like Galen or Hippocrates, but not so much that they didn't feel free to reinterpret their maxims or amend their precepts. Likewise, though sixteenth-century readers bought recipe collections and herbals filled with knowledge that was indebted to a very old textual tradition, thanks to printers' marketing tactics, they often did so under the pretense that these books were novelties. In manuscript and in print, practical books remained popular because they excerpted, revised, and amended authoritative knowledge to suit the needs of compilers and consumers. And because these books blended tradition and innovation, they became important vehicles for introducing truly revolutionary ideas to a popular English readership.

For over two hundred years, from the creation of the first practical manuscript around 1375 to the publication of Bacon's *Advancement of Learning* in 1603, English readers developed their critical faculties through engagement with texts, many of them very old, in practical manuscripts and printed books. Gradually, these interactions with natural knowledge changed English culture, even as shifts within English culture changed how readers interacted with natural knowledge. This book aims to reveal how those changes unfolded over decades and even centuries. I argue that readers' engagement with natural knowledge rooted in a very old tradition—a tradition not particularly well respected in narratives of the so-called scientific revolution—did pave the way for Bacon's ordinary practitioners to feel confident generating new knowledge for themselves. Yet they did so not because they wished to reject the old in favor of the new, but rather because they saw the pursuit of natural knowledge as a very old tradition in which they were eager to participate.

Of Readers, Reformations, and Revolutions

Sometime in late 1557 or early 1558, perhaps just before he paid four shillings to the Stationers' Company of London to register the title, Humfrey Baker put the finishing touches on a new English translation of a best-selling guide to astrology, first published in Paris in the 1540s.[7] The original French text was the work of Oronce Fine, famed cartographer and chair of mathematics at the Collège Royal in Paris. Baker hoped to make Fine's "short Introduction upon the Judicial Astrology" available to an English readership, and so he faithfully translated Fine's text—everything except for the book's original dedication to Jean du Val, secretary to the King of France. Baker had a choice to make: would he follow in Fine's footsteps and compose a paean to some well-connected member of the English nobility? No, he would not. Baker dedicated *The rules and righte ample documentes, touchinge the vse and practise of the common almanackes whiche are named ephemerides,* to "the loving Reader . . . daily more inventive, & inclined also to read & to have understanding in arts and sciences."[8] Baker knew that his success depended on these "inventive" English readers who had their pick from dozens of printed almanacs, prognostications, and other practical books by the time he published his own in 1558.

Once upon a time, book historians might have argued that the virtues of a published astrological treatise like Baker's would have been obvious to English readers. After all, Baker's book had been authored by one of the sixteenth century's greatest mathematicians, Oronce Fine. In Baker's printed edition, illustrated with tables and figures, English readers could be sure they were getting access to the very same knowledge Fine had made available in France. Compared to the error-prone astrological tables and vague prognostications collected in fifteenth-century vernacular manuscripts, Baker's book should have been an obvious improvement.[9] Yet, that's not how Baker saw it. Baker didn't assume that Fine's reputation would appeal to English consumers, or that they would choose his book over much older manuscripts. In his dedication, Baker referenced the ancient histories of Josephus, Herodotus, and Diodorus Siculus rather than the mathematical prowess of Oronce Fine, arguing that astrology was an ancient and venerated science—and that his book partook of that ancient tradition.[10] And, as we know, he flattered English readers' curiosity and inventiveness. Finally, as all good salesmen do, he promised English readers that his book would fix all their problems: "this gentle Reader, / I most heartily thee require / To examine well these Rules / And thou shalt have thy desire."[11]

Book historians and historians of science have come to realize what Baker knew all too well in the sixteenth century: a printed book could only be as successful, influential, and authoritative as its readers would allow.[12] Steven Shapin, Adrian Johns, and Sachiko Kusukawa have demonstrated that even landmark publications in the history of science—from Robert Boyle's pamphlet on the air pump to Tycho Brahe's tables of astronomical data—were fashioned as material arguments that readers then interpreted according to all manner of criteria.[13] The books we now recognize as watersheds were not ensured automatic success among early modern readers. Rather, both authors and publishers worked together to appeal to readers and construct the authority of printed books. This is no less true of inexpensive practical books. Sixteenth-century printers, authors, and translators like Baker never for a moment assumed that readers would necessarily buy their edition of an almanac or recipe collection over one from their rivals, especially given that there was such an "infinite multitude" of these books on the English market. They had to create the conditions in which their editions would sell, and this often meant packaging natural knowledge in particular ways to catch the eye or meet the needs of particular readers.

Chapters 4, 5, and 6 investigate how printers navigated a market glutted with "infinite multitudes" of medical recipe collections, books on astrology, almanacs, herbals, uroscopy treatises, dietary regimens, anatomies, husbandry manuals, agricultural guides, prognostications, plague tracts, and books of secrets, by comparing hundreds of editions of these books produced over decades by dozens of English printers. In decade-by-decade searches of the *English Short Title Catalogue*, I identified more than 475 editions of practical books published in England between 1485 and 1600, many of them containing texts that first circulated in Middle English manuscripts.[14] As I show in those chapters, very little about the contents of practical books changed over the two-hundred-year period under study in this book—probably because the everyday concerns of English people hadn't changed all that much either. English people needed recipes to cure a fever or directions for calculating auspicious days for bloodletting in 1400 just as they did in 1600. Though often derivative and formulaic in their contents, the constancy of practical books makes them ideal sources for detecting how the coming of the press altered English readers' relationship to natural knowledge. Though Roger Chartier almost certainly didn't have practical books in mind when he wrote that book historians ought "to understand how the *same* texts can be differently apprehended, manipulated, and comprehended" in different media and by different readers, he might as well have been describing the aims of this book.[15]

Though the same vernacular texts were copied and then printed over and over again in England from 1400 to 1600, the chapters in this book show how readers' attitudes toward these texts changed dramatically, both as a result of the commodification of knowledge within a commercial print market and as a result of sweeping changes to English society. And there was no more sweeping change than the English Reformation, begun in the early 1530s when Henry VIII declared himself head of the new Church of England. For the next two decades, the throne passed to three of Henry's children, and England swung from hardline to moderate Protestantism and then from hardline Protestantism to Catholicism, until the Elizabethan Settlement of 1559 set the English church on a more moderate course that it would sustain for the next fifty or so years. Over the three decades between 1535 and 1565, the English faithful were asked to adapt to quite a lot of religious change. Somehow, they did. England was a Protestant nation by 1600, and it became so without the violent wars of religion that tore apart both France and the German principalities.

Not surprisingly, historians have spilled considerable ink trying to understand how and why Protestantism took root in England, and many have recognized vernacular manuscripts and printed books as important sources for tracking changes to English belief over time.[16] Eamon Duffy's now classic study of the English Reformation centered vernacular religious manuscripts to argue for the vitality of traditional religion prior to the

Reformation, while Tessa Watt's study of Protestant "cheap print" showed how the old genres of traditional religion were remade in printed pamphlets and ballads to fit the new Protestant order.[17] Yet neither historian could adequately compare Middle English religious texts with Protestant ones because the Reformation did away with genres like Books of Hours, Psalters, and miracle stories that had structured fifteenth-century religious practice. Practical books allow for a kind of comparative analysis that isn't possible with religious books. Throughout the tumultuous English Reformation, before and after the break with Rome, English readers continued to copy recipes in manuscripts, purchase almanacs in bookshops, and borrow herbals from their friends and neighbors. These books of natural knowledge posed little threat to reformers or court ministers. They weren't censored at the printer's shop, and when Henry's son, Edward VI, condemned all sorts of older religious books to the fires in the early 1550s, practical books escaped those too. English people had to learn new prayers to replace the old, but they didn't have to learn new medical recipes.

And yet, though practical books were never a direct target of censorship or reform, that isn't to say that they were left unaltered by England's Reformation. Fifteenth- and sixteenth-century readers alike knew that the natural world was a manifestation of God's power and a reflection of his divine will. Everything in the changeable sublunar sphere had been put there at the time of creation for the benefit of mankind, just as the stars and planets in the spheres of the heavens were set in their regular motion by God. For premodern readers, this basic set of beliefs set the whole of the natural world alight with interpretive possibility because everything from the smallest herb to the most distant star had a divine purpose directly related to human experience. Natural knowledge, in whatever form it took, was reflective of the deeply held belief that God's power and order could be accessed in nature, through the harvesting of herbs or the reckoning of celestial time. Indeed, a recurring theme throughout nearly every chapter of this volume is that practical books were in fact vehicles for the transmission and expression of religious doctrine, whether in manuscript or in print. This book argues that by tracking changes to the presentation and reception of that doctrine in practical books, whether in the removal of saints' days from a printed almanac or the censoring of Latin charms in a manuscript, we can get a better sense of how deeply and thoroughly the Reformation was realized among ordinary English people.

Of course, the only way to track gradual shifts in English attitudes toward nature and the divine—whether those shifts resulted from the introduction of the printing press or the schism of the English Reformation—is to begin this study much earlier than most histories of early modern English medicine and science do: in the fifteenth century, when the English laity were only just beginning to access natural knowledge in manuscripts. Hundreds (if not thousands) of Middle English manuscripts featuring medical recipes, herbal lore, prognostications, and the like have survived, all of them rich sources for understanding English people's relationship to their bodies, to books, and to the divinely ordered natural world. And yet, historians of early modern England have paid them little attention. Though the field of recipe studies is booming, with important works like Elaine Leong's *Recipes and Everyday Knowledge* claiming a pivotal role for manuscript recipe collections within the history of English medicine and science, that history has centered on the later sixteenth and seventeenth centuries.[18]

There are good reasons why historians of medicine and science tend to ignore fifteenth-century England. From roughly 1455 to 1485, England was racked by the Wars of the Roses, a dynastic squabble that pitted rival factions of the royal family against one another and resulted in decades of civil war. As a result, English kings had nowhere near the financial means or the political capital to support the kind of large-scale projects that would have facilitated "trading zones" between artisans and philosophers, which Pamela O. Long has identified as critical to the emergence of the New Sciences within fifteenth-century Italian courts.[19] Moreover, though Danielle Jacquart has shown that fifteenth-century medical faculties at continental universities began to embrace experiential knowledge and question scholasticism, such was not the case in England.[20] In fact, English universities trained only a very few physicians between 1400 and 1500, and those few who did matriculate from Oxford or Cambridge seem to have been little affected by the new humanist learning taking Italy and France by storm.[21] Finally, while fifteenth-century German, French, Italian, and Flemish painters, metalworkers, glassmakers, and goldsmiths pioneered new manufacturing techniques, refining their senses to develop what Pamela Smith has described as a close attention to the properties of natural matter, fifteenth-century English artisans clearly lagged behind their continental counterparts in manufacturing expertise.[22]

In short, the history of fifteenth-century England does not reflect the intellectual or technical developments that have been credited with giving rise to the New Sciences, and for the most part, neither do fifteenth-century English practical manuscripts. Though the Middle English collections discussed in this book do very often reference ancient authorities, they are hardly reflective of the new humanist attention to authenticating ancient sources. Moreover, only very rarely do English practical manuscripts reveal practitioners translating experiential knowledge into writing—acts of translation that Pamela Smith has shown to be crucial to the valorization of observation and experimentation as epistemic practices.[23] Most do not contain what we would identify as proprietary expertise, but rather reflect their readers' desire to access an established and authoritative textual tradition. Perhaps for that reason, most scholarly interest in practical manuscripts has centered on their value *as texts*.[24] Though historians have neglected these sources, the same cannot be said of the literary scholars and manuscript specialists who have worked for decades to catalogue, transcribe, and analyze Middle English medical and scientific writings.[25] This book would not have been possible without their pioneering work, and most especially that of Linda Ehrsam Voigts and Patricia Deery Kurtz, whose database of *Scientific and Medical Writings in Old and Middle English* now contains entries from over 2,000 English manuscripts.[26]

My search for practical manuscripts began with the Voigts-Kurtz database and proceeded through the catalogues of major libraries in the United States and United Kingdom, from which I identified over 300 fifteenth-century manuscripts that I hoped would reveal something about non-elite, lay readers and their relationship to natural knowledge. Armed with this list, I set off on nearly ten years of research trips to London, Oxford, Cambridge, New York, Washington, DC, New Haven, CT, and San Marino, CA. Only after viewing hundreds of manuscripts and taking thousands of photos at libraries across the United States and in the United Kingdom did I begin to see a pattern emerge. I had gone looking for manuscripts filled entirely with vernacular texts related to medicine and science, and I found that the same genres reappeared over and over again in the manuscripts that met this criterion. Mostly absent were vernacular treatises relating to the kind of medicine practiced by learned physicians. Instead, these manuscripts were filled with some combination of medical recipes, herbals, prognostications, plague treatises, or uroscopy tracts, short treatises on bloodletting, astrology, or veterinary medicine, and instructions for textile or ink-making, planting or grafting, and hunting, hawking, or fishing.[27] Though seemingly disparate, these genres of instructional or didactic writing were all related to the *practices* of healing or of manipulating natural matter. I came to believe that these texts comprised an epistemological category in fifteenth-century England, one whose contours closely mirrored the "mechanical arts" as described by scholastic natural philosophers.[28] My contention is that fifteenth-century compilers recognized that useful, hands-on, natural knowledge belonged in its own sort of manuscript, and that the "practical manuscript" was a genre all its own.[29]

To be clear, not every Middle English manuscript with medical recipes or an almanac qualifies as a "practical manuscript" by this definition. A great number of fifteenth-century English medical and scientific manuscripts combine learned Latin theory *and* Middle English instructional texts. Though these manuscripts certainly tell us something valuable about the circulation of natural knowledge in fifteenth-century England—namely, that the learned were not at all opposed to collecting vernacular recipes alongside Latin treatises—these collections suggested a more educated readership than the one I was after.[30] My aim was to understand how readers who had been kept outside of long-standing hierarchies of knowledge (in other words, those without much Latin) engaged with a book of natural knowledge when they could finally own one. That intention guided the selection of the 182 originally separate fifteenth-century manuscripts in my corpus, each of them filled entirely (or mostly so) with Middle English texts explicitly meant to guide everyday practices like preparing medicines, harvesting herbs, tending wounds, caring for animals, producing textiles or ink, forecasting the weather, or diagnosing illness.[31]

And so, even though English practical manuscripts do not resemble Renaissance manuals on mining or metallurgy, and even though they do not contain much that we would call expert or proprietary knowledge, they are nonetheless important sources for understanding English people's relationship to authoritative knowledge, to books, to their bodies, and to the wider, natural world—if we know how to read them. In this book, I read them as epistemic objects whose meaning lies not just in their textual contents but also in their physical characteristics. Instead of focusing on individual Middle English texts, in chapters 1, 2, and 3, I analyze practical manuscripts according to the materials with which they were made, their manner of composition, the order of their textual arrangement, and most important of all, the patterns of use evident in their pages.[32] Readers used these manuscripts in a variety of ways, only some of which can be tracked in reader additions, notes, and deletions.[33] And yet, though necessarily incomplete as a record of English readers' engagements with their books, these reader marks do show English people integrating the written word into their daily lives, using manuscripts as sites for cultivating their own knowledge and expertise.[34] By comparing manuscripts and printed books across more than two centuries, this book aims to reconstruct how English readers became, in the words of Humfrey Baker, "daily more inventive, & inclined also to read & to have understanding."

A Recipe for Reading This Book

The chapters in this book are arranged roughly chronologically, dealing first with manuscript and later with printed practical books. Chapter 1 opens with Peter Cantele, a village priest from Toft Monks in Norfolk, whose practical manuscript serves as our road map through fifteenth-century England. We follow Cantele on an imagined journey to a London stationer's shop to learn how a manuscript was made in fifteenth-century England and then look closely at its contents to trace how natural knowledge moved through networks of monks and clerics until its eventual translation into Middle English in the fifteenth century. This chapter is intended as a primer on manuscripts and medieval medicine and science for non-specialists, and as such, manuscript scholars and medievalists may wish to skip ahead to chapter 2. In that chapter, we meet another member of the clergy, a Benedictine monk who composed what appears to be the first English practical picture book, filled with icons, pictures, symbols, and numbers for keeping track of the agricultural year, the liturgical year, and cycles of celestial time. I explore the visual and ideological influences that bore on the creation of this unusual practical manuscript, before reconstructing how the image cycles in that book were meant to predict the future. I show that the pictures developed by one Benedictine in Worcester were recopied by other Benedictines throughout England and eventually by members of the English laity, too, all of whom shared an appreciation for a visual language intended to make the natural world legible.

Chapter 2 tells a story reflective of broader trends in the transmission of vernacular natural knowledge in fifteenth-century England: knowledge that had once been exclusive to the clergy was no longer so. In chapter 3, we follow the spread of natural knowledge among the fifteenth-century English laity to reconstruct the attitudes of those who commissioned or created practical manuscripts. At the beginning of the fifteenth century, few besides the most elite had ever seen a book of natural knowledge. By the close of the fifteenth century, hundreds if not thousands of practical manuscripts had been commissioned or created by lay English readers. In the intervening years, the form and function of these manuscripts changed dramatically, in no small part because of who was writing them. Only by paying close attention to how natural knowledge got on the pages of these manuscripts—whether through the help of trained scribes or at the hands of determined amateur writers—can we understand what fifteenth-century people hoped to do with that knowledge once it was written. Whether in the inscription of simple charms, the hasty addition of new recipes in the margins, or the copying of a whole book of natural knowledge, this chapter argues that writing was a powerfully transformative act for the laity in fifteenth-century England.

Chapters 4 and 5 take us into the heart of London, where the earliest

English printers began publishing versions of the Middle English medical recipes and herbals that fifteenth-century readers had treasured in manuscript. Chapter 4 explores how the commercial print market exerted pressures that gradually reshaped how very old natural knowledge was presented to English readers. With so much competition in the market for practical books throughout the first half of the sixteenth century, but with very little innovation within the genre itself, early modern printers had to work harder and harder to distinguish their editions from those of their rivals. They did so by innovating marketing strategies that convinced readers of the novelty or proven value of very old natural knowledge. When the Stationers' Company finally did begin to regulate the English print market, putting an end to publishers' reprinting of the same old texts, readers were already primed to seek out knowledge that was novel, original, and had been tried and tested.

Yet not all genres of natural knowledge were so easily translated into print. Though almanacs and prognostications were mainstays of the early English press, in the early decades of the sixteenth century, they were largely the product of continental astrologers. Only in the 1520s did English printers finally begin to reproduce the pictorial prognostications common to manuscript almanacs and calendars—just in time for the English Reformation to wholly transform how English readers marked the passage of time and how they thought about the role of images in books. Protestant reformers were united in their disdain for what they saw as medieval superstitions, and some of them included astrology within this category. But others, like Leonard Digges, believed astrology to be a means of comprehending God's truth as writ in the stars. Protestant insistence that anyone should be able to access God's word, either in scripture or through attention to the natural world, led to a revival of perpetual prognostications in the 1540s and 1550s—but ultimately not a return to the visual language of prognostication that had meant so much to fifteenth-century readers. Chapter 5 thus argues that what the English *stopped* reading had as much to do with their changing attitudes toward the natural world as what they continued to purchase and read in print.

Chapter 6 straddles both manuscript and printed sources to follow the fortunes of women's knowledge among English readers both before and after the coming of the press. Though charms, prognostications, and other rituals of healing were widely read in practical manuscripts, in print they were sometimes associated with women's "secrets." Yet by the mid-sixteenth century, it had become harder for readers to determine whether women's secrets were especially promising—as books of secrets often assured their readers—or whether they were dangerous, as the new Protestant authorities of the Church of England insisted. When these same readers began to look for secrets in fifteenth-century practical manuscripts, some did not like what they found. Charms and reproductive recipes that had circulated without issue in the fifteenth century were censored by early modern readers who took care to blot them out with ink or scrape them from parchment pages.

Chapter 7 picks up with early modern readers where chapter 6 leaves off. By the latter half of the sixteenth century, those with an interest in collecting natural knowledge had begun to see fifteenth-century manuscripts as especially valuable. Years of pandering from publishers hoping to gain advantage in a commercial market had convinced Elizabethan readers that medieval manuscripts contained knowledge that was superior to that available in print—or at least less tainted by commercial interests. Although the collectors of practical manuscripts don't enjoy nearly the reputation of serious antiquarians like Matthew Parker, John Stow, or John Foxe, this chapter argues that their motives were not so different: locating the sources of English medicine went hand in hand with efforts to locate the sources of English history. In both cases, documentary evidence from the medieval past would supposedly confirm that England had been blessed with God's special providence. These efforts were especially necessary in light of reports of marvelous *materia medica* from the New World, which suggested that the Spanish or Indigenous Americans might be recipients of God's divine blessing, too.

Together, these seven chapters show how English readers embraced new roles as adjudicators and purveyors of natural knowledge through their encounters with hundreds of books: manuscripts, then printed books, and then manuscripts once again. By reconstructing their everyday exchanges and interactions with almanacs, recipe collections, prognostications, herbals, husbandry manuals, and books of secrets, this book shows ordinary people learning *how* to think about their bodies and the world around them—but not so much *what* to think in a world cluttered with information. And it's the *how* and not the *what*, I argue, that was so important for cultivating curiosity and inquisitiveness within English readers. Indeed, as the chapters in this book demonstrate, the critical skills that English readers developed in their interactions with these many books were the same ones prized by the likes of Francis Bacon: they learned to be skeptical readers, to relate what they had read to what they had seen or done, and to compare the claims of competing authorities. Which is all to say that practical books taught ordinary English people *how* to claim their place alongside physicians, natural philosophers, and experimenters, united in the pursuit of natural knowledge.

✷ 1 ✷

The Making of a Practical Manuscript

Sometime after he was appointed rector of the parish church at Toft Monks, Norfolk, in 1459, Peter Cantele got his hands on a very nice book.[1] Cantele's manuscript, now British Library MS Sloane 1764, contains a number of medical texts: a popular treatise on the plague; a collection of medical recipes arranged by ailment from head to foot; a series of recipes for manufacturing waters, oils, and "entretes"; and finally, a treatise on the medical properties of herbs. Each of these texts is written in a neat and professional cursive script, with ornamented initials and the occasional pen-and-ink curlicue in its margins. At the back of the manuscript, the scribe who copied those many medical texts added a "concordance of the book," a two-page index annotated with folio numbers set alongside promising recipes.[2] Armed with such a stylish and functional manuscript, Peter Cantele might have treated all manner of ailments and illnesses afflicting him or his parishioners. And perhaps he did—after all, parish priests often served as medical practitioners, tending to the physical and spiritual needs of their flock.[3]

All these many years later, however, we cannot know whom Cantele treated with the many medical texts that fill his manuscript. All we can know for certain is that Cantele consulted his manuscript closely, and even amended it to suit his needs. In its modern binding, Sloane 1764 opens with two paper pages filled with notes and recipes in Cantele's own hand. On the first of these pages, Cantele composed a few lines of formal Latin legalese, testifying that on April 1, 1463, he heard the confession of two of his parishioners, Robert and William. Following this record, he added instructions for how to judge the qualities of a horse by its color; a recipe for size, a kind of glue; and another set of several recipes for making different "waters," or distilled medicines.[4] Finally, Cantele wrote little notes to remind himself of particularly good recipes, like a "precious water for burning and scalding," which was the "fourth water" in his collection.[5] His careful notes also appear in the index at the back of the manuscript, where he added glosses to unfamiliar medical terms: "alopecia," for instance, he defined as "falling of hair."[6] These additions show Cantele to have been a man interested in the collection and assessment of natural knowledge—exactly the sort of discerning reader and writer who, this book argues, used practical books to gain a foothold in an expanding culture of knowledge exchange in fifteenth-century England.

But can we be sure? Cantele's manuscript offers more questions than answers: how did Cantele come to possess his manuscript, and who made it? Did he or someone else choose the medical texts that fill its pages, and why did they choose what they did? And perhaps most central of all, what did he do with this manuscript once it was in his possession? None of these questions are easy to answer—not just in the case of Peter Cantele, but for each of the other 181 practical manuscripts examined for this book. Ownership marks from fifteenth-century readers like Cantele are in fact fairly rare. When they are present, they typically don't tell us much besides a reader's name and the few recipes he or she most appreciated. Moreover, fifteenth-century names are often hard to trace in the archival record. Before the Reformation, the church did not require priests to keep records of their parishioners' births, baptisms, marriages, and deaths, as would be the case after 1535. As a result, fifteenth-century individuals tend to turn up in the archive only after contact with the courts: in land disputes, in quarrels with neighbors, or in the registration of their wills. Though these records do occasionally testify (albeit faintly) to the social and economic status of manuscript owners, this is scant evidence from which to reconstruct how fifteenth-century readers valued and used their practical manuscripts.

For all of these reasons, telling the story of the hundreds, if not thousands, of people who owned practical manuscripts requires thinking creatively, and occasionally, speculatively. The approach I take in this book is to focus on those few readers about whom we know a little something and then to use those few to stand in for the many about whom we know next to nothing. In this chapter, Peter Cantele stands in for any number of fifteenth-century readers who perhaps visited stationers' shops throughout England to commission a practical manuscript. In later chapters, I imagine the lives of creative monastic book artisans, itinerant medical practitioners, and inveterate manuscript collectors. Yet historical imagination can only take us so far. Though it can put flesh on the bones of my narrative, the muscle of my argument—and my second tactic to mitigate the problem of scarce biographical evidence—comes from the manuscripts themselves. As bespoke objects, created by one or more individuals to meet the needs of another, every aspect of a practical manuscript's composition can tell us something about the abilities of its maker and the interests and attitudes of its owners. We can note what sort of material these manuscripts are made from (paper or parchment), the style of their handwriting, the presence or absence of illustration, the size of their pages, the thickness of their spines, and of course, we can analyze their contents. Each of the 182 practical manuscripts in my corpus can be read closely to reveal something about the nature of English people's relationship to books and to natural knowledge.

But before we can begin to assess the relationship between those manuscripts and their readers, we must learn to recognize them as epistemic objects made in particular ways at a particular time with particular attributes. This chapter will offer a primer on how to make and how to read a practical manuscript, with Peter Cantele and his manuscript as our guides throughout. First, we meet Cantele in fifteenth-century England, where established traditions surrounding reading, writing, and book ownership were rapidly changing. We accompany Cantele on an imagined journey to a stationer's workshop, where book artisans worked with parchment, paper, pigment, and ink to create his manuscript. Next, we turn the pages of Cantele's manuscript and examine its contents, investigating the origins of each of its medical texts. We follow natural knowledge as it moved from Arabic to Latin to Anglo-Norman French, and finally, to Middle English, winding through university lecture halls and monastic libraries to reach the pages of practical books.

A Visit to the Stationer

Peter Cantele was a priest, and like generations of clerics before him, he possessed a medical manuscript. We cannot say for certain whether Cantele was responsible for commissioning that manuscript, or if he inherited it upon arrival at the rectory in Toft Monks in 1459, but by 1463 he was jotting notes in Latin legalese and recipes in Middle English on its front flyleaves. Cantele could do so because he was literate in both Latin and Middle English. Perhaps, like many other boys in fifteenth-century England, he first learned his letters at home, studying the Latin prayers in a Book of Hours and setting them to memory. After mastering the alphabet Cantele might have been sent to one of England's primary or "reading schools," where he would have learned basic English literacy. These institutions were fairly common in fifteenth-century English towns and cities, many of them run by parish churches, chantries, hospitals, almshouses, and, in some cases, by guilds.[7] In Cantele's day, a basic education at one of these schools cost only around four pence per quarter, or about a day's wages for a laborer.[8]

But a boy who would grow up to be a village priest needed Latin, too. Cantele may have studied his cases and declensions at Eton College, founded in 1440 as one of several new "grammar schools" explicitly intended to help train boys for clerical office. Some of these new schools—Eton included—set aside a few places for England's poor and needy boys, who would receive an education on scholarship.[9] Cantele may have been one such needy boy. Though his name doesn't appear in Eton's (largely incomplete) fifteenth-century register, it was Eton's provost, William Wayneflete, who appointed Cantele as rector of Toft Monks in 1459.[10] Wayneflete had the right to do so because the revenues from Toft Monks were part of Eton's endowment.[11] If Eton reserved a few scholarships for boys from the parishes in its endowment, as was often the case, it is not an enormous leap to imagine Cantele as a local boy who made good and returned from school to a living at his hometown church.

Cantele's education will have to remain a matter of speculation, but the trajectory I have outlined here was not unheard of in fifteenth-century England. Records show that scores of grammar schools and reading schools were founded over the latter half of the fifteenth century.[12] Indeed, as early as the late fourteenth century, some guilds required boys to be functionally literate in English before they could train as apprentices, indicating that vernacular literacy had become common among the merchant and artisan classes.[13] No doubt rising rates of vernacular literacy contributed to the emerging status of Middle English as an official legal and literary language. After 1362, litigants in English courts could officially plead their cases in Middle English; around 1376, an unknown scribe composed the first legal deed in Middle English; and shortly thereafter, an English customs agent named Geoffrey Chaucer began to compose the most celebrated work of Middle English literature, *The Canterbury Tales*.[14] It was around this same time, between 1375 and 1382, that Henry Daniel, a Dominican friar, authored the very first Middle English medical treatise: the *Book of Uroscopy* (*Liber Uricrisiarum*), a treatise on urine analysis and prognosis.[15] Though Latin was still the language of the church and of the learned, the rise of Middle English meant that many more in fifteenth-century England could read a poem, a writ, or a recipe than ever before.

And yet, it is impossible to say with precision what percentage of men, women, and children in fifteenth-century England did read writs or recipes. Even in the later sixteenth century, a hundred years or more after Peter Cantele made notes in his manuscript, estimates suggest that only around sixty percent of English tradesmen and craftsmen were literate.[16] However, these estimates account for those who could read *and* write, and not those who could only read. As Margaret Spufford demonstrated years ago, there were likely many more who could read in premodern England than could even sign their name.[17] Moreover, even those who might not qualify as strictly literate by our definition certainly possessed what Michael Clanchy termed a "literate mentality," or an awareness of the utility and authority of writing.[18] Though literacy was hardly widespread, Adam Fox's pronouncement that no one in sixteenth-century England lived beyond the reach of the written word could just as easily be extended to the fifteenth century.[19] Even the poorest peasant could still expect to hear poems or prayers read aloud and writs and charters proclaimed in the village square.[20]

This is all to say that though Peter Cantele was more literate than most in fifteenth-century England, the ability to read a book of medical recipes and treatises in Middle English was not so unusual by the time Cantele took up his position at Toft Monks in 1459. By the mid-fifteenth century, it appears that many more in the minor gentry and "middling sort" could engage with texts written in Middle English—and, not coincidentally, many more in these same classes also began making or buying books.[21] In their studies of later medieval wills and inventories, Kathleen Scott, Anne Sutton, and Caroline Barron have identified over 200 fifteenth-century English merchants or artisans who owned manuscripts.[22] Their manuscript commissions were part of a wave of increased consumer spending in the fifteenth century, when living standards rose considerably.[23] Surviving inventories and wills of city-dwelling merchants and artisans show them spending money on textiles, silver plate, and other luxury goods.[24] Outside the cities, members of the up-and-coming gentry classes, like the Yorkshire gentleman Robert Thornton (a well-known owner of a Middle English medical manuscript), were also eager to participate in a burgeoning culture of book ownership.[25] Though Thornton's taste in Middle English literature may have exceeded that of his peers, his interest in book ownership did not. Other landed families, like the famous Pastons, who commissioned a scribe to produce a "little book of physick," or the Tollemaches of Suffolk, who produced a household compendium of medical and agricultural knowledge, made practical manuscripts part of their growing libraries.[26]

Rising literacy rates and improving economic circumstances explain why so many in fifteenth-century England might want to own a practical manuscript, but not *how* these people got their hands on them. Truth be told, there is no single answer to the question of how a practical manuscript got made. Because all it required to make a manuscript was the ability to write and a supply of writing materials, a would-be book owner might employ a professional to copy recipes or try their hand at it themselves. They might use paper, or they might choose parchment. They might have the book embellished with penwork and ornamented initials, or they might not. Manuscript production was bespoke and too decentralized to allow for broad-brushed statements. And yet, because each manuscript was a custom creation, because it was the product of individual artisans with individual styles and abilities, we can say something about how *one* practical manuscript got made. We can, for instance, travel with Peter Cantele and imagine what might have transpired in the making of British Library MS Sloane 1764.

Cantele's parish of Toft Monks was in East Anglia, a region now appreciated as a center of later medieval English manuscript production. From the Toft Monks rectory it would have been just about a day's ride to the city of Norwich, one of England's cathedral towns, where as early as the late thirteenth century book artisans had set up shops in the lanes under the shadow of the great church.[27] In the years following his appointment at Toft Monks, Cantele almost certainly made his way to Norwich to attend markets and to conduct business, during which time he might have conversed with scribes (those who wrote the text of a book), limners (those who produced decoration and illustration for a book), parchminers (those who produced parchment from the skins of cows, goats, or sheep), and bookbinders (those who stitched loose packets of folded sheets of parchment or paper together to form a manuscript), asking about their rates or discussing the possibility of a commission. Fifteenth-century Norwich wasn't unique for having a neighborhood of book artisans. By Cantele's day, residents of most English urban centers, and certainly all of its cathedral towns, could expect to benefit from the services of scribes, limners, and binders—though some towns needed more of these artisans than others. The towns of Oxford and Cambridge, filled with university students constantly in need of books, could support a considerable number of scribes, as could London, with its bustling law courts, and Westminster, with its veritable army of government bureaucrats.[28]

Literary scholars have identified only a few of the scribes and workshops responsible for making Middle English manuscripts of medicine and science, and unfortunately, Cantele's manuscript is not one of those.[29] Though analysis of scribal dialect can sometimes pinpoint a manuscript's locus of origin, we cannot say for certain where Cantele's manuscript was made.[30] We can say, however, that it was made by a professional artisan. We know this because the manuscript tells us so: only a professional scribe

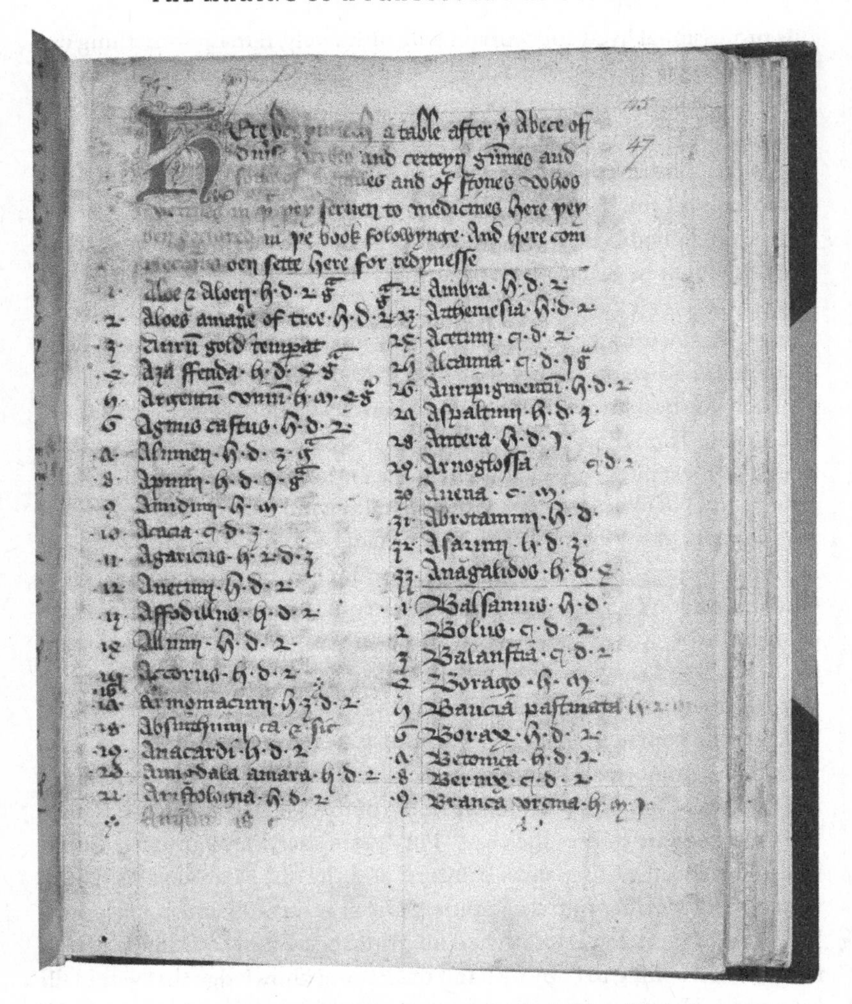

FIGURE 1.1. The opening of the *Circa instans* herbal in a fifteenth-century medical manuscript owned by Peter Cantele, featuring double-column ruling in lead point, as well as a large red initial with pen-and-ink flourishing. © The British Library Board, MS Sloane 1764, f. 47r.

would have pricked the edges of its paper pages at regular intervals and marked out their margins to ensure that its rows of text remained straight and centered on the page. Only a professional scribe could have written the manuscript's recipes in a neat and orderly cursive script. And only a professional scribe would have left blank spaces at the start of each entry in the manuscript for the addition of large red initials and, on the first page of the manuscript, for cascading curlicues of black ink (fig. 1.1). Where this professional lived and worked will, ultimately, remain something of a mystery. But, if we assume that Cantele (or whoever commissioned Sloane MS 1764) followed the usual steps to get a manuscript made in later medieval England, and if we draw carefully from the wealth of scholarship on English manuscript production, we can exercise our historical imagination to reconstruct the making of Sloane MS 1764.

If Cantele had been a much wealthier man, he might have sent off to Paris or Ghent or Bruges for a beautifully illuminated manuscript, as the English aristocracy so often did in the fifteenth century. But Cantele was a village priest without extravagant means, and his ultimate aim was a book of medical knowledge rather than, say, a beautiful Book of Hours, and so probably the best he could afford was a journey 130 miles from Toft Monks to London. Though we cannot be sure that Sloane MS 1764 was produced in England's capital city, by the fifteenth century, London was the undisputed center of manuscript production in England, where many of the 182 manuscripts in my corpus were likely made. The nature of the industry in London was different from that in continental centers like Bruges or Ghent. London's scribes and limners were not beholden to the patronage of an exclusive circle of courtiers as those on the continent often were, and so they catered to a much wider clientele, to include village priests like Cantele, members of the landed gentry, and even merchants or artisans.[31]

Over the course of the fifteenth century, London was home to at least 136 scribes, limners, and binders, all of whom lived in the lanes surrounding St. Paul's Cathedral, just to the north of the River Thames and within the walled square mile of the city.[32] Fifty years later, the same lanes would house nearly all of England's printers, and the shops around St. Paul's would be filled with printed pamphlets as well as secondhand manuscripts. But in 1459, Cantele's London had no print shops, and so Cantele had to find a scribe willing to copy by hand the natural knowledge that would fill his practical book. It wasn't necessary to travel from shop to shop to find a scribe, limner, *and* bookbinder for his proposed project, however. Cantele needed only to seek out a master stationer: a sort of agent or middleman who oversaw the collaborative process of manuscript production. The figure of the stationer was an invention of the very earliest lay scribes working for students at Oxford, and one that was then replicated by the artisans who set up shop in London in the thirteenth century. In 1403, London's scribes, limners, and bookbinders incorporated formally as a guild, which from the 1440s became known as the Mistery of Stationers.[33]

No doubt the stationer who organized the making of Sloane MS 1764 had connections with parchminers who did the smelly work of scraping, drying, and stretching animal hides into thin sheets of parchment. But those connections weren't needed in the case of Cantele's manuscript. For a fraction of the cost, the stationer chose to use paper, a material that had first entered western Europe via Italy in the twelfth century but wasn't commonly used in England until the fourteenth century. By the fifteenth century, paper had become the medium of choice for most vernacular manuscripts produced in England, and for good reason.[34] In 1350, the standard price for a dozen rolls of parchment sold in London was between three shillings and three shillings six pence. Those dozen skins would yield, on average, around ninety-six leaves of the size commonly used for practical manuscripts. The same money could buy five quires of writing paper, which would yield over five times as many equivalent-sized leaves.[35] By 1450 the price of paper had been reduced again by half, and by 1500 the price of a quire of paper was a quarter that of a skin.[36] Peter Cantele's manuscript was one of many created in the second half of the fifteenth century to adopt this less expensive material. Indeed, over two-thirds of the practical manuscripts in my corpus created between 1450 and 1500 use paper as their medium (see table 1.1).[37]

With a fresh supply of paper procured from one of London's Mercers, Grocers, or Spicers, the stationer responsible for Sloane MS 1764 needed only to parcel out the work of copying the medical recipes, herbal remedies, and plague treatise that now fill his manuscript.[38] Today, Cantele's manuscript runs to 114 folios, or 228 front-and-back pages. Originally, however, it appears to have been much longer, probably 165 folios in length. We know it was once quite a lot longer because someone in the later sixteenth or seventeenth century numbered its folios. Fourteen folios are missing from the front of the manuscript, another two leaves are missing between what are now folios 32 and 33, and a whole section of the manuscript (probably thirty-one folios in total) has gone missing prior to what is now folio 47. The ghost folios of Cantele's manuscript remind us that a lot can happen in the intervening centuries between the creation of a practical manuscript and its preservation in a modern library—and, that

Table 1.1. Practical manuscripts made from parchment, paper, or parchment and paper, arranged by approximate date of composition

Approx. date of manuscript	Parchment	Paper	Parchment & paper	Total # of MSS
Pre-1400	4	0	0	4
1400–1449	50	25	11	86
1450–1500	13	71	8	92

deciphering the material clues in a manuscript requires some guesswork. Because the folios that remain in Cantele's manuscript were all copied by the same scribe, it's likely that the missing folios were, too. But, barring discovery of those long-lost pages, we can never know for sure.

At 165 (original) folios, Cantele's manuscript required more than forty sheets of chancery-sized paper (the most common size of paper imported into England), which probably cost somewhere around a shilling in total. Upon procurement of these forty or so sheets, the scribe would have folded the large rectangles of paper in half, bringing the shorter sides together, and then in half once more to form four leaves connected by a fold at the top. This fold would then be cut to form two bifolia of eight quarto-sized pages.[39] Most practical manuscripts are in this "quarto" format, meaning that each page is one-fourth the size of the original sheet of paper or parchment. A few are in smaller "octavo" format, with an additional fold to form eight small leaves. Once folded and cut, bifolia made from paper or parchment were nested together to form a quire, a small grouping of pages that a scribe could work with easily as he began copying a text. Slightly thicker sheets of parchment were often grouped into quires of three to four bifolia, or six to eight leaves, whereas in paper, five to six bifolia (ten to twelve leaves) were common. To prepare each quire of paper for writing, Cantele's scribe used a knife, awl, or spiked wheel to prick the outside vertical edges of each page. The small holes left by these instruments at regular intervals (still visible on Cantele's manuscript) served as guide marks to keep lines of text running straight across the page. After pricking the pages, the scribe next took a lead point and a straight edge and drew faint lines to mark out the central writing space of each page (known as the "text block"), leaving wide margins above and below and at the outside edges to allow for marginal notes.

Only after folding, cutting, pricking, and ruling several quires of paper could this scribe begin to copy the medical contents that fill the manuscript, the most painstaking part of his job. Yet, it was less tedious than it could have been: the scribe responsible for Cantele's manuscript employed a cursive script, which made writing much faster. If he had been working even a century earlier, in the 1350s, he might have used a "book hand" to write out Cantele's medical recipes—a style of handwriting composed of regular, distinct letterforms, each of which required multiple strokes of the quill pen. For a long time, scribes' professional identities were defined by the style of script they used. Scribes who copied manuscripts used "book hand" because it was more legible, while scribes who copied legal documents used cursive because it was so much faster to write when taking dictation. By the fifteenth century, however, these distinctions had dissolved. Nearly every professional writer in England had adopted cursive, and that alone probably cut the cost of Cantele's manuscript considerably.[40] Time was money, then as now. The use of a faster script meant a less expensive book.[41]

Not surprisingly, most of the practical manuscripts examined in this book were written in a cursive hand: either Anglicana, developed in thirteenth-century English courts, or Secretary, a handwriting style that originated on the continent and was adopted in England in the later fifteenth century. Generally speaking, Anglicana and more formal book hands predominate in practical books of the first half of the fifteenth century (including Peter Cantele's), while Secretary begins to appear regularly in practical manuscripts after about 1450. Secretary hand would remain popular throughout the 1500s in England, though it underwent extensive changes to certain letters, like *h, s,* and *e.* These changes in handwriting style make it possible to roughly date the (mostly undated) manuscripts I've analyzed for this book, and to confidently assert in chapter 7 that it was sixteenth- and even seventeenth-century readers who added many of the notes we find in their margins.

With a steady hand and a ready supply of iron-gall ink to fill the quill pen he cut for himself from the feather of a goose, the scribe responsible for Sloane MS 1764 wrote somewhere around 330 front-and-back pages of text in a neat Anglicana cursive, leaving blank spaces here and there for a limner to add a large red initial like that visible in figure 1.1. Sensibly, he broke the work up into chunks, copying out one group of texts at a time and then passing that section off for decoration. Folios 1 through 30, filled with a plague treatise, a section of an herbal, and numerous medical recipes, were one unit. The seventy or so folios containing the entirety of the *Circa instans* herbal were another. Both of these units got the full decorative treatment from the limner, who added red initials and filigree penwork where appropriate. But folios 30 through 47, once containing the missing ghost folios as well as directions for distilling waters and preparing oils and ointments, were either overlooked or hurriedly completed. That section of the manuscript never made it to the limner. Blank spaces remain where bright initials might have been, testifying to the collaborative labor of artisans who lived nearly six hundred years ago.

Though clearly it didn't always go off without a hitch, the piecemeal production of manuscripts was quite common among London stationers, who often hired multiple scribes to work on smaller units of short texts. These "booklets" of a few quires, each one a self-contained unit of text, might be combined into larger manuscripts, or sold as is to circulate as small, unbound pamphlets.[42] Practical manuscripts lent themselves quite readily to "booklet" production because recipe collections or herbals were often fairly short, and because they could be combined in a variety of ways to meet the needs of individual readers.[43] Linda Ehrsam Voigts has identified a cluster of fifteenth-century scientific manuscripts that seem to have been created from booklets prepared by a single scribe or workshop that perhaps specialized in publishing natural knowledge.[44] Though the evidence is still debated among manuscript scholars, London's scribes may even have prepared "booklets" on speculation to be sold ready-made to buyers.[45]

Once the scribe and limner had finished with the text and illustrations—or, at least, once they thought they had—it was time for the stationer to retrieve the manuscript and take it to the binder, who then stitched through the center fold of every one of sixteen or more quires, sewing each one onto cords or strips of parchment or leather placed at intervals down the spine of the manuscript.[46] Following that, it was up to Cantele (or whoever commissioned Sloane MS 1764) to determine what sort of cover he wanted for his manuscript, if any at all. Wear and tear on the manuscript's first pages (where Cantele jotted new recipes and notes) suggests it may not have had a cover until a sixteenth-century owner wrapped the manuscript in a piece of old parchment. This sort of recycled cover was probably common among practical manuscripts, though they do not often survive.[47] Only the most expensive manuscripts or printed books were bound with leather-wrapped boards and brass clasps.

After all was said and done—the texts copied and the book stitched up and bound—what did all this labor cost? It's hard to say. In 1468, John Paston II was charged twenty pence by his scribe, William Ebesham, for "writing of the little book physic."[48] By contrast, John Green, a customs agent for the port of Bristol, charged more than one hundred shillings for the copying of twenty books in the early sixteenth century—a rate that he recorded in the form of a receipt in his own practical manuscript.[49] Yet Green's fee of one hundred shillings was equivalent to a year's pay (not including room and board) for Thomas Elbrigge, a manuscript illuminator working in London in 1425.[50] The range of fees recorded by these scribes highlights the difficulty historians face when tasked with valuing a manuscript. Just a handful of records detailing the itemized costs of manuscript production survive from later medieval England, most of which relate to lavish manuscripts created for religious institutions or universities (in other words, very unlike most practical books).[51] Without scribal accounts, we can only extrapolate from the valuations made by manuscript owners themselves. For example, Thomas Spenser, brought before the Court of Common Pleas in London in 1405, estimated the value of a "book called 'Troylus'" (perhaps Chaucer's *Troilus and Criseyde*) at twenty shillings.[52] Thomas Marlebrugh brought John Yorke before the same court in 1425 to enforce payment for a book called "Policonicon" (probably John Trevisa's Middle English translation of Ranulph Higden's *Polychronicon*), which he valued at forty shillings.[53] However, the very same text was valued by a different claimant in another case before the Court of Common Pleas at one hundred shillings, more than double the price.[54] The value of a manuscript varied considerably according to whether it was elaborately illustrated or plain text, whether it was made on parchment or paper, or whether it was covered in limp vellum or gilded leather-wrapped boards. Manuscripts were always luxury items, but just *how* luxurious depended on the quality of their materials and the labor involved to produce them.

We don't know what the original commissioner of Sloane MS 1764 paid for the manuscript, but the scribe who copied those 330 pages of medical texts did a fine and careful job that probably took him weeks or perhaps even months. That labor must have cost quite a lot, which explains why many in the fifteenth century chose to write their own manuscripts rather than pay a skilled artisan to do so—so many, in fact, that the bibliographer Curt Bühler described the fifteenth century as a period defined by an "every man his own scribe movement."[55] By no means was *every* man his own scribe in fifteenth-century England, but the spread of English literacy and the availability of cheaper book-making materials transformed what had once been a skill exclusive to clerics, and later to professional book artisans, into a practice enjoyed by country gentlemen, medical practitioners, or even a few simple farmers.[56] We will meet two amateur scribes in later chapters: Nicholas Neesbett, a mid-fifteenth-century Yorkshireman who carefully curated medical recipes to compose a homemade surgical manual for an acquaintance, and William Aderston, a London surgeon who compiled and copied out hundreds of medical, craft-related, astrological, and veterinary recipes in a manuscript intended for his own personal use.[57] For both Neesbett and Aderston and the many other anonymous amateur writers who composed practical manuscripts, a book of medical recipes or prognostications required no visit to the stationer. All that was needed was a few sheets of paper, a quill pen, and some iron gall ink.

Cantele wasn't the maker of his medical manuscript, but he, too, was an amateur scribe. He took up a pen and jotted notes at the front of his manuscript and then left annotations in the manuscript's index, highlighting recipes that were especially valuable. Even if he wasn't the figure responsible for commissioning a professional artisan to work for weeks on end copying the medical recipes, plague treatise, and herbal that now fill Sloane MS 1764, it is clear that these texts meant something to Cantele. We cannot now ask Cantele why he valued certain recipes in his manuscript more than others, but we can learn a bit more about the knowledge he referenced when he or others in his parish were sick or injured. To do that, we must travel from a London stationer's workshop back even further in time, to follow the transmission of natural knowledge as it wound its way through monastic libraries and university lecture halls and into the pages of Cantele's practical book.

Translating Natural Knowledge

Peter Cantele was hardly the first English churchman to own a medical manuscript. In early medieval England, as elsewhere in Europe, Benedictine monks collected and preserved bits and pieces of ancient medical writing in manuscripts they created in monastic scriptoria.[58] They did so because they were beholden to the Benedictine *Rule*, which mandated that they should care for the sick and wounded. In monastic kitchens and infirmaries, they brewed plants and animal byproducts into medicaments; they prescribed regimens of food and drink to maintain health; and at set times of the year, they cut patients' veins to release blood and restore balance to the body. These practices were the foundation of medieval medicine, and they were as relevant in Peter Cantele's day as they had been in early medieval England. And yet, even though early medieval English monks produced the earliest vernacular medical writing in Europe, and even though little had changed in the day-to-day care of sick bodies over the intervening centuries, the medicine in Cantele's manuscript has little in common with that in Old English collections.[59] Only a few charms and a few more prognostications from early medieval England remained in circulation in the fifteenth century.[60] Instead, Cantele's manuscript is the product of an enormous period of cross-cultural exchange that began around the turn of the twelfth century when Latin Europeans were introduced to scholarship from the Islamic world. In translations of Arabic manuscripts, they rediscovered the medical theory of Galen and Hippocrates and became newly acquainted with the insights of their medieval commentators: Avicenna (Ibn Sīnā), Rhazes (al-Rāzī), and Albucasis (Al-Zahrawi), among many others.[61] Through their writings, medieval physicians learned that the human body could be understood holistically, its different states monitored through observation of pulse and urine or through a close reading of the stars.

Over the course of the twelfth and thirteenth centuries, Latin clerics inspired by Islamic scholarship produced a whole new corpus of medical texts—texts that form the basis for much of the vernacular medicine collected in fifteenth-century practical manuscripts.[62] Take, for example, the longest entry in Peter Cantele's book, the *Circa instans* herbal, originally composed in Latin in twelfth-century Salerno, Italy, the center of medical education in Latin Europe at that time.[63] Its author, Matthaeus Platearius, composed the text as a guide to the therapeutic properties of 270 herbs, plants, stones, and animal products. At the time of its composition, the *Circa instans* was a cutting-edge medical text. Only a few decades before Platearius authored it, a north African monk named Constantine had arrived in southern Italy with a trove of Arabic manuscripts, one of which contained a treatise on the Galenic theory of pharmacological "degrees."[64] According to this theory, every "simple" ingredient like an herb or plant had some "degree" of hotness, coldness, wetness, or dryness. Prepared correctly, each of these "simples" could restore the body by balancing the four humors (yellow bile, black bile, phlegm, and blood), which were also believed to have qualities of hotness, coldness, moistness, or dryness, owing to their relationship to the four natural elements (fire, earth, water, and air). Treating illness thus became a matter of identifying the simple that would counteract, to a lesser or greater "degree," the complexional imbalance in the patient.[65]

We know that Cantele's manuscript contains a Middle English translation of a text composed in twelfth-century Italy, based on a manuscript by an eleventh-century North African monk, founded on the doctrines of a second-century Roman physician. But Cantele may not have known any of that history or understood the theory of pharmacological degrees. He didn't need to. The point of Platearius's *Circa instans* was to distill the complex workings of the natural world into a text that was useful and easily referenced.[66] This was the premise of all herbals found in Middle English manuscripts, some of which were translations of older Latin texts—like the eleventh-century verse herbal *On the powers of herbs* (*De viribus herbarum*), also briefly excerpted in Cantele's manuscript—and some of which were more recent compositions.[67] Two of the most popular herbals in practical manuscripts were authored in Middle English: the *Agnus castus*, so named for the first herb in the collection, and another that began with the incipit "Here men may see the virtues of herbs, which been hot and which been cold."[68]

As table 1.2 illustrates, herbals were enormously popular among fifteenth-century compilers of natural knowledge. Even more popular, however, were medical recipes, found in 160 of the 182 practical manuscripts in my corpus, including Peter Cantele's. As a genre, medical recipes long predate the twelfth-century Renaissance. Early medieval monks had been capable healers, and their manuscripts were filled with practical texts

Table 1.2. The most common categories and texts found in my sample of 182 fifteenth-century practical manuscripts, omitting texts that only appear in one manuscript witness

Medical recipes, 160 MSS total, including:	
"The men that will of lechecraft lere"	13
"Good leeches hath drawen out of books"	11
Reference to *Thesaurus pauperum*	3
Gilbertus Anglicus *Compenium medicinae*	3
Liber de diversis medicinis	3
Herbals, 73 MSS total, including:	
Agnus castus	19
"Here may men see the virtues of herbs"	14
Macer's *De viribus herbarum*	8
Virtues of Betony	11
Virtues of Rosemary	11
Circa instans	8
Latin/English herbal glossary	8
Of herbs twenty-three	5
Prognostications, 75 MSS total, including:	
Canicular/perilous days	28
Whether a sick man will live or die	24
Lunary	13
Dominical letter	12
Wise Book of Astronomy and Philosophy	9
Brontology	7
Nativities	7
Christmas Day	5
Uroscopy, 47 MSS total, including:	
Letter of Ipocras	13
Twenty Jordans	7
Dome of Urynes	5
Judicial of Urines	4
Twenty Colors of Urines	3
Walter of Agilon	2
Surgeries, 47 MSS total, including:	
Various surgical recipes	29
Guy de Chauliac	5
Lanfranc of Milan	5
John Arderne on Fistula	3
Phlebotomy, 36 MSS total	
Dietary, 31 MSS total, including:	
"Galen the good leech teacheth us . . ."	9
"Hippocras the good leche of mete . . ."	2
Queen Isabella's dietary	2
Animal husbandry, 27 MSS total, including:	
Miscellaneous veterinary recipes	24
"Book on the Conditions of Horses"	5
The Booke of Marchalse	5
Plague treatise, 13 MSS total, including:	
John of Burgundy	12
Hunting & fishing, 13 MSS total, including:	
Fishing with an angle	5
"Book of Saint Albans"	2
Coney catching	2
Planting or grafting, 7 MSS total, including:	
Godfridus Super Palladium	5
Nicholas Bollard	4
Craft recipes, 28 MSS total, including:	
Ink, color, or book making	18
Textiles	13
Metallurgical	3

that described how to prepare medicines or heal fevers. And yet, within early medieval monastic education, medical knowledge based on skill or hands-on practice was afforded very little status.[69] A true scholar devoted himself to the seven *artes liberales* (grammar, rhetoric, logic, arithmetic, music, geometry, and astronomy), a curriculum with a suitably classical pedigree. It wasn't until the twelfth century, when Arabic science became available in Latin translation, that medieval clerics began to see that medicine had something in common with these other fields of knowledge—that is, that it had a theoretical basis.[70] In the newly formed universities of the later twelfth and thirteenth centuries, medical faculty embraced a view, first espoused by Galen and reiterated in important works of Arabic medicine, that the art of medicine had two distinct subcategories: internal medicine and operation, or, as one Middle English translator rendered it, "theoryk and practyk."[71] Internal medicine, based on a deep understanding of the imperceptible but nonetheless entirely logical system of influences and elements that governed the human body, would fall under the purview of the university. At Bologna, Paris, Padua, and Montpellier—universities renowned for their medical training—students listened to lectures and pored over textual commentaries to earn their doctorate in the medical arts. The rest of the practitioners of medieval Europe, including surgeons, leeches, and the amateur village healer, could certainly master the techniques of operation, but these practitioners would have none of the learned status of the physician and no grounding in theoretical principles.

Or at least, that was how it was supposed to be. Very quickly, university-educated practitioners saw a need for texts that brought theoretical principles to bear on healing practice. The first to compose new *practica* were the university-educated surgeons of Italy. Their Latin surgical manuals incorporated elements of classical and Arabic surgical theory while also stressing experience as a valuable source of knowledge.[72] Lanfranc of Milan, who trained in surgery in Bologna and authored one of these surgical manuals, brought Italian theoretical surgery to the University of Paris. Later, Guy de Chauliac, a younger contemporary of Lanfranc's trained at Paris, used his mastery of surgical theory to compose perhaps the most widely read surgical manual of the fifteenth century—a manual that probably inspired the several recipes for ointments and "entretes" that appear in Cantele's manuscript.[73]

But it wasn't just surgeons who wrote recipe collections; physicians did, too. Gilbert the Englishman, who may have been educated in Italy, composed his *Compendium of medicine* around the year 1230. Organized by ailment from top to toe, it was the first all-purpose, general guidebook to medical practice (and not just surgical practice) to include both Gilbert's own empirical observations and the learned theory of ancient and Arabic scholars.[74] Just a few decades later, Peter of Spain, an expert logician and would-be Pope, composed his own all-purpose guide to medicine known as the *Treasury of the Poor* (*Thesaurus pauperum*), which again offered both theoretical explanations and practical treatments for a huge array of ailments. Peter's collection was perhaps the most popular general medical guidebook of the later Middle Ages, which probably explains why the recipes that fill folios 7 through 29 of Cantele's manuscript—recipes that have nothing to do with Peter of Spain's Latin originals—invoke his title "*Thesaurus pauperum*" and proceed in order from top to toe, following the model set by Gilbert the Englishman two centuries earlier.[75]

By Cantele's day, the surgeries of Lanfranc of Milan and Guy de Chauliac and the *Compendium* of Gilbert the Englishman had all been translated into Middle English. Once translated into the vernacular, their recipes could be easily excerpted, adapted, and repackaged by scribes and compilers.[76] For that reason, it is not easy to unspool the thread of transmission that connects Chauliac's massive surgical manual with Cantele's brief section of surgical recipes, or that which connects Peter of Spain's *Thesaurus pauperum* with the collection that bears the same name in Cantele's book. As recipe texts moved from manuscript to manuscript, they underwent incremental changes thanks to scribal error or omission, or even as a result of compilers' experience.[77] And yet, there can be no question that Cantele's recipes—those few related to ointments and "entretes" and the larger collection of medical recipes organized by ailment—are an outgrowth of a twelfth- and thirteenth-century revival in the composition of *practica*, manuals intended to bring theory to bear on the hands-on practices of healing. Indeed, most of the hundreds of recipes that fill fifteenth-century practical manuscripts riff on established methods, techniques, and ingredients first elaborated in Latin by thirteenth- and fourteenth-century practitioners. That is why historian Peter Murray Jones has enjoined scholars "to assume that every vernacular text of medicine and science, for which conclusive evidence to the contrary does not exist, was originally translated out of Latin."[78]

Assuming is one thing; proving it is another. Identifying the sources for Middle English recipes is extremely difficult because most recipes circulated without attribution or, just as often, with entirely spurious attributions. The mistitling of Peter Cantele's recipe collection as "*Thesaurus pauperum*" looks like a minor scribal error when compared with the wildly ambitious incipit of another popular Middle English recipe collection: "Here beginneth medicines that good leeches have found and drawn out of books, that is to say Galen and Asclepius and Hippocrates, for those were the best leeches in the world."[79] This recipe collection, found in nearly a dozen manuscripts, is just one of numerous Middle English medical texts to claim direct transmission from the ancients. Another, a popular uroscopy treatise that described how to interpret the color of a patient's urine to diagnose illness, presented itself as a copy of a book that "Hippocrates . . . sent to the emperor Caesar."[80] Two more popular treatises, both of them dietaries advising on which foods and drinks to consume throughout the year, were attributed to "Galen the good leech" and "Hippocrates the good leech," respectively.[81] Though these texts all prominently cite ancient authorities, it is important to recognize that these Middle English treatises have almost nothing to do with authentic Galenic or Hippocratic writings. In Middle English collections, these famous names were markers of authority. If "Galen the good leech" himself directed a reader to "rise up early from thy bed and early eat and drink and use hot meats" in the month of May, a reader could be sure the text was a genuine reflection of sound medical principles.[82] And indeed, Galen had insisted on the regulation of "non-naturals" like food and drink and sleep to maintain health. The ideas in this dietary are Galenic, but the Middle English text that bears his name is of no relation to Galen's authentic writings on hygiene.

To borrow a phrase from the historian Faith Wallis: fifteenth-century compilers cared a great deal about authority, but very little about authenticity.[83] In part, this ambivalence stemmed from the fact that the burgeoning vernacular literary culture of fifteenth-century England did not yet have a notion of authorship akin to our modern one. In medieval Europe, authors were called such because they were "authorities" and authority was synonymous with Latin, the language of the ancients and of the church.[84] Even learned writers like Guy de Chauliac were careful to frame their work as no more than an extension of the writings of the ancients, going so far as to invoke the medieval cliché that he was just a boy, standing on the shoulders of giants.[85] What was to stop vernacular writers from claiming a similar relationship to ancient authorities by slapping the name Hippocrates or Galen on a recipe collection—especially since all authoritative medical knowledge was presumed to derive from those two great men anyway?

The recipes and instructions Cantele added to the first two paper flyleaves of his manuscript show him to be a participant in a long tradition of playing fast and loose with authorities, selecting from established texts and combining them into new sources of knowledge. Yet, the bulk of Cantele's manuscript is better organized than many other exempla, likely because it was copied by a trained scribe who must have referenced an orderly copy-text. Many practical manuscripts composed by amateur writers feature seemingly unrelated texts juxtaposed alongside one another, reflecting their compilers' interests rather than a well-ordered plan to produce a compendium or commentary. These compilers don't seem to have recognized anything unusual about putting recipes for headache or fever alongside recipes to "handle a serpent" or "for biting of a toad," nor were they bothered by redundancy or repetition.[86] A single practical manuscript often contains multiple iterations of the same recipe or prognostication. If we can assume that fifteenth-century people were not bitten by serpents and toads all that often, then a likely explanation for the presence of these unusual recipes and all the redundancy and repetition in practical manuscripts is that fifteenth-century compilers tended to copy wholesale whatever natural knowledge was available to them. Acquiring and possessing knowledge was often more important than organizing or analyzing that knowledge.

Historians have reason to be thankful that fifteenth-century compilers weren't more selective. The wholesale copying of entire collections has enabled us to reconstruct "discourse communities" of textual exchange simply by comparing the contents of practical manuscripts.[87] Clusters of the same texts often appear in manuscripts that share similarities in terms of dialect or style, suggesting that manuscripts passed from hand to hand among friends and neighbors. Only by comparing Cantele's manuscript to other Middle English practical books can we tell that the *Book on the Condition of Horses* (*Liber conditionis equorum*) added to the front flyleaves of his medical manuscript was actually excerpted from a much longer set of instructions for equine care that appears in at least four other manuscripts in my corpus.[88] But we have no idea where he encountered these instructions, nor do we have any idea where those instructions originated. They may have been adapted from one of a number of estate management and husbandry-related texts composed in England in the thirteenth century for the French-speaking nobility.[89] If so, then Cantele's manuscript carries forward traces of another, earlier wave of Latin-to-vernacular translation in England, one reflected in the huge variety of Anglo-Norman practical texts that survive from the thirteenth and fourteenth centuries.[90] A number of these Anglo-Norman practical texts were later translated into Middle English, and many fill the pages of fifteenth-century manuscripts.[91] But did these Anglo-Norman texts on husbandry inspire Cantele's *Book on the Condition of Horses*? What about the recipe for "size for flocks and gold" that Cantele added immediately after these horse-related instructions? Did that recipe originate in Latin or Anglo-Norman French collections of craft instruction?[92] Or, did Cantele learn how to make this glue for flocking gold leaf from an acquaintance, someone who showed him the technique or shared the ingredients in conversation with Cantele?

These questions illustrate why identifying a textual origin for the recipes in a practical manuscript is not the same thing as identifying its immediate source. Even though most recipe knowledge in fifteenth-century collections likely originated in a written source, those written sources could then be transmitted orally. Cantele's recipe for glue may have originated in a Latin collection, but he may have learned it from a friend. The same may be true of a recipe for treating bladder stones in Bodleian MS Ashmole 1443, supposedly "proved for truth by one John Edward Brykyndynmaker the which Master Geram the physician taught him in Saint Margaret in Lothbury."[93] Though this recipe's disclaimer indicates oral transmission, Master Geram's recipe includes standard ingredients and techniques common to many Middle English recipes for expelling unwanted matter from the body, be it a bladder stone or a woman's afterbirth, making it difficult now to ascertain the extent to which Geram's recipe was an original, or simply another riff on a recipe he had read in a book.[94] In the same way that literary scholars have noted the mutability of vernacular verse as it moved from manuscript to manuscript and through oral exchange, so too might recipes transform as they were exchanged both orally and textually.[95] No wonder, then, that we cannot definitively trace the origins of the recipes in Peter Cantele's manuscript, or in most of the other 159 practical manuscripts in my corpus to feature recipe texts.[96]

Fortunately, there is one final medical text in Cantele's manuscript that is easily identifiable: "a noble treatise made of a good physician John of Bordeaux for medicines against the pestilence."[97] John of Burgundy, a physician from Liège, composed the plague treatise in Cantele's manuscript in 1365, just as a second wave of illness hit an already decimated Europe. The original, longer version of the treatise includes John's explanation that he composed the work to address a gap in the existing medical literature. He argued that the ancient authorities had simply never dealt with a disease like the Black Death. Given that John had experience treating the pestilence, he felt obligated to write a practical guide detailing the causes of illness (evil vapors); outlining steps for prevention (eating a good diet and avoiding activities that would open the pores); and offering treatments (involving both a better diet and bloodletting). The work became the most widely copied plague treatise of the fifteenth century.[98]

John of Burgundy's experience of the Black Death revealed the insufficiency of ancient medicine. He sought to address that insufficiency by incorporating his experiences into a new guide to practical healthcare.

In that respect, John of Burgundy shared much in common with other important figures of Renaissance medicine, like John Arderne, surgeon to English royalty, whose Latin surgery known as the *Practica* incorporated Islamic surgical practices alongside descriptions of techniques he had developed treating battlefield wounds. The same instinct is evident in the treatises and case histories of the fifteenth-century Italian physician Michele Savonarola and the French physician Jacques Despars. As Danielle Jacquart has noted, these learned men, trained in scholastic medicine at the best universities, nevertheless recognized that tradition could be amended in light of experience—and, moreover, that it *should* be.[99] Because John of Burgundy, Arderne, Savonarola, and Despars were all educated men (though Arderne wasn't university trained), each could read Latin and had a deep knowledge of the power and influence of texts. It is therefore not at all surprising that each took the time to compose *new* texts to challenge old ideas: a plague treatise, a surgical manual, case histories, and *practica*. Yet many who commissioned or created vernacular practical manuscripts over the course of the fifteenth century did not have Arderne's or Savonarola's experience reading Latin medical books. Their collections of recipes and herbal lore were their ticket into this world of learned tradition—even if that tradition had been diluted by haphazard transmission and translation. And when itinerant practitioners, artisans, merchants, and farmers gained entry into that world, they very quickly intuited what Arderne and Savonarola and John of Burgundy knew: that books could be tools for fitting ancient and authoritative knowledge to suit their own needs. They, too, could amend tradition to reflect their own experiences.

Moreover, where John of Burgundy, Savonarola, and Despars—all professors—had to fit themselves within the rigid hierarchical structure of their universities, and where John Arderne had to cater to the needs of powerful patrons in the English court, the merchants and artisans and country householders who owned practical manuscripts had no such concerns. They were informally trained practitioners, and fifteenth-century England simply didn't have the formal institutions of medical licensure or powerful medical universities to shape or constrain their experience of natural knowledge. Whereas continental European cities had civic bodies to license practitioners, England had nothing of the kind in the fifteenth century. Only briefly, in the 1420s, did the physicians and surgeons of London experiment with creating a regulatory body to organize their trade within the city. It quickly failed, and no other attempt was made until 1518, when the College of Physicians was founded.[100] Nor did England's universities wield much authority over medical practice. Whereas in fifteenth-century Paris, the powerful medical faculty at the university licensed medical practitioners, the faculties at Oxford and Cambridge had no such power. Over the entirety of the fifteenth century, Cambridge and Oxford produced fewer than seventy-five MDs.[101]

When compared to Italy, or even to France or Germany, the institutions of English medicine in the fifteenth century seem outdated, even backward.[102] Looked at another way, however, the same societal factors that depressed learned medicine in England were a boon to the amateur compiler or owner of a practical manuscript. Because expertise was thin on the ground in the fifteenth century—or, at least, what experts there were lacked official status—when learned texts (or derivations from them) were made available in Middle English, ordinary English readers felt very free to pick and choose among them, and in doing so, to take ownership over that once-exclusive knowledge. As we have seen in this chapter, Middle English practical manuscripts contained the residues of "expert" medicine first developed in the twelfth and thirteenth centuries, with much of the rigid formality of Latin authority stripped away. An amateur writer like Peter Cantele could insert his own recipe for book glue and a copy of instructions for selecting a horse in the same manuscript with a complete Middle English translation of a twelfth-century herbal. Textual tradition and authority clearly mattered to Peter Cantele, but then again, so did his own experiences. The marvelous thing about a practical book was that it could accommodate both.

Conclusion

In this chapter, we have visited Peter Cantele's England and imagined him traveling to a stationer's shop. We have learned how and from what a practical manuscript was made, and in the process, discovered why this genre of book was so popular in fifteenth-century England. These manuscripts of natural knowledge were owned by members of the well-to-do middling sort—artisans and merchants and village priests and surgeons—who heartily embraced a new vernacular culture of book ownership and knowledge exchange. By commissioning or creating practical manuscripts filled with medical recipes, prognostications, and herbals, these men (and a few women) were joining an old and learned conversation, whether they knew it or not.

In the second half of the chapter, we listened in on this learned conversation as we traced the origins of the texts that fill Cantele's manuscript from twelfth-century Salerno to fourteenth-century Liège. The links we identified between classical, Arabic, scholastic, and Middle English texts illustrate that there was nothing about vernacular natural knowledge that set it in opposition to learned discourse. Both emerged from the same sources and dealt in the same theories of the body, of natural matter, and of the heavens. Peter Cantele encountered twelfth-century pharmacology and thirteenth-century medical recipes and fourteenth-century plague remedies in the pages of a relatively unremarkable fifteenth-century Middle English manuscript. And so did many other fifteenth-century readers. In the following chapters, we will meet these other readers who made an old and learned tradition their own, shaping it to meet their needs and expanding upon it to arrive at new methods for making sense of their bodies and the world around them.

✸ 2 ✸

Picturing the Natural World

Richard Skires made his living trawling the ships coming and going from the port of King's Lynn, Norfolk. Shortly after the coronation of Henry VI in 1422, Skires had begun serving as official "searcher" for the Customs Office at King's Lynn. His job was to certify that all goods and persons on ships either leaving from or arriving to the port were duly taxed according to the various rates imposed by the English Crown, and occasionally, to seize goods that were either prohibited from export or had been smuggled into England without the necessary custom.[1] In May of 1429, Skires made such a seizure when he uncovered a barrel full of "hardeware and haberdashery" that had been smuggled into England by "a certain alien merchant."[2] This barrel full of odds and ends represented quite a lot of work for Skires—carefully itemizing its contents and recording them in his ledger—with very little hope of return. Searchers like Skires made their money by selling the goods they had confiscated, but the contents of this barrel had been ruined by saltwater. We have all the more reason to be thankful, then, that Skires was nothing if not a committed bureaucrat. He dutifully carried out his job, examining and recording the dozens of tiny items in that barrel: seven dozen thimbles, eleven dozen curtain rings, six dozen harp strings, a dozen inkhorns, two dozen writing tablets, six folios of paper, and—most importantly for this chapter—one dozen "lewdecalenders."[3]

Besides valuing them at just six pence in total, Skires didn't go into much detail about these dozen lewdecalenders. Manuscript scholars, however, believe that Skires was describing manuscripts that looked like the one pictured below in figure 2.1.[4] Made of a single leaf of parchment, folded once lengthwise and then at least six times accordion style, this simple calendar employs icons of the saints, astrological symbols, and illustrations of the "labors of the month" to impart the rhythms of the liturgical year—a rhythm that also dictated the patterns of everyday life.

Eamon Duffy has described how the rituals of the liturgical year in England "provided a means of ordering and perhaps also of negotiating social relations."[5] For example, a tenant could expect to pay his annual rent on Michaelmas, celebrated on 29 September, but he might also expect to receive small gifts from his landlord on the Feast of the Annunciation of the Virgin, celebrated on March 25, which in medieval England marked the beginning of the new year. Major feast days were always reserved as rest days from labor, but minor local saints might be celebrated with fervor too, their celebrations a welcome respite from the toil of agricultural or manual labor.

Lewdecalendars presented information about these important social rituals succinctly, in images. The order of feast days represented by saints' icons is legible on one side of the calendar (fig. 2.1), while the other side displays each month's agricultural labor, astrological sign, and hours of daylight expected every month (represented by the red lines within the circle in fig. 2.2).[6] The saints depicted on this calendar (and two others just like it that survive from the early fifteenth century) tell us that this manuscript was created by Flemish artisans, who appear to have produced these simple calendars for export to an English clientele.[7] Richard Skires clearly stumbled upon evidence of this Flemish export trade in 1429. And when he did, he described the manuscripts in that salt-soaked barrel as "lewdecalendars" because in Middle English, the adjective "lewde" meant uneducated or unlettered.[8] Skires believed that calendars composed of pictures were the sort of book that might be used by readers without advanced literacy. They were, in other words, calendars for "lewde" people.

The impulse to read pictures as a means of marking time lies at the heart of this chapter, which examines eight practical manuscripts composed of a variety of images related to religion, medicine, and the environment. These practical picture books borrowed from several pictorial traditions current in later medieval England, combining diagrams of the human body with saints' icons or depictions of the "labors of the month" with images of solar and lunar eclipses. And in addition to these combinations of well-known images, all eight contain at least one of a set of three pictorial prognostications unattested outside England, intended to help a reader forecast upcoming weather events, crop yields, illnesses, and auspicious days for travel or bloodletting. Each of these was a pictorial translation of a prognostication that had circulated for centuries in Old English or Latin text. It wasn't until the last quarter of the fourteenth century, however, at the same time that the earliest Middle English translations of medical treatises began to circulate among English readers, that a Benedictine artisan was inspired to transform these textual predictions into pictures. The

FIGURE 2.1. Detail of pictorial liturgical calendar with illustrations of saints' icons to mark the feast days for the months of January, February, March, and April. © The British Library Board, Additional MS 70517, recto.

FIGURE 2.2. Detail of pictorial liturgical calendar with "labors of the month" illustrations and zodiac signs to indicate the months of January, February, March, and April. © The British Library Board, Additional MS 70517, verso.

aim of this chapter is to understand why he was so inspired and to explore what these unusual pictorial arrangements can tell us about reading in fifteenth-century England—whether that reading involved manuscripts, the human body, or the whole of the natural world.

We will first inhabit the world of the late fourteenth-century Benedictine who created the earliest of these eight pictorial practical manuscripts, Bodleian MS Rawlinson D.939, to reconstruct the influences that inspired his work. We will then examine that manuscript's contents, including its three pictorial prognostications. Finally, we will follow these prognostications as they circulated in fifteenth-century England, learning in the process how a visual language was passed from teacher to student and reader to reader, most of them probably members of Benedictine communities. Admittedly, the eight practical manuscripts examined in this chapter represent just a small fraction of the manuscripts in my corpus, most of which are filled with recipes or instructions in Middle English text rather than images. Yet, a fifteenth-century reader of pictorial prognostications would understand why these manuscripts warrant this closer look: it is only through attention to anomalies, instances where the pattern breaks or shifts, that we begin to understand the order of the whole.[9] If we can understand why fifteenth-century Benedictines valued the images in these manuscripts, and how they used them, we may be able to reconstruct how they read the signs all around them in the natural world: lunar phases, weather cycles, patterns of sickness or crop failure, even cycles of time. Indeed, as I argue in this chapter, the visual language of prognostication in these manuscripts referenced and reinforced the inscrutable but no less meaningful patterns of the natural world and the human body. These simple books of numbers and pictures were premised on an understanding that the natural world itself was legible if one could accurately interpret signs—from the pictorial ones in their almanacs to the ones manifest in the human body or on display in the starry night sky.

Images and End Times

Sometime around 1389, an anonymous English artisan began to devise an all-purpose practical book to guide him through the rhythms of his day. That manuscript is now Bodleian MS Rawlinson D.939, composed on six long, rectangular parchment leaves of irregular size, each of which folds once lengthwise and then several times accordion style to form a small rectangle of parchment. Like single-sheet lewdecalendars that folded in the same manner, the Rawlinson calendar would have been perfect for tying up with a leather thong and wearing hung from a belt, or for slipping into the pocket of a garment. The panels of parchment created by these folds are filled with images that convey information: about the order of the church year, about biblical suffering and redemption, about the movement of heavenly bodies and their effects on ailing human ones and, most intriguingly, about the future. The artisan who created this calendar—the earliest of the eight surviving pictorial practical manuscripts examined in this chapter—may very well have birthed a new pictorial tradition and, more broadly, a new model for merging the practices of reading the world with those required to read the page.

Who was this artisan? Though his name is lost to us, the manuscript he created holds several clues to his identity. The second long leaf of the manuscript—almost a meter in length when fully unfolded—contains a pictorial liturgical calendar very much like the lewdecalendar shown in figure 2.1, with icons of the saints marking the order of the ritual year. Included in this calendar are several minor saints celebrated only in the region of Worcestershire, as well as St. Maurus, an acolyte of St. Benedict whose cult was only celebrated in Benedictine houses. We can infer, then, that the calendar originated in a Benedictine monastery around Worcester, and possibly at Evesham Abbey, which had a long tradition of innovation in calendrical manuscripts.[10] It is thus reasonable to assume that the manuscript was created by a Benedictine monk, trained in the scribal arts but somewhat less skilled at illustration. A Benedictine would have been accustomed to using manuscripts as guides to his day-to-day routines: Psalters and antiphoners and breviaries marked the passage of the monastic day, month, and year. Perhaps this monastic artisan thought to make a manuscript as a similar sort of guide to a life that melded religious observance with agricultural work and medical care. Yet this artisan did not follow the model established in Psalters or breviaries. Though he could read and write Middle English, Anglo-Norman French, and some Latin, he chose images as his medium for conveying the patterns of the world around him. Why?

The first clues to our Benedictine's intentions emerge from two full-length portraits on the manuscript's first and third leaves. The figure of "Harry the haywarde" stands with his dog, Talbat (shown in fig. 2.3), on what would be the cover of the Rawlinson calendar when folded, while "Peris the pyndar" stands alone on another panel of parchment that served as the cover for the last three leaves of the manuscript.[11] Some have suggested that Harry and Peris were the manuscript's intended users, and therefore that the Rawlinson calendar's images were designed to make the manuscript legible to two "lewde" laborers.[12] Though we cannot entirely rule out this possibility, it would be remarkable indeed if manorial labor-

FIGURE 2.3. Harry the Haywarde and his dog, Talbat, on the front panel of a folding pictorial calendar. The Bodleian Libraries, The University of Oxford, MS Rawlinson D.939, section 1r. CC-BY-NC 4.0.

ers like a hayward or pinder were literate enough (and wealthy enough) to commission and use such a manuscript. More to the point, however, a hayward's job was to mend enclosures to keep livestock from wandering astray, and a pinder's job was to gather up those stray livestock and return them to the flock. Their labor was, quite literally, the work of pastoral care. Indeed, the priest-as-hayward was a common trope in Middle English literature: he appears in John Wyclif's sermons and two of his treatises, as well as in William Langland's poem, *The vision of Piers Plowman.*[13] Harry's dog, Talbat, was also a stock character of Middle English verse, appearing in Geoffrey Chaucer's "The Nun's Priest's Tale."[14] Harry and Peris were thus probably not literal portraits of potential readers. Instead, each image is a visual allegory: together they speak to life on a Benedictine estate, where the work of tending fields and livestock was also the work of devotion. Our Benedictine understood that an image could convey these references all at once in a manner text could not replicate.

He knew as much because he lived in a world thick with images: heraldry banners trailed from the wagons of the well-to-do and wooden signs with painted emblems marked the entrances to shops throughout England. But perhaps nowhere was the language of imagery more developed than within the four walls of a medieval church, where icons of the saints might appear in stained glass or on murals, and where the symbols of Christ's passion were immediate evocations of the sacrifice at the heart of the Christian story. The English faithful, our Benedictine included, learned to parse this iconography while listening to sermons or stories drawn from the *Golden Legend*, the standard corpus of saints' lives in Latin Christendom.[15] They might look around their parish church at wall paintings, stained glass, or sculptures and "read" the stories of the saints through their emblems: Saint Katherine's sacrifice represented as the instrument of her torture, or the miracle of the Annunciation made visible as a peace lily passing between Gabriel and Mary (shown in the third row of fig. 2.1, at right).[16]

This rich visual language of sacrifice, suffering, and redemption had developed over centuries with encouragement from church authorities. Despite the second commandment's condemnation of graven images, from very early the church defended religious imagery as an important tool for inculcating belief among the laity. The sixth-century Pope Gregory the Great famously wrote in a letter to one of his bishops that, in images, "the ignorant may see that which they ought to follow, in them they read the letters they do not know; thus for the people an image is chiefly for reading."[17] Despite periods of iconoclasm in the eastern Orthodox church, Gregory's early medieval defense of religious iconography held up within the Latin church for the remainder of the Middle Ages. In fact, the church's insistence on the didactic power of pictures only intensified following the Fourth Lateran Council of 1215, whose rulings refocused the church's attention on the spiritual lives of laypeople.[18] Given a mandate to minister more directly to the needs of the laity, esteemed theologians like Thomas Aquinas reaffirmed images' power to inspire the faithful by imprinting important truths in their minds.[19]

For their part, the laity responded to the mandates of the Fourth Lateran Council by enthusiastically embracing the prospect of reading images as a means of contemplating the divine. By the middle of the thirteenth century, the very wealthiest members of society had begun to commission beautifully illustrated manuscripts to guide them through their religious practice: fabulously illustrated Bibles, lavish Books of Hours, and illustrated guides to biblical exegesis.[20] By the time our Benedictine began work on the Rawlinson calendar, these illustrated devotional guides had become widely popular among lay and ecclesiastical readers, and they were clearly a source of inspiration for his work. Perhaps he admired lavish Books of Hours, which often contained illustrations of the "labors of the month" and astrological symbols, in addition to depictions of the saints or the Virgin Mary, linking seasonal and celestial time to liturgical time in the same manner as the Rawlinson calendar.[21] These motifs were also adapted in inexpensive lewdecalendars, which were clearly another influence on the creator of the Rawlinson calendar.

Yet it was another illustrated devotional text, *The Mirror of Human Salvation* (*Speculum humanae salvationis*), that seems to have inspired our artisan more directly. The *Mirror*, first composed in the fourteenth century in German lands, was a guide to reading the Bible typologically. Typological exegesis, enormously popular in the later Middle Ages, held that every event in the New Testament was preordained and thus foreshadowed by an event in the Old Testament. In a typical *Mirror* manuscript, every page contained two illustrations—one from the Old and one from the New Testament—underneath which were two columns of Latin verse explaining the connection between the two. As it happens, the Rawlinson calendar includes several illustrations that are very similar to those found in Morgan Library MS M.766, a copy of the *Mirror* created in England around the turn of the fourteenth century, nearly the same time as the Rawlinson calendar.[22] Though these similarities do not mean that both manuscripts were the product of the same limner, or even the same workshop, they can tell us something about our Benedictine's preference for pictures. Clearly the Rawlinson calendar's creator had some familiarity with another manuscript genre that used images to tell the story of God's divine plan for his creation, and clearly, this artisan embraced the premise of the *Mirror*'s composition: that images could and *should* be read as a means of comprehending that plan.[23]

In 1389, when the creator of the Rawlinson calendar started work on his manuscript, the argument that images *should* be read was an especially contentious one in England. Only seven years earlier the Oxford cleric John Wyclif had been condemned as a heretic at the Council of Blackfriars for beliefs that included rejecting the veneration of saints' icons as akin to idolatry. Though Wyclif had endorsed images' power to "rouse, assist and kindle the minds of the faithful to love God more devoutly"—very much as earlier theologians had done—his followers, pejoratively called Lollards, made iconophobia a rallying cry of the movement.[24] Following Wyclif's death in 1384, there were scattered instances of Lollards removing and burning images from churches across England.[25] In response, the English church felt it had to double down on its defense of religious imagery and commission new works of vernacular theology that could explain orthodox doctrine to the lay faithful. In one of these new works, a treatise on the Ten Commandments known as *Dives & pauper*, the problem of the second commandment was addressed in the same language used by Gregory I centuries earlier: "[images] be ordained to be a token and a book to the lewd people that they may read in imagery and painture that clerks read in the book as the law sayeth." But lest these images become idols, English readers were enjoined to "do thy worship afore the image, not to the image."[26]

The late fourteenth-century English church was racked by controversy over Lollardy, orthodoxy, and images, and this climate of controversy is important context for understanding the motivations of a Benedictine artisan who chose images over text for his practical manuscript. Yet Wycliffism was not the only—or perhaps even the most serious—heresy with which church authorities had to contend in the late fourteenth century.[27] The papal schism that followed the conclaves of 1378, which saw first Urban VI and then Clement VII elected Pope in Rome and Avignon, respectively, was an undisputed threat to the church and a constant source of conflict between European powers. But for many, the fracturing of church authority portended something even more significant. There existed a very old tradition within Christianity of describing and predicting the end times, when Christ would come again to bring about the end of this world. The papal schism of 1378 suggested that moment had come. According to a traditional reading of 2 Thessalonians, one of Paul's letters to an early Christian congregation, the end of times would be revealed when a "man of sin" and a "son of perdition" emerged within the church to separate true believers from false.[28] Each side of the papal schism believed the other side's Pope to be that "man of sin," the Antichrist who would bring about the end of days. As a result, the late fourteenth and early fifteenth centuries saw a number of churchmen—from revered authorities like Pierre D'Ailly to disgraced visionaries like John of Rupescissa—prophesying on the likelihood that the end times were near.[29]

Our Benedictine thus lived at a time when the practice of reading images was increasingly championed by church authorities as an antidote to heresy, while at the same time, church authorities described how the end of the biblical cycle of suffering, death, and redemption could be read in the signs all around them. These contemporary events shaped how this artisan imagined his manuscript taking shape, but so, too, did trends in manuscript production. This artisan was clearly familiar with richly illustrated devotional guides like *The Mirror of Human Salvation*, as well as simpler illustrated manuscripts like lewdecalendars. Moreover, as a member of a Benedictine house, he would have ordered the rhythms of his day by other manuscripts, like Psalters and antiphoners and breviaries, all of which marked the passage of the monastic day, month, and year. With all these many influences in mind, we can imagine this artisan thinking about the passage of time and its relationship with divine order. We can picture him speculating about the future, about the predictive nature of signs both biblical and natural. We can even imagine him poring over religious illustrations, eager to make meaning from the juxtaposition of familiar shapes and forms. All of these experiences drove him to produce Rawlinson D.939, a manuscript that merges several traditions of illustration from religious, medical, and astronomical genres to produce something altogether new: a manuscript through which a user could "read in imagery and painture" the cycles of biblical, seasonal, and celestial time in the world around them—and perhaps even read the future.

Borrowing Pictures for Rawlinson D.939

Every one of the six leaves of the Rawlinson manuscript is chock full of pictures, from religious icons to astrological symbols to diagrams of the human body. Most of these images—save for its three pictorial prognostications—were borrowed from other contemporary manuscript genres. For example, the first leaf of the Rawlinson calendar features an illustration of the Angel Gabriel visiting Mary to announce the good news of her pregnancy, very similar to one found in *The Mirror of Human Salvation*.[30] It's a fitting opening for a calendar, of course, as the Feast of the Annunciation of the Virgin, celebrated on March 25, marked the New Year in England. The second leaf of the calendar clearly draws on the tradition of illustration in lewdecalendars. This panel, nearly a meter long when unfolded, contains a liturgical calendar with illustrations of the saints on the verso, and astrological signs and "labors of the month" illustrations on the recto. Together, these leaves present religious imagery fitted to the cycle of the year.[31] The last three leaves of the Rawlinson calendar feature a different cycle of religious illustration related to the cycle of biblical time.

Running the length of the verso of folios four, five, and six are scenes from the story of Adam and Eve and the Fall of Man; scenes of Adam and Eve's death and Christ's "harrowing of hell"; and finally, scenes of Christ's birth, the Adoration of the Magi, and Christ's crucifixion.[32] Together, these illustrations reference a cycle of suffering and redemption that was not yet complete. The promise implicit in the Christian Bible, illustrated in these scenes of the Fall of Man and Christ's crucifixion, was that God would redeem the suffering of mankind, just as Christ had already redeemed the suffering of Adam and Eve when—according to popular medieval belief—he descended into hell prior to his resurrection and brought salvation to those who had been condemned since the beginning of the world. In other words, these illustrations were a visual explication of typological exegesis, a reassurance that what God had already enacted in the past ordained what was to come in the future.

But not all of the images in the Rawlinson calendar are so overtly religious in nature. The final two folios of the calendar contain illustrations and tabular arrangements of symbols for use in medical practice.[33] Like the religious illustrations mentioned above, these medical diagrams and tables were borrowed from contemporary manuscripts, too—in this case, from a popular genre known as the "physician's almanac." These were folding manuscripts that could be carried by medical practitioners to patients' bedsides and used as handy guides to medical practice.[34] The core components of these manuscripts were astronomical tables detailing the conjunctions and oppositions of the sun and moon for four Metonic cycles, the nineteen-year period between the alignment of the solar and lunar calendars. These tables were foundational to the practice of astrological medicine, a discipline that had taken hold in Europe following its introduction from the Islamic world in the twelfth century.[35]

According to Islamic scholars, the movements of the sun, moon, and planets through the zodiac directly influenced the four elements (fire, air, water, and earth) whose qualities were manifest in the human body in the four humors.[36] Because the body was directly influenced by the stars, it stood to reason that a physician could not properly diagnose or treat illnesses without knowledge of the movements of the heavens—knowledge that could only be gained through reference to tables of astronomical data. Tables of planetary positions (called "almanacs" for the first time by the English polymath Roger Bacon) were first produced in England in the mid-twelfth century, but because the calculations required to produce these tables were quite complex, and the positions of the planets and moon were so many, the tables only covered a period of fifty-seven or seventy-six years (three to four Metonic cycles).[37] Moreover, the data in these tables were only relevant for the latitude and longitude at which they were calculated, meaning that a physician in Oxford could hardly use tables created for Paris or Rome. As a result, successive generations of English scholars had to systematically update their astronomical tables in order to practice astrological medicine. One such update (for the years 1301–1366) was the product of a monk who lived at Evesham Abbey, the likely home of our Benedictine artisan, too.[38]

By the late fourteenth century, when that Benedictine set to work on his calendar, the tables of Walter of Odington, former resident of Evesham, had expired. They had been replaced by the *kalendaria* of the Friars John Somer and Nicholas Lynn, calculated for the years 1387 to 1462. Somer completed his tables in 1380 as a gift for Joan of Kent, mother of King Richard II, and just a few years after that, Lynn completed his own set of tables at the request of John of Gaunt, Duke of Lancaster.[39] The new calendars of Somer and Lynn, and the canons they wrote explaining how to use them, were immensely popular among English readers, so much so that their names became synonymous with astronomical expertise.[40] Chaucer himself advertised the tables in part four of his *Treatise on the Astrolabe* as deriving from "the kalendars of the reverend clerks, friar J. Somer and friar N. Lynn."[41] But while Chaucer had astronomical instrumentation in mind when he cited the good Friars Lynn and Somer, most English readers prized their *kalendaria* not so they could measure the movements of heavenly bodies, but rather so they might minister to ailing human ones. Indeed, what distinguishes the *kalendaria* of Somer and Lynn from earlier astronomical tables is that both integrated astrological material with their astronomical data, specifically so their *kalendaria* could be used in the practice of astrological medicine. Both men provided tables showing the moon's position within the astrological houses, lunar and solar alignments and eclipses, and the ruling planets for every hour of the day and night.[42] This was all that most physicians needed to determine auspicious times for administering medicines or bloodletting, particularly when coupled with a *homo signorum* (man of signs) diagram and a *homo venarum* (vein man) diagram.[43] A *homo signorum* illustration featured an outline of the human body overlaid with astrological symbols indicating which zodiac sign influenced which part of the body. A *homo venarum* diagram similarly employed a human figure overlaid with illustrations of veins critical to the practice of phlebotomy, or bloodletting. Whereas neither of these figures were typical in earlier ecclesiastical or astronomical calendars, both were commonly included in the copies of Lynn and Somer's tables that fill the more than two dozen physician's almanacs that survive from England.[44]

The creator of the Rawlinson calendar clearly understood that the movements of the sun, moon, and planets exerted influence on the human body, but beyond that, there is little to suggest that he was a particularly learned practitioner of medicine. The Rawlinson calendar has no tables detailing upcoming solar and lunar eclipses or the monthly alignments of the sun and moon. Even so, the tabular arrangement of information and diagrams in physician's almanacs clearly influenced the Rawlinson calendar. Inspired by these guides to medical practice, our artisan put his own spin on the genre: instead of using text and Arabic numerals in his tables, the creator of the Rawlinson calendar used planetary symbols and astrological signs.[45] Figure 2.4 shows the final four columns of one of these tables, indicating which astrological house the moon would enter on which days of the month. This table would have been used in conjunction with a *homo signorum*, also in figure 2.4 at the right, which links these astrological positions to the human body.[46] Consulting these two diagrams in concert was critical to medical practice because, as one Middle English medical treatise explained, "when the moon is in a sign and one lets blood or is wounded [in a part of the body] governed by that sign, the body will soon be dead or destroyed in this world."[47] In other words, noting the position of the moon within the zodiac was critical for determining when and where *not* to let blood in a patient. The table and *homo signorum* in figure 2.4 tell us that a medical practitioner should be careful not to let blood in mid-January from the vein in the right arm that supposedly treated cardiac ailments. According to the table in figure 2.4, the moon was in Cancer in mid-January, making its influence on the heart, the part of the body governed by Cancer (as seen in the *homo signorum* in fig. 2.4 below), far too intense for bloodletting.

Like so much else in the Rawlinson calendar, then, this table of astrological symbols and *homo signorum* illustration makes eternal, cyclical time relevant at the level of individual, everyday human experience. That act of translation—from divine, eternal scales of time to human ones—was the fundamental purpose of the Rawlinson calendar. It's the reason the calendar features biblical illustrations of cycles of suffering and redemption and charts of astrological and planetary influences. It's also the reason the calendar includes a pictorial history, indicating the years elapsed since important events like the biblical great flood, Christ's birth, Saint Augustine's arrival in Britain, and the murder of Thomas Becket at Canterbury Cathedral.[48] The Rawlinson calendar is a guide to understanding an individual's place within divine order, an order that extended well beyond the scale of one human life and repeated in preordained cycles. And that is why our Benedictine artisan included prognostications within his calendar, too: if time was cyclical and the events of the past could be

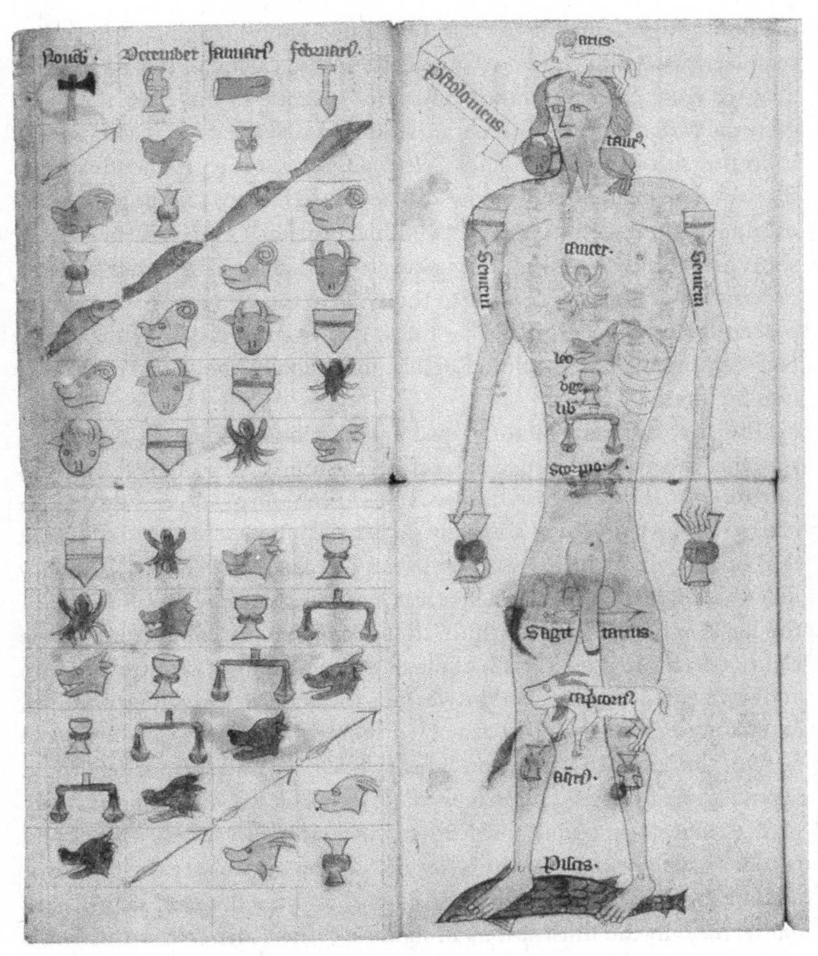

FIGURE 2.4. Portion of a pictorial table detailing which zodiac sign the moon is in at every day of the year, accompanied by a *homo signorum*, or "man of signs." The Bodleian Libraries, The University of Oxford, MS Rawlinson D.939, section 5v. CC-BY-NC 4.0.

read forward, then who was to say the events of the future couldn't be read from the present day?

Prognostications in Pictures

What would it look like to read the future in the pages of a book? It's a question that must have percolated in the mind of our Benedictine artisan as he worked on his manuscript and developed the three cycles of pictorial prognostication that are (as far as I know) unattested in earlier manuscripts. And yet, like so much else in the calendar, these too were adapted from contemporary manuscript sources. In this case, the three pictorial prognostications are translations of texts that circulated widely in fourteenth-century England.[49] Indeed, two of the prognostications, discussed in greater detail below, can be traced back to Old English manuscripts from the eleventh and twelfth centuries.[50] For centuries, these short prescriptive texts describing how to predict future weather patterns or harvests or illness according to the cycle of the sun, moon, or weather passed from manuscript to manuscript in Old English, Latin, Anglo-Norman French, and Middle English, until finally, they were translated into pictures.

The first of these pictorial prognostications, shown in figure 2.5, stretches nearly the length of the recto of the fourth folio of the Rawlinson calendar. On the far left-hand side, a vertical column of letters *A* through *G* runs from top to bottom. These are dominical letters, or letters indicating the date on which the first Sunday of the year fell. For example, if January first was a Sunday, the year's letter was *A*; if January second was a Sunday, the letter was *B*; if January third, then *C*, and so on. Proceeding to the right of each letter are a series of illustrations with brief captions in Latin conveying predictions about the weather, the grain yield, coming illnesses, and natural disasters for that year. If we follow the icons proceeding to the right of the letter *A* (Sunday on January first), we find that readers could expect a warm spring (*printemps calida*), a weak harvest without much grain (*frumentum carum*), the death of the young (*juvenes morientur*), and battles (*pugna erunt*). For the letter *B* (Sunday on January second), one could expect a wet summer (*estas aquosa*) and the death of kings (*reges morientur*). In the illustrations in figure 2.5, the same colors and forms repeat across each of the seven rows of the prognostication to indicate weather and crop yields, forming a pictography that could be read by a user, even if that user could not read Latin to decipher the brief captions that accompany each illustration.

The premise of this prognostication—that the dominical letter was determinative of patterns that would emerge throughout the year—reflects a central principle of Christian timekeeping. In order to determine the date of Easter, celebrated on the first Sunday after the first full moon after the vernal equinox, early medieval churchmen needed to align the lunar calendar with the Julian solar calendar. According to the definitive time-reckoning treatise followed by medieval Europeans, authored by the eighth-century English monk known as the Venerable Bede, the first step in this calculation was to figure out which dates throughout the year

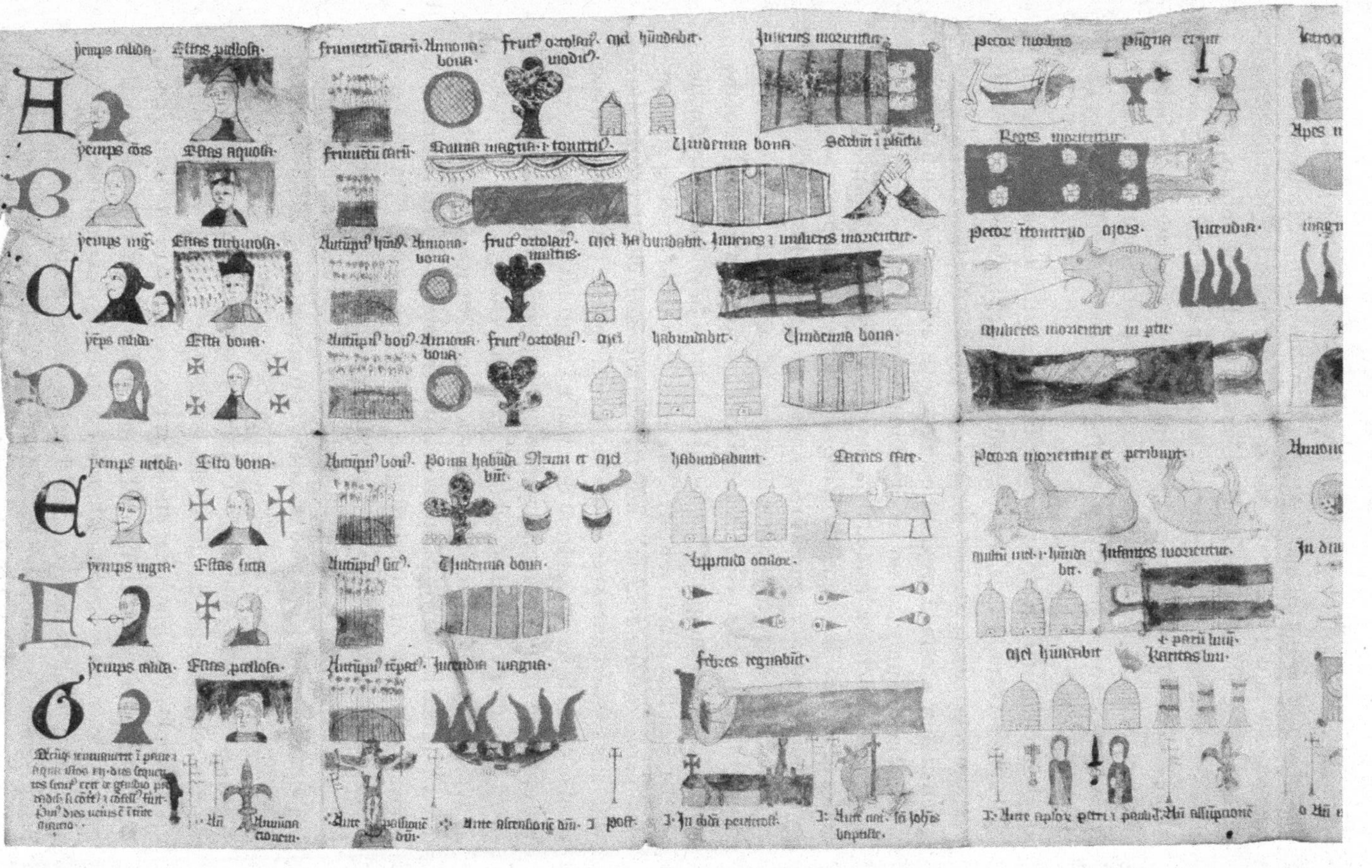

FIGURE 2.5. Pictorial prognostication according to dominical letter, featuring heads in profile to indicate seasonal weather and various icons indicating crop yields, disease, and warfare. The Bodleian Libraries, The University of Oxford, MS Rawlinson D.939, section 4r. CC-BY-NC 4.0.

would be Sundays.[51] That is where the dominical letter was useful: if the first of January was a Sunday (letter *A*), then the eighth of January would be a Sunday, and the fifteenth, and so on. If the second of January was a Sunday (letter *B*), then the ninth of January would be a Sunday, and the sixteenth, and so on. In other words, knowing the date of the first Sunday of the New Year had important predictive power in determining the future of the church calendar. It is not hard to imagine why some might believe it could also prove critical for predicting the weather, the harvest, or the death of kings.[52]

The second of the pictorial prognostications in the Rawlinson calendar again predicts crop yields, weather patterns, and political events.[53] In this case, however, the prediction depends on whether one hears thunder in a given month, a form of prognostication known as *brontology*. In figure 2.6, we see the prognostication arranged in two columns of six monthly entries, all of which are headed by a short Latin verse and composed of a series of images. The same illustrations from the annual prognostication by dominical letter (pictured in fig. 2.5) are repeated in this monthly prognostication to represent weather, illnesses, and crop yields. Here, however, the pictography is also supplemented by illustrations of the "labors of the month." So, for example, if one heard thunder in July, represented by the scythe used that month for threshing grain, one could expect a good annual harvest (*bona annona*), indicated by the same circular illustration with crisscrossed lines seen in the top row of figure 2.5 (probably a sieve used to separate the wheat from the chaff). Unfortunately, that same year the offspring of livestock, represented by the figure of a boar seen in row three of figure 2.5, would perish (*fetus pecorum peribit*).

Finally, the third pictorial prognostication in the Rawlinson calendar, stretching half the length of the sixth leaf of the manuscript, offers predictions for every day of a standardized thirty-day lunar cycle.[54] In this case, there are no Latin captions or descriptions to help us decipher the symbols that fill figure 2.7. Thankfully, however, we can guess at their meaning, because these icons also translate a prognostication widely popular in Middle English verse and prose.[55] On the far left of the image in figure 2.7, dots, semicircles, and circles (representing units of one, five, and ten, respectively) indicate the day of the lunar cycle.[56] To the right of those are symbols indicating lunar phases as the moon waxes and wanes over the course of the month. The full moon, at day fifteen, is visible exactly halfway down the prognostication. The rest of the symbols in the calendar are more difficult to interpret, but their meanings can be at least partially recuperated when compared with another pictorial prognostication of the same type in Bodleian MS Ashmole 8.

FIGURE 2.6. The brontology, or "Book of Thunders" prognostication, predicting weather, crop yields, disease, or warfare depending on whether one hears thunder in a given month. The Bodleian Libraries, The University of Oxford, MS Rawlinson D.939, section 3r. CC-BY-NC 4.0.

The Ashmole manuscript is yet another folding, single-sheet manuscript from the early fifteenth century, but its pictorial prognostications have been annotated with Middle English captions by a very helpful fifteenth-century reader. Thanks to these annotations, we learn that the four columns of symbols in the manuscript represent, in order: auspicious days for bloodletting; good days for particular agricultural activities; horoscopes for children born on that day of the lunar cycle; and prognostications about the outcome of illnesses that onset on that day.[57] Many of the

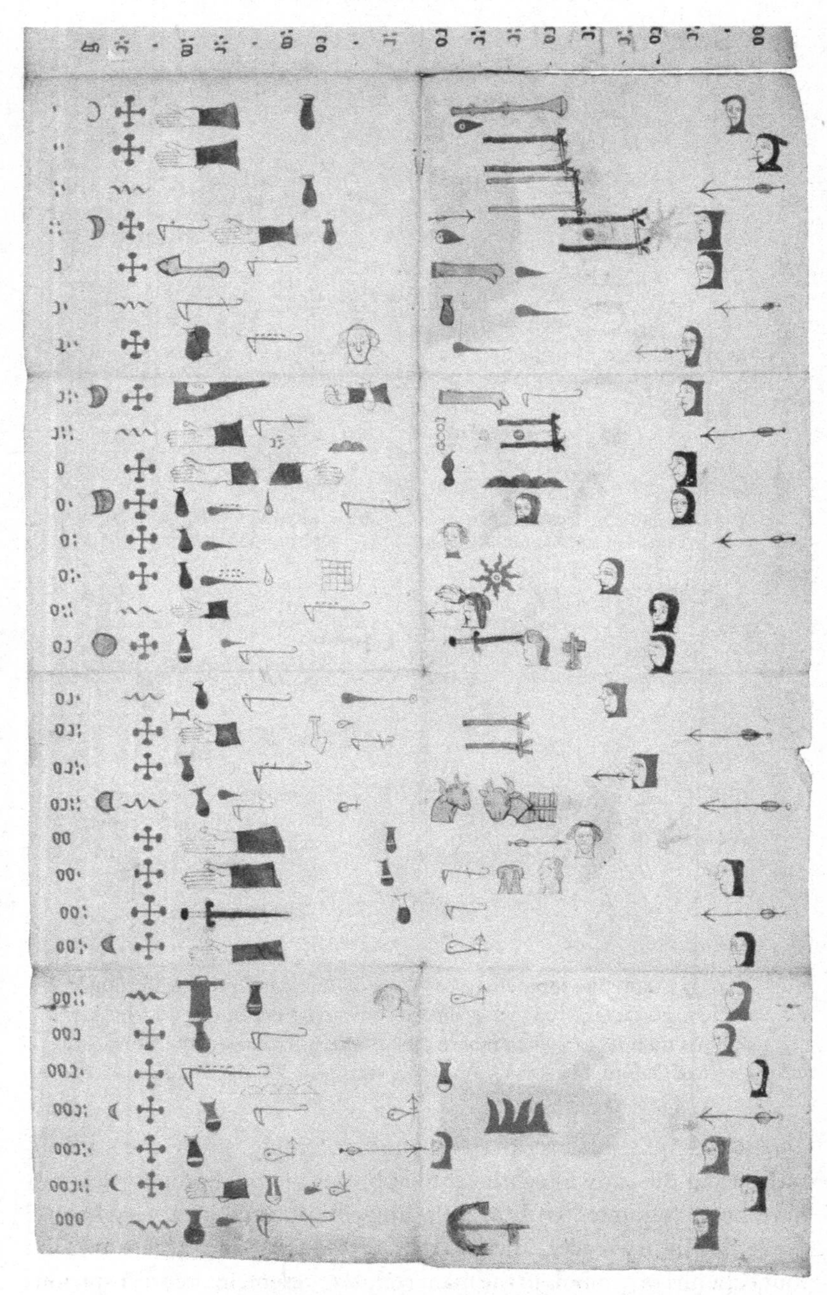

FIGURE 2.7. Pictorial lunar prognostication, indicating auspicious days for bloodletting, taking a journey, outcomes for illness, and horoscopes according to the waxing or waning of the moon. The Bodleian Libraries, The University of Oxford, MS Rawlinson D.939, section 6v. CC-BY-NC 4.0.

symbols in the Rawlinson calendar, above, are the same as those in the Ashmole version. So, we can surmise that the cross-marks and squiggles to the right of the symbols showing the lunar phase indicate good (cross-marks) and bad (squiggles) days for bloodletting. The symbols to the right of the central fold in the figure above offer horoscopes for children born on those days. For example, children born under the full moon (day fifteen) were doomed from the start: the sword striking a skull in that day's prognostication reveals that the child would become a "man killer," as the Ashmole version puts it.[58] To the far right of the prognostication, the hooded busts indicate probable outcomes for illnesses. Our Benedictine wouldn't have wanted to come down with an ague (or fever) on the third day of the lunar cycle, for example, because the arrow to the far right of that day's prognostication forecasts a very poor outcome: death.

Together, the three pictorial prognostications in the Rawlinson calendar predict the future at different scales of time: the year, month, and day, according to the cycle of the solar calendar, the weather patterns in a given month, or the waxing and waning of the moon. In their pictorial form, they fit neatly within a manuscript that provided visual representations of other significant cycles—the cycle of the church year, the cycle of suffering and redemption in the Bible, and the cycle of celestial movements—all of which exerted influence on the human body and spirit. In the context of the Rawlinson calendar, filled as it is with other established cycles of religious, astrological, and medical illustration, these three prognostications supply yet another visual language for making sense of the world. Yet, what separates these three prognostications from the other pictorial elements in the Rawlinson calendar is their independence from other known traditions of manuscript illumination. The scenes from the Bible in the Rawlinson calendar share similarities with illustrations in a contemporary English copy of *The Mirror of Human Salvation*. The arrangement of saints' icons and "labors of the month" in the Rawlinson calendar are imitative of lewdecalendars. The *homo signorum* and tables of planetary and astrological influences in the Rawlinson calendar draw their format and function from physician's almanacs and the *kalendaria* of Friars Lynn and Somer. But the three pictorial prognostications, so far as I know, have no similar antecedent. They appear to be a novel arrangement of images meant to translate popular Latin or Middle English prognostications into visual form.

But though they are novel, the prognostications in the Rawlinson calendar were clearly not created in a vacuum. Our Benedictine drew inspiration from the legible pictures in *The Mirror of Human Salvation,* the pragmatic repurposing of saints' icons in lewdecalendars, and even the tabular and spatial arrangement of information in Lynn's or Somer's *kalendaria*. These other traditions for representing cycles of time and the interplay between human experience and the natural world gave rise to a new program of prognostic illustration. Art historian Sonja Drimmer has described this mode of creativity—invention through allusion to convention—as a unique phenomenon of later medieval English manuscript production. Surveying the cycles of illustration that limners developed to accompany new Middle English verse, Drimmer notes that artisans developed "new images through referential techniques—assembling, adapting, and combining image types from a range of sources in order to answer the need for a new body of pictorial matter."[59]

In the case of our Benedictine artisan, a new body of pictorial matter arose from a need to read and interpret nature in order to understand his position within a divinely ordained cycle of time. Understanding weather cycles or predicting a failed harvest or diagnosing a sudden illness was about more than exerting control over capricious and unpredictable forces. It was about comprehending the patterns of God's power at work within the world. God's will was manifest in the regular movement of the sun and moon through the astrological houses. That cycle's influence on the four elements (water, earth, air, and fire) and thus on the four humors (blood, black bile, yellow bile, and phlegm) was comprehensible with the help of an astrological table and a *homo signorum*. The birth, death, and resurrection of Christ, as depicted in the illustrations of *The Mirror of Human Salvation* and the Rawlinson calendar, was another cyclical manifestation of divine will. According to these typological images, what God ordained for mankind's salvation through Christ had already been revealed in the events of the Old Testament. One had only to see the correlations between past and future, and the images in *The Mirror of Human Salvation* allowed a reader to do just that.[60] Yet the cycle preordained by scripture was not yet complete, as later fourteenth-century churchmen knew all too well. In the midst of a papal schism that seemed to mark the beginning of the end days, returning to the past (in scripture) in order to predict the future was especially critical, just as it was equally important to read the signs manifest in the heavens or in nature.[61]

Reading the cycles of illustration in the *Mirror* or aligning astrological tables with a *homo signorum* suggested that God's plan was orderly. And yet, there were so many unpredictable manifestations of God's power, too: earthly plagues, failed harvests, and sudden storms, not to mention devastating illnesses. Might these also be predictable if only a different sort of pattern was uncovered in lunar phases, in summer thunder, or in the cycle of the calendar year? Prognostications were that pattern. They provided a loose framework into which one could fit an experience of death, disease, or famine and render it comprehensible, making order from disorder. And crucially, as our Benedictine artisan realized, that loose framework for ordering experience was much better communicated by images than either Latin, Old English, or even Middle English text. Pictures were far better suited as a medium for finding the patterns that made sense of the natural world, because reading a picture entailed locating patterns, too. One needed to notice similarities of color, line, or form in order to make meaning from repetition. An image of a head in profile in red (indicative of a warm spring, according to the prognostication in fig. 2.5) could only have meaning when arranged alongside other nearly identical heads in profile colored in white or black. One could only "read" that head in profile with an awareness that the whole program of imagery was interconnected. Readers of pictorial prognostications had to have the ability to draw significance from slight alterations within an expected pattern—a skill that was especially well suited to a genre that offered readers a way to decipher the meaning behind a summer thunder shower or a waxing moon. Presumably, our Benedictine was already a practiced reader of lunar cycles or of ominous thunderclouds, as were most in late fourteenth-century England who had to anticipate bad weather or plan their evening rituals of prayer. All that was required to make new meaning from these everyday habits was to translate those observational practices from the natural world to the pages of a book.

Reading the "Boke of Paynture"

What did it look like in practice to read a book of pictures? At Evesham Abbey, the likely birthplace of the Rawlinson calendar, it may have looked something like this: our Benedictine consulting with others in his community, describing the visual language he developed in the pages of his calendar, demonstrating how a brother could reference its symbols to determine the best days for planting or for bloodletting. Monastic life was, after all, communal life. The work of tending the crops and livestock of the estate, preparing food, nursing the sick, reading, and praying was shared by many. The contents of the Rawlinson calendar address nearly every aspect of this shared work, and we must assume that the calendar itself was shared among the brothers of Evesham, too. Moreover, if we assume that our Benedictine passed his calendar among his brothers, then it is much easier to understand how a manuscript like the Rawlinson calendar could indeed function as a guide to practice.

Pictures might certainly be "books for laymen," as Pope Gregory once put it, but they could only serve that function for those well versed in their visual language.[62] We know that the language of saints' icons and astrological symbols, also present in the Rawlinson calendar, was constantly reinforced for late medieval readers in encounters beyond the pages of a manuscript: in wall paintings, in sculptural facades, or in the stained glass of churches. But, as far as we know, the pictorial prognostications developed in that calendar benefited from no commensurate reinforcement in everyday exposure, nor was their meaning stabilized by well-established narrative. Instead, we can only assume that our Benedictine artisan explained its pictorial vocabulary to a fellow monk or perhaps even a novice, pointing out the cross-marks and squiggles indicating auspicious days for bloodletting in figure 2.7, for example, while treating a sick brother in the monastery's infirmary. Or perhaps, at the start of the new liturgical year on January 1, he sat with a different brother to note the forecast for the coming year, remarking upon the likelihood of a good harvest.

We have no way, now, of reconstructing the oral and aural traditions by which the system of meanings attached to those pictorial prognostications was shared, and yet we know that they must have been shared. That is because the same visual language of prognostication developed in the Rawlinson calendar found new life well beyond Evesham Abbey and Worcestershire in at least seven other manuscripts.[63] Every one of these manuscripts adheres generally to the contents of the Rawlinson calendar, with one notable exception: none contains the cycle of biblical imagery relating the temptation and suffering of Adam and Eve with the suffering and redemption of Christ. The remaining seven are devoted solely to natural knowledge, as are the other practical manuscripts in my corpus. Moreover, because the contents and format of these seven manuscripts are so similar, we can assume—given the incredibly limited survival of medieval manuscripts—that these pictorial calendars and almanacs represent the fragmentary traces of a relatively widespread and well-established subgenre of practical manuscripts. They were produced in England by artisans who understood themselves to be participating in a particular tradition of manuscript production that had well-defined parameters as set out in the Rawlinson calendar.[64] They understood this tradition because they, too, were members of Benedictine communities.

Take, for example, the anonymous artisan who produced British Library MS Egerton 2724 around the year 1430. No doubt this artisan had seen a manuscript similar to the Rawlinson calendar (if not that manuscript itself) before creating his own copy. The Egerton calendar is composed on several strips of parchment folded lengthwise and then accordion style to form a small square, just like the earlier Rawlinson calendar.

It also features icons of the saints, a *homo signorum* figure, and a copy of the pictorial prognostication by dominical letter, though without any of the Latin captions to clarify its meaning.[65] Without those captions, the calendar would have been indecipherable to the uninitiated, its iconography totally incomprehensible. Except, this artisan wasn't uninitiated. The liturgical calendar in the Egerton manuscript includes an extra icon for the month of September not found in other similar calendars: an icon of a cathedral with a Latin caption, "*Dedicatio*," above which is written in much smaller letters "norwyc," or Norwich.[66] We can assume that this calendar's artisan lived close to, if not within the grounds of, Norwich's cathedral, whose twelfth-century dedication was celebrated with a feast on September 24. And unlike most urban cathedrals, Norwich's was staffed not by secular priests, but by Benedictine monks.[67]

So, it seems, a network of Benedictine knowledge transmission carried pictorial prognostications from Worcester to Norfolk, and from there, the tradition continued to spread. The Egerton calendar, created in Norfolk, was the model for yet another manuscript created just three years later, in 1433. That manuscript, now in the Morgan Library in New York, is missing the leaves of parchment that once contained most of its liturgical calendar, making it difficult to ascertain the locus of its creation.[68] What is clear from the remaining leaves is that whoever made the copy (perhaps a novice at Norwich?) was much less skilled in the arts of manuscript production than the creator of the Egerton calendar. Moreover, it seems he had real difficulty deciphering the abbreviated Latin captions that accompany many of its illustrations.[69] Despite these difficulties, however, the same artisan faithfully reproduced the pictorial prognostication by dominical letter. Reading Latin was a problem; reading pictures was not.

This isn't to suggest that pictorial prognostications were inherently simple, or that they only appealed to the relatively unlearned. To the contrary, in the north of England, the pictorial prognostications in the Rawlinson calendar found new life in manuscripts that presented relatively learned astronomical data alongside icons, symbols, and images in a traditional codex format. The creator of British Library MS Harley 2332, completed around 1412 at yet another Benedictine foundation near York (perhaps St. Mary's Abbey), was clearly familiar with the pictorial conventions developed in the Rawlinson calendar. The Harley manuscript reproduces many of the same elements found in that earlier calendar: a liturgical calendar of saints' days with labors of the month and astrological signs (fig. 2.8), a *homo signorum* and accompanying pictorial table of astrological influences, and finally, a pictorial version of the prognostication by dominical letter (fig. 2.9).[70] Yet the Rawlinson calendar was not the most direct

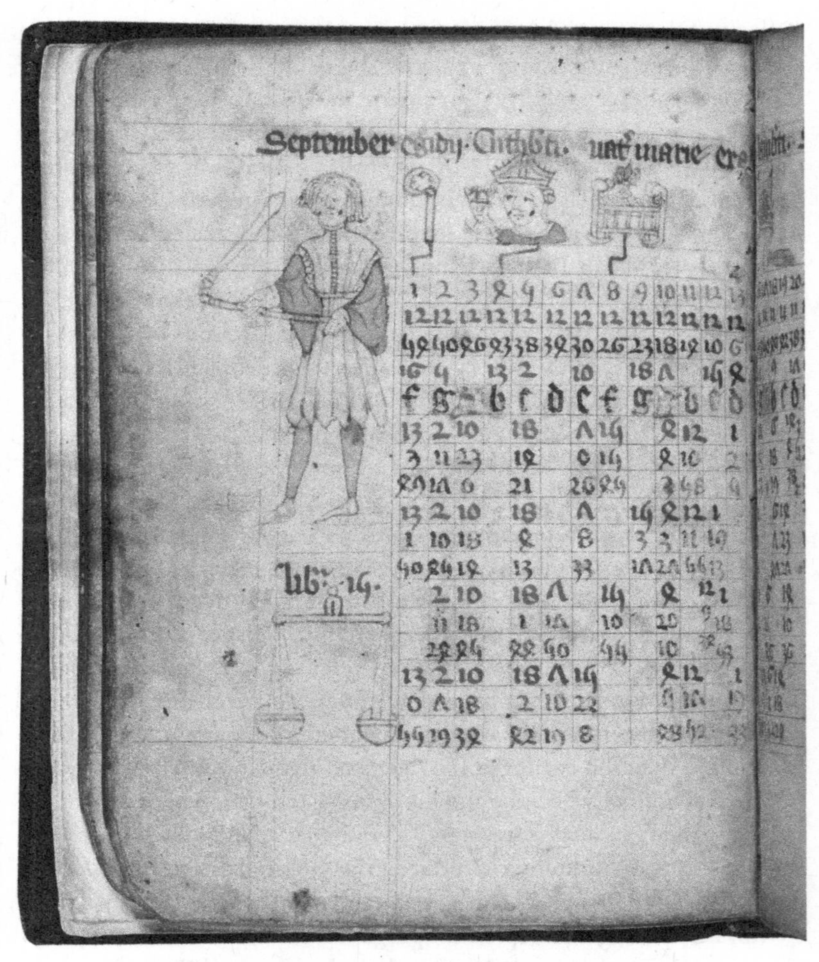

FIGURE 2.8. Liturgical calendar and astronomical table for the month of September, with data from John Somer's *kalendaria*. © The British Library Board, MS Harley 2332, f. 9v.

influence on Harley 2332. The Harley manuscript is in fact modeled very closely on a Latin almanac created in 1408, probably at the same Benedictine foundation.[71] Not only is this Latin almanac roughly the same size, with around the same number of pages as the Harley manuscript, it also loosely follows the same order of entries. And most telling of all is that the tables of upcoming solar and lunar eclipses in the Harley manuscript are an exact match to those in this earlier, Latin almanac.[72] So, it seems, the creator of the Harley manuscript merged two traditions of Benedictine manuscript production—one, a learned Latin almanac calculated for the latitude of York, and the other, a pictorial calendar and prognostication created far away in Worcester—into a new hybrid manuscript genre: the pictorial almanac.

Again, it is difficult to explain the existence or function of a manuscript like the Harley almanac without fitting it into its monastic context, within a community of readers who could explain its meaning to the uninitiated. Not only is the pictorial prognostication within the Harley almanac (shown in fig. 2.9 below) totally indecipherable without foreknowledge

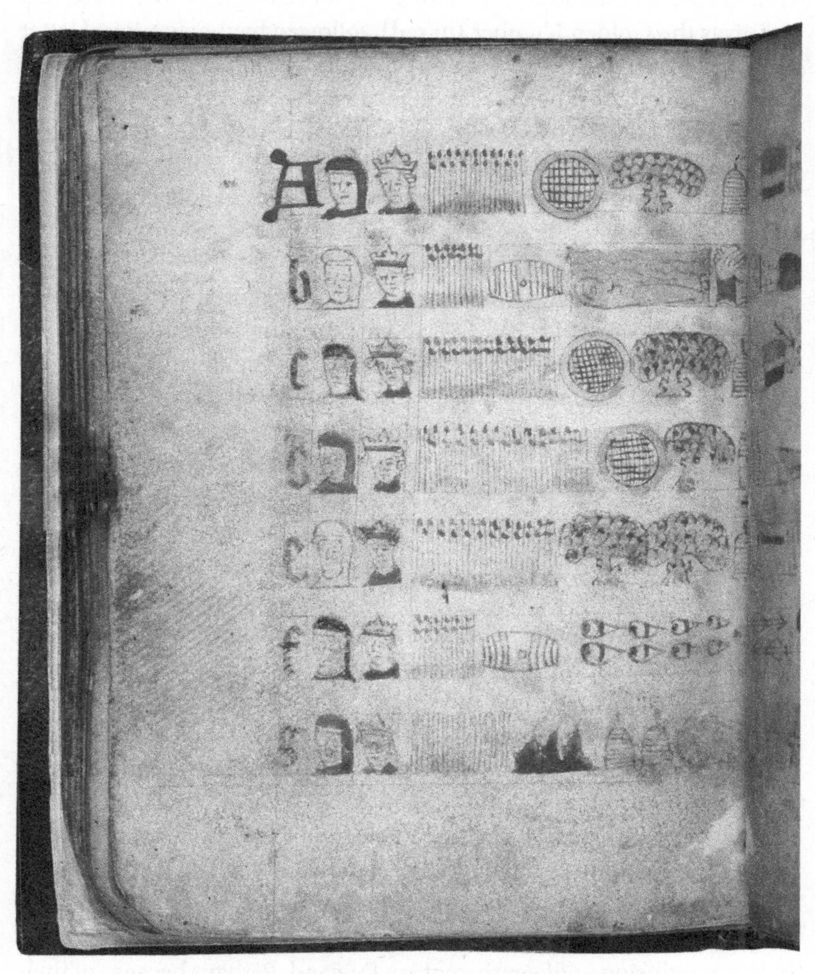

FIGURE 2.9. Pictorial prognostication by dominical letter in a pictorial manuscript almanac created around 1411 in Yorkshire. © The British Library Board, MS Harley 2332, f. 19v.

of its meaning, so too are its tables of astronomical data (shown in fig. 2.8 above). Most almanacs from fifteenth-century England, whether in Latin or Middle English, contained explanatory canons detailing how a user should interpret the rows and rows of Arabic numerals in tables for every month of the year. For example, Wellcome Library MS 8004, a codex-style Middle English and Latin almanac created in 1454, offered line-by-line instructions for deciphering a table like that in figure 2.8: "upon the first line . . . are the number of the days in the month [in black]. And in the second line is set the hours [in red] and the minutes [in black] of the day from the rising of the sun to the setting."[73] Proceeding downward, the next row gives the Golden Number (in red), followed by the dominical letter (in black). The rest of the table's red and black numbers give the precise hour and minute for the new moon for every year from 1405 to 1481, arranged by Metonic cycle. A user could consult these tables to accurately calculate the moon's phase within the calendar month, which was requisite information if one was to determine auspicious days for collecting medicinal herbs, administering treatments, or letting blood, as outlined in the lunar prognostication in figure 2.7, above.[74]

But the Harley calendar contains no such instruction. Without a guide to its astronomical tables (fig. 2.8), and without any captions accompanying its pictorial prognostications (fig. 2.9), the Harley almanac's user must have relied on significant training in astronomy as well as prior instruction in the visual language of prognostication. In short, the ability to use this manuscript would have depended on a secure position within a community like that at St. Mary's Abbey in York, Evesham Abbey in Worcester, or Norwich Cathedral in Norfolk—a community that had a shared understanding of signs and symbols and the relationship between the natural world and the pages of books. It was those communities and their networks of exchange that allowed for the proliferation of the illustrated prognostications in the Rawlinson calendar, just as these networks also led others to copy the format and contents of the Harley almanac. In 1420 and again in 1425, two more pictorial almanacs were made in the vicinity of York (if their astronomical tables are any indication), both of which are very close copies of the Harley almanac.[75] One of these manuscripts eventually ended up at Luffield Abbey, another Benedictine foundation nearly 150 miles to the south in Buckinghamshire, where it was still in use when Thomas Rowland was appointed prior in 1487.[76]

But almanacs aren't the end of the story of pictorial prognostications and their diffusion in fifteenth-century England. Perhaps because of their popularity within monastic communities, the iconography of practical picture books moved beyond Benedictine networks.[77] A manuscript featuring an enormously popular Middle English history of Britain known as

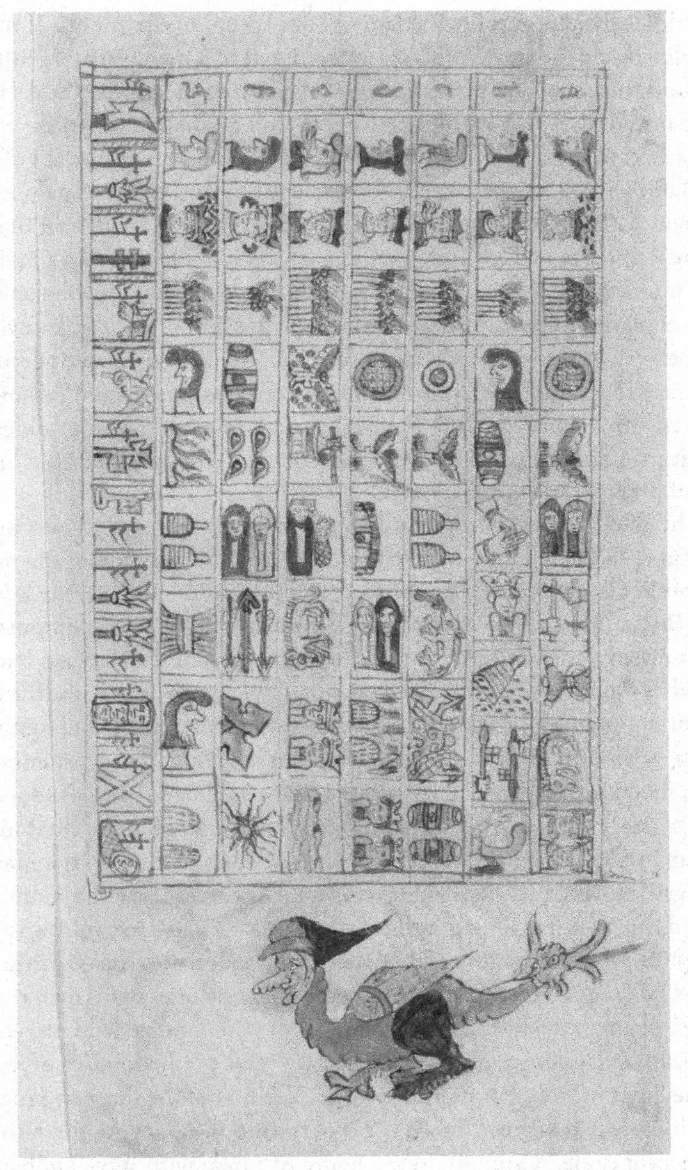

FIGURE 2.10. Pictorial prognostication by dominical letter on the frontispiece of a prose *Brut* manuscript from the early fifteenth century. Cambridge, Harvard University, Houghton Library, MS Richardson 35, f. 1v.

the prose *Brut*, now held at the Houghton Library at Harvard University, contains a compact version of the pictorial prognostication by dominical letter on its opening folio (fig. 2.10).[78] Not only that, but the margins of folios 14 through 25 of this *Brut* manuscript feature exact copies of the astrological signs and illustrations of the labors of the month from the Rawlinson calendar, right down to the detail of a man warming just one unshod foot over the fire as January's "labor of the month."[79] As far as we can tell, the *Brut* manuscript has nothing whatsoever to do with a Benedictine foundation. The manuscript includes an armorial guide to England's nobility, with illustrations of heraldry and captions naming important earls and dukes of the mid-fifteenth century, suggesting that the manuscript was created for a member of the lay elite and not a member of a monastic community.[80] But, as we will see in the following chapter, clerics played important roles in the transmission of natural knowledge to the laity, particularly as scribes-for-hire. The *Brut* manuscript with its labors of the month and armorial illustrations was almost certainly owned by a wealthy member of the English laity, and perhaps even a nobleman, but it may have been copied by a Benedictine who knew a thing or two about pictorial prognostications.

The same might be true of one of the manuscripts bound into British Library MS Harley 1735, owned by John Crophill. The first of the manuscripts in Harley 1735 is a beautiful copy of the Middle English *storia lunae*, or "Thirty Days of the Moon" prognostication.[81] Though composed in verse rather than pictures, the *storia lunae* prognostication gives advice on travel or bloodletting for every day of the lunar cycle, much like the third pictorial prognostication in the Rawlinson calendar, shown in figure 2.7. Though we cannot say where Crophill got his copy of this prognostication, it certainly looks like something that could have been produced in a monastery by trained scribes. It opens with a large historiated initial, and many of its pages feature red pen-and-ink flourishing in the margins alongside rows of very neat cursive script. However, it wasn't a monk who added the marginal illustrations that now adorn Harley 1735; it was John Crophill, who was bailiff of the Benedictine foundation of Wix Priory in Essex from 1455 to 1477 and also a medical practitioner in his own right.[82] Crophill amended his practical manuscript with figures that he sketched in the margins alongside each of the thirty verse predictions for every day of the lunar month. For example, beside the first of the month's prognostications, which begins "The first day of the moon Adam our forefather to this world came," Crophill drew a figure of a man with shovel in hand (a common symbolic representation of Adam in medieval visual culture).[83] Crophill also drew an illustration of a millwheel beside the prognostication for day four, on which it "is good mills to begin," as well as numerous other icons alongside other daily prognostications.[84]

In addition to these sketches referencing the contents of each day's prognostication, Crophill annotated the top margin of each leaf of the text with short summaries of what the day portended for those who were sick and ailing: "soon he shall rise," for day one, or "soon in few years to be dead," for day three. And, most intriguingly, under each of these captions he drew a squiggly line, a single arrow, or a cross—exactly the same symbols used in the lunar prognostication in figure 2.7 to indicate the outcomes of illnesses. Did the brothers at Wix Priory where Crophill served as bailiff teach him the symbols he used to annotate his verse prognostication? Did Crophill see a pictorial calendar like those discussed in this chapter? We simply cannot know. What is clear is that Crophill recognized a relationship between reading the body, reading the natural world, and reading pictures to read the future, as did every one of the artisans who produced the pictorial calendars and almanacs discussed in this chapter.

Conclusion

For the later medieval interpreter of a pictorial calendar or almanac, repetitive patterns were everywhere: in the movements of the constellations, in the rising and setting of the sun throughout the year, and in the mathematics of the church calendar, with its dependence on both the solar and lunar cycles. Comprehending these patterns in order to understand their influence on human experience was the primary purpose of the pictorial practical manuscripts that proliferated in later fourteenth- and fifteenth-century England. Because all of God's creation was interconnected in the macrocosm of celestial and earthly spheres, it stood to reason that attention to anomalies or affinities within the order of that creation would produce meaning. A medicine that cured one patient but failed in another, or a crop that flourished one year but withered on the vine the next—these events could not be random. They required explanation through attention to the patterns of the seasons, the weather, the day of the month, week, or year. To make sense of the capriciousness of the natural world, one had to remain attuned to one's surroundings, hone one's observational skills, and integrate that experiential knowledge within a loose framework that would ultimately give it meaning. The symbols and icons in pictorial prognostications presented that loose framework in a medium that reinforced these habits of observation and interpretation already practiced every day by Benedictine monks, but also by lay medical practitioners like John Crophill. Images were both fixed and flexible enough to accommodate a belief in divine order and the experience of unpredictable natural disasters, illnesses, or crop failures.

It was this belief system that ordered the lives of the Benedictine monks who developed a visual language for interpreting the natural world, and who spread that visual language across fifteenth-century England. The pictorial calendars and almanacs created at Evesham Abbey, Norwich Priory, and St. Mary's Abbey demonstrated their users' participation in a community of like-minded believers, one that was bounded and exclusive, but also deeply creative—a community that felt perhaps especially important in the chaotic atmosphere that saw the emergence of Lollard heresy and papal schism. Many of the illustrations in the Rawlinson, Egerton, and Harley manuscripts would have been indecipherable to the uninitiated, which was part of what made those manuscripts so meaningful to their owners. They carried forward traditions that were nurtured through personal exchanges. It wasn't enough for the artisans who created the Rawlinson calendar or the Harley almanac to possess knowledge in the pages of any old book. They had to make that knowledge their own, choosing pictures as their medium for conveying information about the interplay between cycles of eternal, celestial time and individual, human experience.

3

Writing Recipes, Wrangling the Power of Nature

At some point in the middle decades of the fifteenth century, Robert Taylor of Boxforde, Suffolk, got the idea to commission a practical manuscript.[1] That book, now Huntington Library MS HM 1336, contains more than 200 medical recipes—among them, recipes for "jaundice," for "stinking breath," and for "a drink for a wounded man."[2] But that's not all. It seems Taylor wanted more than just a litany of cures for whatever ailed him. The final two pages of his manuscript feature Middle English versions of the same prognostications discussed in the last chapter (predictions for weather and crop yields according to the year's dominical letter or according to whether there was thunder in a given month), followed by a dietary instructing Taylor on what to eat and drink throughout the year.[3] In addition, scattered among the medical recipes at the start of the manuscript are recipes for making colors, black ink, wax for sealing letters, and glue for torn parchment—all of which suggest that Robert Taylor knew how to write.[4] And yet, Taylor was not responsible for copying the hundreds of medical recipes and instructions in his practical manuscript. Rather than do the work himself, he outsourced it. He was fortunate to live in Boxforde, Suffolk, only about a day's ride from Cambridge, where students might be hired to write for cheap. Taylor was lucky to find Symon Wysbech, a law student at Cambridge, who took care with this scribal work: he marked out the margins of twenty-five parchment leaves with black lines; he copied hundreds of recipes in a neat, cursive Anglicana script; and finally, he finished his work by titling each recipe in red ink. Occasionally Wysbech's handwriting ran into the margins as he copied recipe after recipe, but all in all, he did a fine job at a scribal commission that was outside the purview of his usual routine listening to lectures in canon law. Justifiably proud of his work, Wysbech signed his name on the last folio of the manuscript, and then, not able to resist the temptation of a blank section of parchment, added a joke for posterity: "Now I have written everything, for Christ's sake give me a drink."[5]

Wysbech's joke about the hard work of writing raises an important point: any history of the circulation and influence of natural knowledge in fifteenth-century England must also be a history of scribal acts. Every one of the 182 practical manuscripts discussed in this book was the product of at least one and in many cases several writers. Though we cannot always identify those writers by name, the choices they made about the style of script to use, the arrangement of text on a page, or the contents selected for a manuscript shed light on fifteenth-century attitudes toward both books and the natural world. And those attitudes were hardly static. At the beginning of the fifteenth century, few besides the most elite laypeople had ever seen a book of natural knowledge. By the close of the fifteenth century, hundreds if not thousands of practical manuscripts had been commissioned or created by lay English readers. In the intervening years, the form and function of these manuscripts changed dramatically, in no small part because of who was writing them.

The first section of this chapter examines the earliest practical manuscripts, created for the wealthiest members of the laity by professional book artisans. These lavishly decorated creations reflect a tradition of book-making that had long centered on religious manuscripts. Though practical books were not prayer books or psalters, they did tell a story of God's provision for mankind. A popular medieval adage held that God had endowed his "virtues" in herbs, stones, and words, and practical books offered English readers access to all these sources of divine power: recipes and herbal texts described how to prepare plants to harness their "virtues," while charms invoked the divine power of words. As I argue in the chapter's second section, charms are in fact key sources for understanding how fifteenth-century readers grew comfortable using writing as a response to experiences of illness or injury—even if those responses entailed no more than scrawling a few holy words onto paper, parchment, or even a sage leaf. Over time, as the English laity became more capable writers, their written responses to illness or injury grew more sophisticated. The final section of this chapter details how amateur writers used practical books to engage with a very old and learned textual tradition, interjecting their own words alongside those of revered authorities. In marginal notes or recipes scribbled in sloppy cursive, we can glimpse these writers observing natural or bodily phenomena and capturing those experiences on the page.

The previous chapter examined how reading natural knowledge was especially meaningful for members of the clergy, while this chapter explores what it meant to write natural knowledge for English laypeople. As we will see, that meaning shifted over the course of the fifteenth century as many more among the English laity learned to wield a pen. And yet, one truth remained constant: whether a practical manuscript was a deluxe creation commissioned by a noblewoman or the product of an amateur healer whose scribbled recipes fill every inch of the page, all who sought to own a practical manuscript in fifteenth-century England did so because they knew that in writing, they might harness something of the power of nature.

Patrons and Their Books

By the time Robert Taylor commissioned Symon Wysbech to produce his medical manuscript, sometime around the middle of the fifteenth century, scores of texts related to the practices of preparing medicines, healing wounds, tending plants, making textiles, or predicting the future were available in Middle English. Broadly, these texts were the product of growing interest in vernacular composition and translation across later fourteenth- and early fifteenth-century England, but specifically, a number of these texts were also the direct result of commissions from noble patrons. Two of the most notable were John, Duke of Bedford, and his brother Humfrey, Duke of Gloucester, both of whom acted as regents for their nephew, Henry VI, after their brother, Henry V, died in 1422. Bedford was regent of France, while Gloucester was Lord Protector of England. From these positions, both men were able to secure extraordinary collections of manuscripts. Not only did Bedford purchase wholesale the library that had been established by the French king Charles V at the Louvre, he also commissioned new works, including a number of medical, astrological, and mathematical treatises. He asked a member of his household medical staff, Jean Tourtier, to produce a French translation of Galen's commentary on Hippocrates's *Aphorisms*, as well as a French translation of a popular book on astrology.[6] It appears that Bedford also commissioned one of several Middle English translations of Guy de Chauliac's *Chirurgia magna*.[7] Humfrey, Duke of Gloucester, also found time to patronize Middle English didactic literature. He commissioned a translation of the classical treatise *The Work of Agriculture* (*Opus agriculturae*), a practical guide to land cultivation, sometime around 1442–1443.[8] And by the time of his death in 1447, Humfrey had become renowned for his learning throughout Europe, in large part thanks to his magnificent manuscript collection, featuring Latin medical texts by Avicenna, Bernard of Gordon, Galen, Gilbert the Englishman, Dioscorides, and Constantine the African.[9]

Though few in England could match Bedford or Gloucester for wealth or learning, there is good reason to assume that other members of the elite were eager to emulate the dukes' patronage of Middle English letters. Several lavish and expensive manuscripts featuring Middle English translations of learned medical texts survive from the early fifteenth century, each one a testament to its patron's recognition that books were important markers of status—and that status in fifteenth-century England had as much to do with intellectual pursuits as it did with pursuits on the battlefield. Like the Dukes of Bedford and Gloucester, whoever commissioned the first English translation of Lanfranc of Milan's *Chirurgia magna* in the latter decades of the fourteenth century knew that the finished manuscript, composed of 276 parchment folios adorned with red and blue initials and marginal decoration, would convey an interest in surgical medicine *and* affluence.[10] Bodleian MS Ashmole 1505, created for Henry VI's master distiller, is a huge manuscript featuring the Middle English translation of Bernard of Gordon's *Lily of Medicine* (*Lillium medicinae*). Its illustrations and decorative flourishes are fittingly sumptuous for a member of the royal household.[11] British Library MS Sloane 277 likewise reflects its anonymous patron's learning and largesse. That manuscript, composed in the first quarter of the fifteenth century, features the Middle English translation of a surgical manual by William of Parma and John Arderne's *Treatise on Fistula in Ano*, copied in double columns on parchment folios and augmented with exquisite diagrams of Arderne's surgical instruments.[12] Similarly, the translations of Guy de Chauliac in University of Glasgow MS Hunter 95 and British Library MS Sloane 1 were each copied onto parchment in the early decades of the fifteenth century by well-trained scribes for obviously wealthy patrons.[13] But most magnificent of all is an early fifteenth-century copy of Guy de Chauliac in the New York Academy of Medicine, copied on 181 large folios of parchment, with floriated initials and borders of red, blue, green, and gold.[14] Though none of these exorbitantly expensive manuscripts has been linked to a well-known collector like Bedford or Gloucester, the survival of so many lavish and learned volumes of Middle English medicine from the first half of the fifteenth century suggests that other members of the elite shared the dukes' intellectual proclivities.

What is perhaps more surprising is that even those who did *not* share the same interest in learned medicine as men like Bedford and Gloucester also saw something of value in owning a lavish practical manuscript. Though not quite as extravagant as those thick surgical manuscripts cited above, early fifteenth-century medical miscellanies full of recipe texts, herbals, craft instructions, and prognostications were often ornamented, too. In British Library MS Sloane 2584, one of the earliest surviving Middle English recipe collections, a trained scribal artisan prepared every one of its parchment leaves with marginal boundary lines drawn in lead point and pricking to keep its twenty-six lines of text straight and neatly centered on the page. Though the scribe didn't leave space for large colored initials, the top lines of the first several pages of text are decorated in their own special way: the scribe lengthened the ascenders of certain letters (the vertical lines that extend upward in the letters *b, d,* or *h,* for example), extending them into the top margin where they were flourished with red ink. This decoration ornaments a collection containing directions for ink- and color-making and for catching coneys and for taking deer, as well as scores of medical recipes, a series of prognostications, a short primer on apothecaries' weights and measures, a simple table of multiplication facts, and a treatise on veterinary medicine related to horses.[15] Wellcome Library MS 406, created just at the turn of the fifteenth century, contains directions for bloodletting, medical recipes, and a treatise on the virtues of rosemary, each entry ornamented with a beautiful initial decorated with red and blue flowers (fig. 3.1, below).[16] Likewise, British Library MS Additional 34210, also created around the turn of the fifteenth century, features elaborate red pen-and-ink flourishing in the margins of its scores of medical recipes, many of which open with large blue initials.[17] Trinity College Cambridge MS R.14.51, composed around 1425, has pages of recipes ornamented with blue initials and curlicues of red ink.[18] And in Wellcome Library MS 5262, another practical manuscript of 129 medical recipes from the late fourteenth century, an artisan sketched fanciful animals around each of the manuscript's catchwords, adding a bit of beauty and whimsy to a collection of medical recipes and charms.[19]

Of all of these beautiful recipe books, however, British Library MS Harley 2320 is perhaps the most ornate. Composed around the turn of the fifteenth century on seventy-four parchment leaves neatly ruled with wide margins, the manuscript contains a book of "nativities," predicting the disposition, health, and wealth of men and women born under each zodiac sign; a "Thirty Days of the Moon," or *storia lunae,* verse prognostication; and finally, a series of directions for braiding silk threads into trim to adorn clothing.[20] Silk braiding, or "silk-throwing" as it was sometimes called, was women's work in fifteenth-century England.[21] Between 1300 and 1500, at least 123 silkwomen worked at their craft in London, and court records reveal that women in Coventry, York, and Nottingham were also engaged in the trade.[22] The inclusion of silk braiding directions in Harley MS 2320 thus indicates that its patron was a woman, though she was probably not a member of the trades. We may in fact catch a glimpse of her in the manuscript: each of the entries in Harley MS 2320 opens with a large historiated initial, and within the large historiated *I* at the

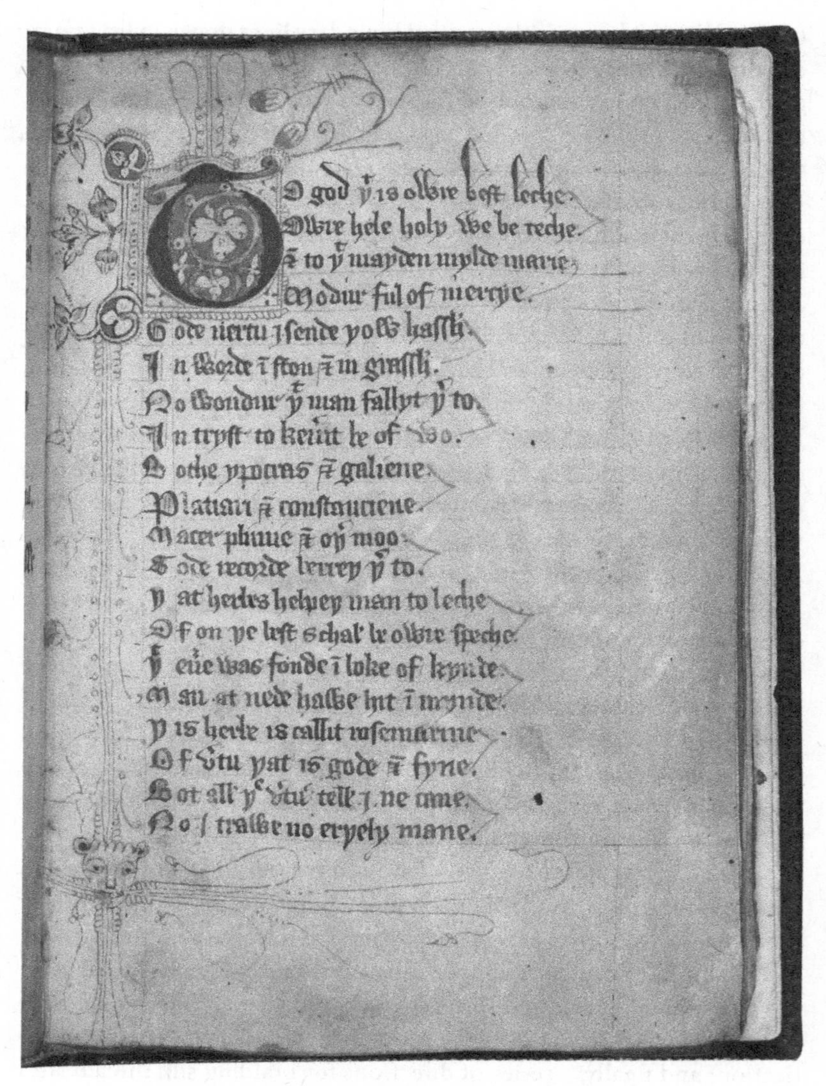

FIGURE 3.1. The verse prologue to a collection of medical recipes with a historiated initial *T*. Collection of medical tracts in verse and prose, mostly in Middle English (Leech-Books, III). London, Wellcome Library, MS 406, f. 14r. Wellcome Collection. Public Domain Mark.

start of the directions for silk braiding is a portrait of a well-to-do woman seated with a distaff, spinning thread.[23] Though the craft of silk braiding didn't require a distaff, the limner responsible for ornamenting Harley MS 2320—probably a man—may not have known as much. That anonymous limner may simply have done his best to illustrate Harley MS 2320 in a style similar to other genres popular with elite women readers, most of which were religious manuscripts like Books of Hours. These beautifully illustrated guides to daily prayer were the most popular genre of manuscript in later medieval England, and a few even featured portraits of their female patrons.[24] In fact, the owner of Harley MS 2320 seems to have been inspired by the format of these devotional manuscripts: not only did she enlist a limner to adorn her practical manuscript with historiated initials, she also tacked a liturgical calendar to the front of the manuscript, just where it would be in a Book of Hours.

The format of Harley MS 2320 may have been directly inspired by a Book of Hours, but in fact, all practical manuscripts were at least indirectly influenced by these ubiquitous guides to daily prayer. In many ways, the Books of Hours that were commissioned for wealthy lay patrons beginning in the later thirteenth century and increasing in popularity throughout the fourteenth century were the first "practical" manuscripts: they taught English readers that books could be private and personal, and they were manuals for religious *practices*, just as remedy books were manuals for healing.[25] And so, even though most practical manuscripts do not mimic the format of a Book of Hours quite so closely as Harley MS 2320, they are nonetheless a product of that tradition. The pen-and-ink flourishes in Wellcome MS 406 (fig. 3.1, above), for example, or the fanciful catchwords at the bottom margins of Wellcome MS 5262 are expressions of a manuscript culture that recognized books as precious objects—a recognition that came from readers' encounters with religious books in church or at home.

Yet we should remember that while early practical manuscripts may look like prayer books, and while both genres did serve as guides to practice, their contents were not intended to inspire the same reactions in readers. A prayer book was meant to invoke reverence for the divine, and so it makes sense that they would be beautifully illustrated. Thanks to Kate Rudy's pioneering work with a densitometer, we know that readers were overawed by their encounters with the images they found in their psalters and Books of Hours.[26] Some readers even went so far as to kiss the pages from which saints' serene faces looked back at them.[27] In other words, the splendor of these books encouraged reader responses that were appropriate given their contents. But what purpose did illustrations serve in a manuscript like Harley 2320? Did the initials in that manuscript inspire its reader to more fervent work as a silkthrower? Did its marginal flourishing give her pause as she consulted the manuscript to determine auspicious days for letting blood or taking a journey?

Unfortunately, there is little evidence in the Harley manuscript of its original owners' interactions with its contents, nor in fact is there much evidence to reconstruct how any early fifteenth-century reader used their practical manuscript. First of all, very few early fifteenth-century readers signed their names or added new recipes in the margins of their books, probably because very few could write. No doubt the commissioner of the Harley manuscript felt comfortable reading Middle English verse, but that does not mean she could write much more than her name—if she could even do that. Men like the Dukes of Bedford and Gloucester and the wealthy patrons who commissioned lavish translations of Latin surgical manuals probably did know how to write, but like Peter Cantele or Robert Taylor, they weren't keen to undertake such work themselves. In the late fourteenth and early fifteenth centuries, writing was a skill practiced by experts.[28] And second, besides a lack of written evidence to indicate readers' responses to recipes or instructions, there is also very little material evidence to suggest that early owners of practical manuscripts consulted them regularly in kitchens or gardens while making medicines or tending herbs. We might expect to find drips or smudges from pots of simmering ointments or residues from grinding herbs on their pages, but those non-textual signs of use are just as scarce as early fifteenth-century reader marks. Though Pamela Smith has argued that premodern recipes communicated a kind of tacit knowledge that practitioners acquired through hands-on manipulation of matter, there are few traces of that hands-on manipulation in early English practical manuscripts.[29]

And yet, the evidence assembled in this section suggests that early practical manuscripts were prized as objects with material, social, and intellectual value. They were commissioned by members of the nobility, like the Dukes of Bedford and Gloucester, and by less renowned fifteenth-century readers, who sought to imitate the practices of England's most elite.[30] But even though would-be readers commissioned artisans to write useful natural knowledge in beautiful books, it is not at all clear how this knowledge was put to use. Early fifteenth-century English practical manuscripts would thus seem to challenge the assumptions that underlie a number of recent projects emphasizing recipe reconstruction as a means of recovering premodern attitudes toward nature.[31] Some of the recipes in Middle English manuscripts, like the instructions for silk braiding in Harley MS 2320, record the kinds of hands-on practices prized by historians of science. Yet even so, these written directions represent just a small fraction of the tacit practical knowledge that must have circulated in premodern England by other means, including through oral transmission or observation. By contrast, the majority of recipes recorded in early fifteenth-century English manuscripts seem to have been valued by their readers *as texts*. Rather than seeking evidence of emergent experimental practice in these collections, perhaps we ought to ask instead why so many lay readers in fifteenth-century England—readers who no doubt had tacit knowledge of healing practices and local herbal ingredients—were so enamored by the prospect of possessing natural knowledge *in writing*.

Writing the Power of Nature

The trained scribe who in 1462 had just finished copying the Middle English translation of Gilbert the Englishman's *Compendium of Medicine* in Wellcome Library MS 537 might have had an answer to that question. Though he probably felt the same relief that Symon Wysbech did upon completion of Robert Taylor's manuscript, this scribe resisted the temptation to fill what blank space remained after 262 folios of medical recipes with a joke. He had completed a gargantuan task, but he didn't make light of it, perhaps because the very last line he copied into that compendium was this: "To three things God giveth virtue: to words, to herbs, and to stones. *Deo gracias*."[32] If this scribe had any doubts about the consequence of his work, this popular medieval adage should have put them to rest. God had imparted divine qualities (virtues) into words and into natural matter. In the act of copying the many, many words of the thirteenth-century physician Gilbert the Englishman into Wellcome MS 537, this anonymous scribe had done his part to make those divine qualities accessible to others.

Nature and words, words and nature: these were the tangible and sensible manifestations of God on earth. Every parishioner who attended Mass in fifteenth-century England knew that God *was* Words, as the first verse of the Gospel of John insisted.[33] Every time a priest recited the *Pater noster* or the *Ave Maria* or uttered the ritual blessing that transformed bread into the body of Christ, he was calling down God's presence or power. So, too, did fifteenth-century readers recognize that God's power could be felt in nature, and specifically, in the natural ingredients of medicines.[34] As a prayer in one fifteenth-century manuscript put it, God had "created and marvelously formed mankind," and made natural matter into "medicine for restoring the health of the human body."[35] God's providence, manifest in nature, was the promise underlying every medical recipe that required the use of herbs, plants, stones, and animal byproducts. To prepare an ointment or concoct a healing drink from these ingredients was to harness the divinely appointed healing properties of nature, amplifying their "virtues" in order to heal, as God had intended.

Medical recipes and herbals were essentially primers for understanding how these divine qualities might be harnessed most effectively. One very popular Middle English herbal promised as much in its incipit: "Here may men see the virtues of herbs, which been hot and which been cold."[36] As this incipit suggests, the language of divine "virtues" could be made to align quite nicely with ancient pharmacological theory, which held that the whole of the universe was composed of four elements (air, water, fire, and earth) whose qualities of hotness, coldness, wetness, or dryness were manifest in all natural matter. In the human body, it was the balance of the four humors (blood, black bile, yellow bile, and phlegm) that determined the body's combination of hotness or coldness, wetness or dryness. Phlegm, for example, was cold and wet, and thus, a body with too much phlegm would have a cold and wet complexion. The phlegmatic patient would need a remedy whose "virtues" were hot and dry to counteract this imbalance. If this patient were lucky enough to have a copy of the *Circa instans* herbal lying around, a quick glance would reveal that aloe was "hot and dry in the second degree," and was thus an ideal treatment.[37] Without this text as a guide, however, no amount of touching, tasting, or smelling aloe would have revealed its elemental qualities. They were intrinsic properties that were neither sensible nor tangible, only comprehensible through recourse to texts.

The words collected in practical manuscripts were thus the keys to understanding the hidden qualities of the natural world, put in place by God for the benefit of mankind. For this reason, early fifteenth-century readers valued their practical manuscripts tremendously, even if they never took them into a stillroom or herb garden to make a medicine or harvest an herbal ingredient. Like the prognostications we examined in the previous chapter, the theory of divine virtues referenced in medical recipes and herbals imposed order on a world that was unpredictable and difficult to interpret. But because this order was not immediately obvious—because the healing properties of natural matter were only explicable through recourse to theory—fifteenth-century compilers quite reasonably preferred practical texts attributed to well-known authorities over those without a venerable pedigree. They had to trust these texts to tell them what their eyes could not. In that situation, who wouldn't value a remedy collection that supposedly originated with "Ypocras [Hippocrates] the good surgeon, and Socrates & Galen," over one compiled by the local village healer?[38] Surely Hippocrates and Galen could be trusted to explain the order of the natural world when observation was insufficient.

Sometimes, however, even the theories of Hippocrates and Galen fell short. Sometimes a recipe's efficacy couldn't be explained according to theories about the divine "virtues" of herbs or plants or the complexion of the human body. Sometimes a recipe harnessed divine power more directly, bypassing the usual order of elemental qualities. These recipes are what we would call charms, the vast majority of which employed words from Christian ritual (or words that evoked Christian ritual), repurposing them to direct divine healing toward a specific ailment or injury. For example, one popular charm for a woman in childbirth directed a reader to write a list of holy mothers and their offspring (Mary and Jesus, Anna and Mary, Elizabeth and John, Cecilia and Remigius) on parchment or paper and then bind it to the woman's right thigh.[39] Another charm "for the fevers" instructed a reader to write holy words (*pater est alpha, filius est vita,* and *spiritus sanctus est remedium*) on three communion wafers, or obleys, and ingest them over three days.[40] Still another charm, also "for the fevers," had a user write holy words on three sage leaves, which again would be ingested over three days.[41] All three of these charms were included within the popular Middle English recipe collection known by its incipit, "The man that will of leechcraft lere [learn]," quoted above invoking the authority of Hippocrates and Galen.[42] Whether inscribed on parchment as part of a healing charm or attributed to ancient authorities, in this recipe collection and many others in Middle English, words were the key to accessing divine power in nature.

Nevertheless, healing with holy words was a contested practice in the eyes of the fifteenth-century church. St. Augustine, one of the founding fathers of the early church, condemned the use of amulets and charms as pagan superstition, laying the foundations for medieval canon law, which officially categorized healing charms and amulets as illicit magic.[43] According to church doctrine, it was perfectly appropriate to recite the *Pater noster* as a prayer to call down God's blessing on a sick patient, but it was another thing entirely to recite the *Pater noster* over an herb to transform it into medicine. Yet, because the line between licit and illicit healing practices was so fine, even learned clerics like Gilbert the Englishman and Bernard of Gordon found reason to include charms in their Latin compendia of medical recipes.[44] They did so because God's power was magnificent and mysterious, and who could say that the right words spoken over herbs didn't call down divine healing?

As Thomas of Cobham, a fourteenth-century English bishop and theologian, put it: "we know something of the virtue of herbs and stones, but of the virtues of words we know little or nothing."[45] Because the power of healing words couldn't be explained through medical theory, learned medical writers like Gilbert the Englishman classified charms as *experimenta* or *empirica*, a designation that indicated their power had been proven through experience.[46] This classification meant that charms could be more easily reconciled with other medical texts whose authority depended on textual tradition. Charms that were designated as *experimenta* had value precisely because *someone* had witnessed their efficacy and put them into writing. Indeed, Gilbert the Englishman defended the inclusion of charms in his thirteenth-century recipe collection, despite being unable to explain how or why they worked, on the grounds that they reflected the wisdom of the ancients.[47] Not surprisingly, Middle English practical manuscripts whose recipes were derived from earlier Latin collections followed the example set by Gilbert, Bernard, and Thomas and often included charms, too. Of the 182 manuscripts surveyed for this book, seventy-four contain at least one charm and many of those contain numerous charms within their medical recipe collections.

In all cases, whether recipes were the powerful words of the ancients or simply words powerful enough to harness divine power directly, fifteenth-century readers of practical manuscripts, like the owner of Cambridge University Library MS Dd.4.44, were keenly aware that words were vehicles for the transmission of authority. First copied in the second quarter of the fifteenth century, MS Dd.4.44 opens with three treatises on horse medicine, each of them written in a neat cursive hand, with red and blue capitals. The first veterinary treatise in the collection, called "The Boke of Marchalsi," is composed as a dialogue about equine medicine between a "Master" of horsemanship and his student.[48] The other two veterinary treatises in MS Dd.4.44 are attributed to a named author: the *Practica of William Marshal* (*Practica Willelmi Marescalli*) and "The marchalsie of Piers Moris."[49] Though perhaps not so illustrious as Galen or Hippocrates, all three treatises invoked the authority of an expert whose words could reveal the sometimes hidden workings of nature. But MS Dd.4.44 also contains charms that use words to harness the power of nature more directly. "The Boke of Marchalsi" contains two charms, one for the "farcine" (now known as glanders, a bacterial infection still fatal to horses), and one to stanch blood. The cure for "farcine" directs its reader to "charm" a sick horse with *Pater nosters* and crosses of herbs, lead, and leather—a combination of acts and ingredients made powerful because, as the recipe explains, Jesus Christ "granted virtues to be in words & in stones & in herbs."[50] The treatise of William Marshall contains another charm attributed to a "Saracen of Spain," which nonetheless invokes the figures of Saint Peter and Saint Martin as interlocutors.[51] And finally, the two folios following the treatise of Piers Moris contain various medical recipes for human and horse ailments, including multiple charms: two more for the "farcine" and one to make a horse stand still.[52]

Perhaps because this manuscript made very clear that words were sources of power and authority, the mid-fifteenth-century owner of MS Dd.4.44 felt inspired to try his hand at writing some of those powerful words on blank folios at the back of the manuscript. Figure 3.2 below shows his hasty and amateur writing in attempts to copy a "charm to make a horse stand still" as well as a charm "for the trenches." Perhaps most intriguing in figure 3.2, however, is the section of parchment cut away from the bottom right corner of the page. The rectangle of parchment now missing from this page has the same dimensions as the charm for "trenches" to its left. Though we have no way of knowing what was once written on that small rectangle of parchment, the size and position of the excised section certainly invites speculation that it once held another copy of the charm. Perhaps, after failing to write those healing words as neatly or accurately as he had hoped, this amateur writer and manuscript owner crossed out the charm on the left and tried again on the right. Satisfied with the second attempt, he cut that charm from the manuscript so that it could be used as a cure for his afflicted horse.[53]

Of all the many recipes in MS Dd.4.44's collections relating to horse medicine—many of which required the selection of natural ingredients and the preparation of ointments and powders—the cures that were the easiest for this amateur writer to attempt were charms like the ones in figure 3.2, which merely required inscribing words on parchment. The owner of MS Dd.4.44 didn't need to collect herbal ingredients or boil greasy ointments to cure "trenches" in his horse; he needed only to procure a quill and some ink. Perhaps these simple scribal acts gave this fifteenth-century reader a sense of agency when his or his master's horse took lame. Perhaps that is why he chose to add another twelve charms to the final folios of MS Dd.4.44, including four more charms for the "farcine," a charm to "stanch blood," a charm for worms in a horse, a charm for toothache, and, finally, a "medicine for the axes," a corruption of the Anglo-Norman *ague*, or fever:[54]

> Take a sage leaf that is not pierced and write this on with a pen with ink: *In principio erat verbum angelus nunciat.* And then give it to the sick to eat and let the sick say first five *pater nosters* in the worship of the five wounds of our lord Jesus Christ and five *aves* in the worship of the five joys of our Lady and then in the second day take another leaf and write this on: *Et verbum erat apud deum Johannes predicat.* And say the prayers aforesaid and the third day take another leaf and write this on: *Et deus erat verbum Cristus tonat.* And give it to the sick and let him say the prayers aforesaid and by God's grace he shall be healed.[55]

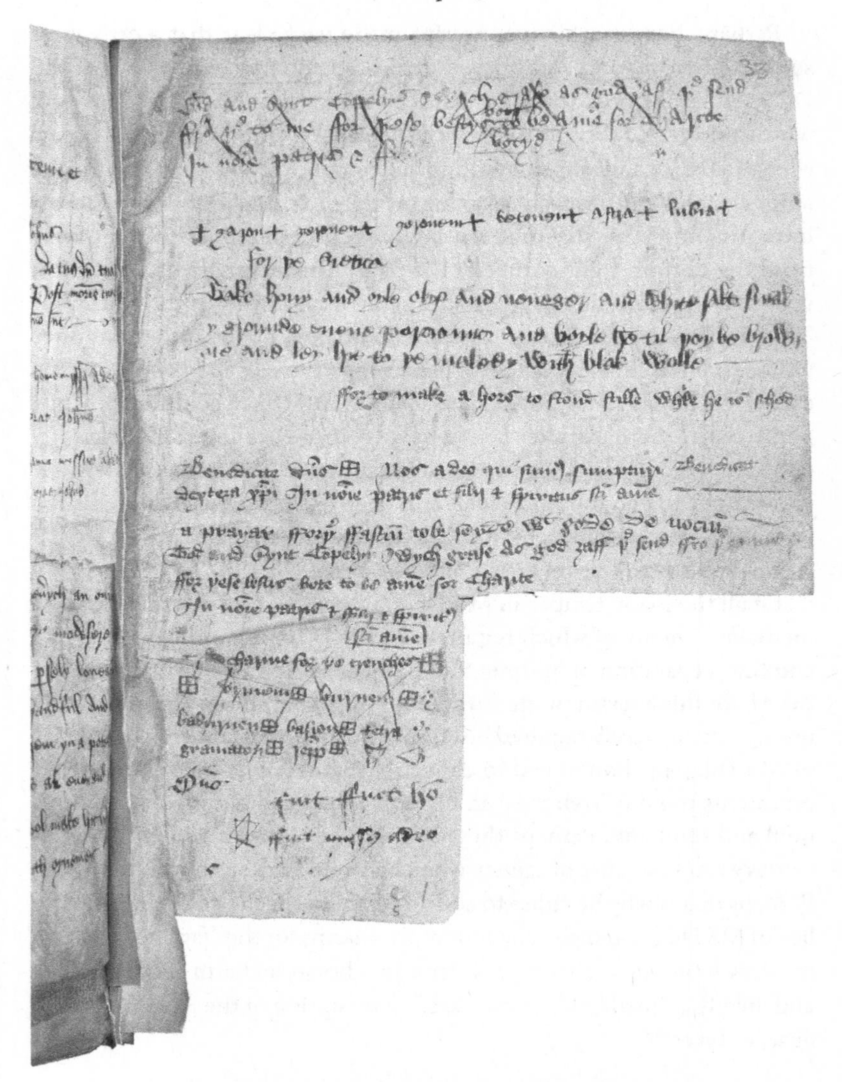

FIGURE 3.2. Charms and recipes added by a mid-fifteenth-century reader to an earlier collection of recipes for horse medicine. Cambridge University Library MS Dd.4.44, f. 33r. By kind permission of the syndics of Cambridge University Library.

Of all the charms added to blank pages in MS Dd.4.44, this sage leaf medicine makes explicit the relationship between healing words and divine power. It invokes the most well-known formulation of the power of words in medieval Christianity, excerpted from the first verse of the Book of John: "In the beginning there was the Word and the Word was with

God and the Word was God."[56] This verse was one that our writer would have heard at Mass, at the moment of the ritual blessing of the holy bread, and at Rogationtide, recited as a means of protecting the parish from evil.[57] Its words were invoked in rituals of prayer, for the sanctification of consumable matter, and as protection—the same functions this edible, textual medicine would perform. For this writer, who clearly recognized the power of words to heal, there could be no verse more representative of that power, which may explain why this copy of the popular sage leaf charm includes the verse from the Book of John when none of the other seventeen copies of the charm that I have examined in practical manuscripts do.[58] The sage leaf medicine in MS Dd.4.44 is the only exemplar with this variation.

If the owner of MS Dd.4.44 was responsible for adding these Latin phrases to a charm already popular in mid-fifteenth-century England, it was because the addition made sense: if God is words, then his Word on a sage leaf should, logically, produce powerful curative effects. Yet we should not overlook the significance of this scribal act. This writer's insertion of a verse from scripture was an act akin to experimentation. In this recipe, words are a key ingredient in the production of a medicine. Adding a new combination of powerful words was thus a bit like adding an extra herbal ingredient in a more traditional recipe for curing fevers. Indeed, if we were to omit the holy words from the sage leaf charm above, we would be left with a recipe that, in every other respect, adheres to the conventions of any other herbal remedy. The sage leaf cure is just as exacting in its instructions for selecting ingredients ("a sage leaf that is not pierced"), in directing a user how long to prepare the cure ("the third day"), and in explaining what to do with the herbal preparation when it was completed ("give it to the sick to eat") as most recipes for drinks or salves or ointments. Like all herbal remedies, the sage leaf medicine contains specific directives for the physical manipulation of natural matter in order to render its God-given virtues sensible and to direct the power of those virtues in a particular manner. In most herbal remedies, those physical manipulations involved chopping plants, boiling them in water, or simmering them in animal fat. In the sage leaf charm, the process that transforms natural matter into a powerful cure is writing. Only writing could call down the divine power of words and at the same time activate the divine power inherent in the created stuff of the natural world.

If fifteenth-century readers thought of words as ingredients and writing as a transformative process—a process not so very different from grinding, soaking, or boiling—then we can better appreciate why charms so often appear right alongside traditional recipes in Middle English remedy collections. In MS Dd.4.44, for example, our amateur writer copied the sage leaf charm just above a recipe for curing hemorrhoids with leeks and another for the mixing of painter's oil, mastic, red lead, yellow ocher, and frankincense. On the verso of that folio, he copied still more instructions for grinding pigments to make dyes for cloth.[59] When the sage leaf charm appears in other practical manuscripts, it is similarly set right alongside herbal or craft instructions. In BL MS Sloane 3160, for example, the charm appears immediately following the herbal that begins "Here may men see the virtue of herbs."[60] Indeed, all of the charms cited thus far in this chapter are found embedded within larger collections of traditional recipes. The charm to assist a laboring woman in CUL MS Additional 9308, mentioned above, appears just after a more traditional recipe for removing a thorn from a wound.[61] In yet another manuscript, the charm for fevers involving "obleys," or communion wafers, appears amid recipes for facial cosmetics, for treating clouded vision, and for cleansing the breast.[62] And in still another manuscript, a charm for toothache requiring the inscription of vaguely Latin nonsense words onto "virgin wax" is the last recipe in a string of more traditional remedies for tooth pain.[63]

Fifteenth-century compilers simply did not perceive charms to be categorically different from traditional recipes. Perhaps they did not because these compilers recognized that words were a component of a divinely ordained system of therapy, which included herbs and stones as well. And perhaps fifteenth-century owners of practical manuscripts realized that both charms and herbal remedies instructed users to manipulate natural matter with precise physical processes. Though it may be difficult for us now to recall the physical work of forming letters by hand, it would not have been for a fifteenth-century reader. Indeed, for those who could only marvel at the skill with which expert scribes cut a feather into a quill and mixed ink from oak galls to prepare a parchment leaf for writing, the physical act of writing *would* seem transformative. Drawing a quill pen across parchment and moving the nib this way and that to form tiny letters was painstaking work that was difficult to perform—to say nothing of the difficulties posed by the furred surface of a sage leaf. Much as it might require special instruction in the art of brewing or fermenting to produce an herbal medicament or ointment, it took patience, dexterity, and practice to render God's word onto something so fragile, and in so doing, transform a leaf into a powerfully curative medicine. And yet, for those few who could form clumsy letters, like the owner of MS Dd.4.44, recipes that invoked the power of words were opportunities to attempt the practice of healing. With seemingly insignificant scribal acts, they could respond to instances of illness or injury, leaving their mark on paper, on wax, on parchment, or on leaves.

Writing Experience

It was no small thing to leave one's mark, as the scribe responsible for Huntington MS HM 19079 no doubt knew. Sometime in the mid-fifteenth century, at the request of his "brother in Christ," he undertook the work of copying hundreds of medical recipes from Gilbert the Englishman's *Compendium of medicine* into what is now a thick manuscript of 244 parchment leaves.[64] But unlike the scribe we met in the previous section, who left no jokes or notes at the end of the same text copied in Wellcome MS 537, this monastic scribe concluded the manuscript with a few addenda clarifying difficult medical terms and methods of preparation. Regarding the use of violet in "electuaries or in syrups," this scribe reminded his reader to "take the flower of violet and not the leaves." Next, he corrected an error in his transcription (or perhaps in the copy text he was using), explaining that when the words "cytr" or "cytre" appear in recipes for electuaries, "it should be saffron." Finally, he added some advice on how to source the "rust of iron" called for in several recipes: "it is not of old rusted iron," but rather rust that has just formed and can be "beaten away" into a powder that should then be "sodden in vinegar thrice or four times."[65]

Because this scribe was probably a monk (a "dear brother in Christ"), and thus both a trained writer and healer, he could leave his mark on a collection of recipes that, according to its prologue, had been "drawn out of good leech's books."[66] He could participate in a written conversation among learned authorities and contribute knowledge garnered from his own experience. But this wasn't so unusual for a churchman like our scribe. Gilbert the Englishman had done very much the same in the thirteenth century when he composed the original Latin text of the *Compendium of medicine*. Indeed, for centuries, churchmen like Gilbert and the scribe responsible for Huntington MS HM 19079 had had a monopoly on the composition of medical texts, simply because they were the only figures in medieval English society who knew something about healing *and* had the skills to write that knowledge down. But of course all that had begun to change by the time Huntington MS HM 19079 was composed in the mid-fifteenth century. As we learned in chapter 1, rising literacy rates combined with a rapid growth in vernacular translation had begun to crack open the church's monopoly on learning. And as these cracks began to show, informally trained or amateur writers and healers found they could push their way into learned conversations from which they had once been excluded. A few, like the Yorkshire medical practitioner Nicholas Neesbett, even found they could leave their mark on this learned textual tradition in the pages of a practical manuscript.

Nicholas Neesbett was an experienced healer. He knew how to apply ointments and salves for cuts or bruises, how to stanch a bleeding wound, and how to brew special drinks to cure ailing patients. He understood the "virtues" of various herbal ingredients growing near his home in Yorkshire, and he also knew where to locate pricier chemical ingredients at the local apothecary's shop. But Neesbett wasn't a university-educated cleric, nor does he appear to have had guild affiliation. There is no trace of Neesbett in the register of the York Barber-Surgeons or in university records or court documents.[67] If not for his collection of surgical recipes, composed on a booklet of twelve paper pages around the middle of the fifteenth century, there would now be no trace of Neesbett's skill as a medical practitioner.[68] Yet if Neesbett's manuscript is a testament to his skill, it is also a testament to the power of a textual tradition. Like most practical manuscripts, nearly all of the surgical recipes in Neesbett's collection derive from versions common to Latin surgeries. Neesbett may have adapted these recipes after reading one or several of the learned surgical manuals available in Middle English translation by the middle of the fifteenth century. Or, alternatively, he may have learned the tried-and-true recipes in his collection while apprenticing with another practitioner. He may even have gathered his recipes from a variety of Middle English manuscripts, most of which didn't retain information about their original sources.

However Neesbett learned of the recipes he copied into his manuscript, it is clear that he took care with its composition. Though Neesbett was an amateur writer, and though his manuscript's pages are neither ruled nor pricked and his handwriting frequently runs into the margins, it seems he was familiar with the materials and methods of professional writers.[69] Indeed, in one medical recipe in his manuscript involving beaten egg white, or glare, Neesbett reminded his reader that glare is what "writers make red ink with."[70] Though there is no red ink in Neesbett's manuscript—no rubricated headings or red initials—Neesbett did try to emulate professional manuscript artisans in other ways to give his recipe book the look of an authoritative text. He decorated the opening initial in his recipe collection with black ink flourishing, and he gave his collection a title (*Sururgia*) and signed his name to it as author, right on the collection's first page (top right, fig. 3.3 below). With this front-page claim to authorial status, Neesbett claimed a place for himself among learned surgeon-authors like Guy

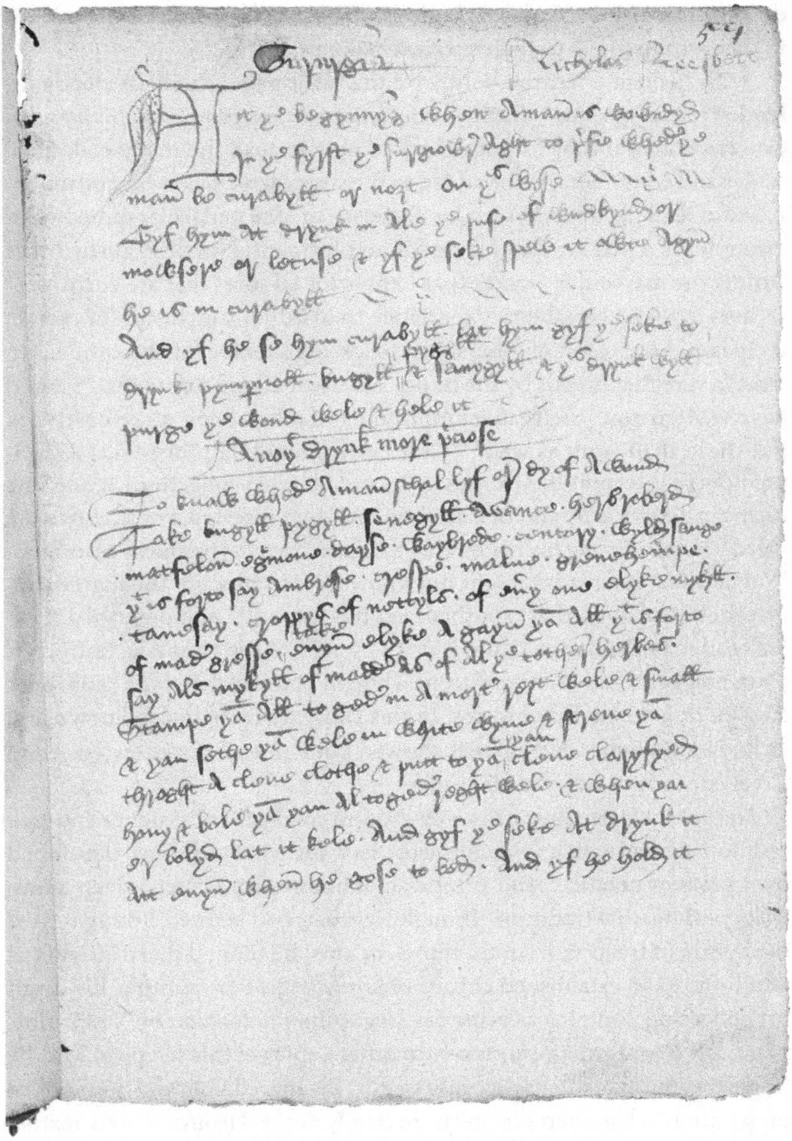

FIGURE 3.3. The first page of the *Sururgia* of Nicholas Neesbett, written in his own hand and featuring his signature at top right. The Bodleian Libraries, The University of Oxford, MS Ashmole 1438 pt. I, p. 57. CC-BY-NC 4.0.

de Chauliac or John Arderne, both of whose names were also prominent in the introductions to their surgical manuals.[71]

With his name written boldly on the front page, Neesbett clearly intended readers to know that his manuscript was the product of his *personal* expertise. And if they forgot, they had only to read the many colloquial asides and commentaries he added to the surgical recipes in his collection. A few of these asides reflected his enthusiasm for a particular remedy. For example, as a preface to a recipe for an ointment, Neesbett gushed that "this is the marvelous worker that I know for all sores that are corrupt."[72] Others invoked Neesbett's experience to justify his inclusion of certain recipes and exclusion of others. Just following a recipe for an ointment to treat "a sore that breeds by the bone," Neesbett added a note that "I could have written you [the reader] ointments that had been more costly . . . but these shall work as well."[73] Still more offered helpful advice regarding how to interpret the recipes he included in his collection. When one recipe called for a "dram" of camphor, Neesbett inserted a note above the word "dram" alerting his reader that it equals "two d. [pennies] of weight." Worried that his reader might not know how to procure the ingredient, Neesbett explained that "apothecaries has it to sell" and described it for his reader as similar to "alum but it is soft."[74] As it happens, both rock alum and camphor are translucent white in color and do look quite a lot alike in their crystallized forms. In this comparison between known and unknown ingredients, Neesbett showed both an expert's knowledge and a real concern to educate his reader.

Neesbett clearly knew his way around medical ingredients and had opinions about which remedies were best for which injuries, developed over years of practice. And yet, these opinions also reflect a deep familiarity with textual tradition. Though Neesbett had learned a thing or two over years of treating wounds, burns, or cuts, he framed these insights as additions to an established corpus of knowledge. For example, his surgical collection contains a recipe for *Unguentum judeorum*, or "Jew's ointment," that conforms very closely to other copies of this recipe in Middle English surgical manuals. In other words, the ingredients and methods of preparation in Neesbett's recipe were clearly derived from a shared, textual source.[75] But Neesbett was not content to leave this standard recipe as is. Like the reader of MS Dd.4.44 who amended the sage leaf charm with the verse from the Book of John, Neesbett added to this recipe based on his experience. At the top of the recipe, he noted his own recommendations for application: "this is the ointment that I told you of & a man were pricked with a thorn . . . take a black wool lock & wet it in this ointment & lie thereto & he shall feel no work nor no disease after that."[76] Neesbett interjected his voice and his experience within an established text. The recipe for *Unguentum judeorum* remained the same, but Neesbett offered his own developed technique for application.

In the midst of a different recipe for a less expensive, "general" ointment for cleansing festered wounds, Neesbett again amended the established text to explain that woodbine—the main ingredient in the ointment—was a "worthy" ingredient that could heal all by itself, making it ideal for "poor" men unable to pay for the costlier ingredients of other ointments. But though Neesbett gave this advice willingly, he also recognized that his experiential knowledge was valuable, and that it distinguished him from other practitioners. He cautioned his reader that it is "a precious thing that I let you wit [understand]" and warned that he should keep Neesbett's tricks of the trade "privy" to himself.[77] Even though Neesbett's collection often presents personal, experiential knowledge as mere addenda to established recipes, this direct address to his reader shows that Neesbett was aware of the value of these additions. That same awareness drove Neesbett to reflect, near the midpoint of his collection, "what mysters [compels] me to write you more when these are sufficient"?[78] If knowledge born from experience was the currency that set Neesbett apart from other practitioners, why did he continue to put it in writing so that others could read it?

Though we cannot now answer that question with certainty, one thing is clear: for Neesbett, writing a practical manuscript was a means of shaping a legacy that he hoped would inform future generations. It was Neesbett's abilities *as a writer* that allowed him to craft a recipe collection that moved beyond the formulaic repetition of authoritative texts to instead reveal something of his individual expertise. Because Neesbett was comfortable with a pen in hand, because he had the means to procure a few sheets of paper, he could begin to make tiny adjustments and additions to the textual tradition that governed how he saw his role as a medical practitioner. And as the fifteenth century wore on, there were many more in England who found, like Neesbett, that they could use pen and paper to craft a record of their expertise. William Aderston, a London surgeon who practiced in the latter decades of the fifteenth century, was able to copy hundreds of medical recipes, prognostications, charms, and craft recipes into a personal collection of practical knowledge, now Bodleian MS Ashmole 1389.[79] The unruled paper leaves of his manuscript are filled to the edges with his sloppy cursive script, while partially filled leaves and blank spaces throughout the manuscript indicate that Aderston recognized he would accumulate more knowledge as he treated patients in the bustling environs of London—knowledge that he hoped to set down in writing. Not only could Aderston fill those blank spaces with recipes copied from established and authoritative manuscript collections, he could also collect tips and instructions from his contemporaries. And, like Neesbett, Aderston could incorporate his own expertise as a practitioner alongside the knowledge he gathered from elsewhere. For example, in a note added to a recipe for "Pellet of Antioch," Aderston wrote: "This I have much used and loved, for with it I healed the sheriff of Bristol."[80] Under another recipe for the pox, he added a mark of approbation with his signature: "approved by me W Aderston."[81]

Aderston and Neesbett were amateur writers whose skills with a pen changed how they thought of their roles as both consumers and purveyors of natural knowledge. In the early fifteenth century, they would have been unusual. By the later fifteenth century, they were in good company. Whereas only fifteen of the ninety-one practical manuscripts that I have dated prior to 1450 look like the work of an amateur writer, thirty-five of the ninety-one practical manuscripts that I've dated to the second half of the fifteenth century look like the work of an untrained scribe. The numbers in table 3.1 below suggest a greater than twofold increase in the number of amateur writers who felt they could compose their own practical manuscript over the course of the fifteenth century.[82]

Not that these amateurs were always able to carry off the job perfectly. Robert Thornton, a member of the Yorkshire gentry whose manuscripts have attracted considerable attention from scholars of Middle English, did his best to copy a collection of Middle English recipes known as the *Book of Diverse Medicines* (*Liber de Diversis Medicinis*) around the middle of the fifteenth century, but unfortunately for Thornton, the tedium of copying got to him. His eye skipped over words when copying recipes, and he misspelled or misunderstood the names of a number of ingredients.[83] Nor could the amateur writer who copied the many recipes that fill a late fifteenth-century manuscript known as *The Tollemache Book of Secrets* quite understand (or correctly spell) the Latin words in the many charms that fill that manuscript.[84] Yet even though these writers weren't expert scribes, they were still able to create textual sources that more directly reflected their needs and desires. It wasn't necessary to have the resources of the Thorntons or Tollemaches (both landed families with country estates) to use pen and ink to engage with natural knowledge, either. Later fifteenth-century readers who may not have been able to create their own manuscripts felt perfectly comfortable amending established collections with added recipes, notes, and simple signatures. Most of these reader marks are quite simple—an added recipe here, a "*probatum est*" there—and do not tell us much about their writers. But some, like those

Table 3.1. Number of practical manuscripts copied by trained, semi-skilled, and amateur writers, arranged by approximate date of composition

	Pre-1400	1400–1449	1450–1500	Total MSS*
Trained scribe (Neat handwriting, ruled pages, bounded margins, often with ornamentation)	4	57	37	98
Semi-trained scribe (Inconsistent letter forms, some ruling, some marginal lines, no decoration)		21	26	47
Amateur scribe (Sloppy handwriting, no ruling, no marginal lines, no decoration)		15	35	50

* Thirteen practical manuscripts feature both trained and semi-trained hands, both trained and amateur hands, or both semi-trained and amateur hands and are thus counted twice in this table.

of John Tyryngham, show a writer putting his experience—in this case, not as healer but as patient—on the page.

Tyryngham was an attorney and not a medical practitioner, but, like so many in fifteenth-century England, he was preoccupied with maintaining his health, and he consulted medical manuscripts as one means of doing so. Tyryngham owned all six of the manuscripts now bound together as Bodleian MS Ashmole 1481, three of which can be classified as practical manuscripts.[85] But Tyryngham was not just a passive reader of medical knowledge. He was a writer, too. On several blank pages in one of those practical manuscripts, Tyryngham made a list of "fesicions" and surgeons in his neighborhood and noted troubling symptoms of illness. On October third (of what year, we do not know), he recorded the emergence of a "swelling," after which he copied excerpts from John of Burgundy's treatise on the pestilence and a short list of the "principal veins" in a man's body.[86] Certainly any one of the physicians or surgeons listed by Tyryngham might have treated his symptoms, but in an era when illness very often proved fatal—and indeed, when emergent "swellings" were especially ominous—Tyryngham clearly felt the need to take an active role in his own healthcare. And the first step in doing so was to pick up a pen and write, listing his symptoms in the blank pages of his notebook and drawing up a list of physicians whom he could consult.[87]

By the close of the fifteenth century, there were even a few women who used pen and paper to manage their household or craft a record of their expertise. *The Tollemache Book of Secrets*, mentioned above, may or may not have been written *by* a woman, but it was certainly intended for a woman's use. Like British Library Harley MS 2320, created nearly a century earlier and discussed in the first section of this chapter, *The Tollemache Book of Secrets* contains instructions for silk braiding, that "craft of women" since time immemorial.[88] But where Harley MS 2320 was composed on parchment with historiated initials and ruled pages, giving it an appearance somewhat like a devotional manuscript, the *Tollemache Book*'s paper pages are totally unadorned, filled with recipes written in a rough cursive script. By the close of the fifteenth century, a woman's book might look very much the same as a surgeon's hastily scribbled collection of recipes. Moreover, when written down, women's medical knowledge might even circulate in networks more often dominated by learned men. This was precisely the hope of John Paston III, who wrote a letter to his wife Margery sometime in the last decade of the fifteenth century requesting that she send him a "plaster of your *flos unguentorum*," as well as written instructions detailing how to apply the common ointment. John was away in London on business where his powerful friend, the King's Attorney, was suffering from knee pain. John had faith that his wife's experience as a healer could help his friend, but he needed her to communicate that expertise in writing.[89] Though Margery's response does not survive and we cannot now read her recipe for *flos unguentorum*, John's letter alone is evidence that the act of writing medical knowledge could generate real authority in later fifteenth-century England—even among friends in high places.[90]

In the early fifteenth century, when practical manuscripts were ornate books with historiated initials and few women could write their name—much less a recipe—there were surely many matrons like Margery Paston who knew how to make special ointments. Some probably sent them off to their husbands and powerful friends, too. Likewise, there were certainly itinerant practitioners like Nicholas Neesbett with considerable experience stanching bleeding wounds or tending broken bones. But in the early fifteenth century—and indeed, for the many centuries preceding—those men and women had no ability to set down their hard-earned knowledge in writing. That they could do so with regularity by the later fifteenth century represents a tremendous advancement in English medicine and science. Women like Margery Paston and men like Nicholas Neesbett,

William Aderston, and John Tyryngham appreciated that their abilities as writers mattered. They composed recipe collections and made written records of their symptoms because putting their knowledge into writing was an important step in making it authoritative. And on some level, they knew that writing was power. Writing made the wisdom of the ancients accessible; it made the workings of nature legible; it made the force of divinity tangible; and finally, it made the hard-won but fleeting knowledge of experience durable, so that others might learn from it, too.

Conclusion

In this chapter, we have followed readers and writers as they engaged with practical manuscripts whose form and function changed over the course of the fifteenth century. As we have seen, those changes paralleled shifts in the way English people thought about writing. In the early fifteenth century, practical manuscripts carried with them the legacies of older book-making traditions centered on devotional practice. Following the lead set by royal patrons of learning like Humfrey, Duke of Gloucester, or John, Duke of Bedford, affluent members of the laity commissioned beautifully ornamented recipe collections as symbols of their wealth, status, intellectual ambition, and reverence for the divine. Maybe they didn't take them to the garden or hold them over simmering pots of ointments, but they did understand the recipes and instructions in these collections as keys to unlocking God's power as it was harnessed in herbs, stones, or even words. Indeed, the charms in these manuscripts, like the recipe for writing the first verse from the Book of John on a sage leaf, helped early fifteenth-century people appreciate writing both as a conduit for the divine and as a transformative process, capable of rendering natural ingredients into powerful cures.

But charms like the sage leaf medicine could only work if readers developed their own writing skills. Over the latter decades of the fifteenth century, as readers grew more comfortable wielding a pen, they also grew more capable of contributing to the authoritative knowledge copied on the pages of their practical manuscripts. Amateur writers like Nicholas Neesbett drew from established textual sources and arranged and commented on this extant knowledge to present a new, and presumably more thorough, textual source. Though still very much anchored in a textual tradition of the medieval past, later practical manuscripts show non-elite fifteenth-century people beginning to assess that tradition, to interject their voices alongside those of the ancients. For the first time, itinerant practitioners like Nicholas Neesbett, or guild-trained surgeons like William Aderston, or skilled household healers like Margery Paston, could record their experiential knowledge, and in doing so, render it authoritative.

As this chapter has demonstrated, written words had real power as far as fifteenth-century readers were concerned. Owning a book of authoritative words by Hippocrates or Galen meant owning the keys to the hidden workings of the natural world. If that book contained divine words with the power to heal directly, even better. It was the perceived power and authority of the written word that drove the proliferation of practical manuscripts in the early fifteenth century, well before we find evidence of fifteenth-century readers consulting them while making medicines or treating wounds. As the fifteenth century progressed, and greater numbers of English people at different levels of society found they could make and own a practical manuscript, the authority of the written word never diminished. All that changed is that many more in England realized they could claim that authority for themselves in the pages of their very own practical manuscripts.

4
Marketing Natural Knowledge

In September 1485, just weeks after King Richard III was killed at the Battle of Bosworth Field and the young upstart Henry Tudor claimed the English Crown for himself, Londoners began succumbing to a mysterious and terrifying illness. In just a matter of hours, a man might go from minding his business, to burning with fever, to dead. It was unlike anything the English had ever experienced. Some called this new pestilence "the sweat" or the "hot sickness" after the fever that struck the afflicted. Others coined the name "stopgallant" for the speed with which this disease killed the young and wealthy. Thomas Le Forestier, a French physician living in London at the time, reported that 15,000 people died during the first outbreak of "the sweat" in September 1485, including two mayors of London who died within eight days of each other. Things were so bad that the would-be Henry VII had to postpone his coronation until the pestilence passed.[1] That same year, William Machlinia, a Belgian-born printer who had set up shop in London, published the first medical book in English: a plague treatise attributed to a fourteenth-century physician from Montpellier.[2] Though the tract concerned the bubonic plague and not "the sweat," there is reason to speculate that Henry VII commissioned the publication, perhaps because he wished to dispel rumors that this new plague was God's punishment for the regicide at Bosworth Field. Machlinia's pamphlet explained that the true portents of plague were signs in nature—a falling star or thunder and lightning, for example—and not, as it were, the death of a king on the battlefield.[3] But then again, Machlinia may simply have been inspired by publication trends on the continent, where pamphlets on the symptoms, origins, and treatments for the plague made up the vast majority of the earliest printed vernacular medical books in Italian and German.[4] In fact, the very same plague tract printed in English by Machlinia in 1485 had been published just about a decade earlier in France by Guillaume le Roy.[5]

The first medical book printed in English was thus very likely a translation of a pamphlet published in France. Yet the reason for its publication—the appearance of a new pestilence—was uniquely English. Those who snapped up the treatise at Machlinia's shop in London likely had no idea that French readers had done the same just a decade earlier. The tract was new to English readers, and it was a hit. Within the year, Machlinia published two more editions, tinkering with the presentation of his best-selling treatise each time (though none of the editions are dated, making it difficult to surmise which came first).[6] In one edition (perhaps the first), the book's very long title appears as the first paragraph in the main body of text. In another (perhaps the second), Machlinia gave his treatise a short title—*Here begynneth a litill boke necessarye & behovefull agenst the pestilence*—which he set just above the first paragraph of text on the first page of the book. Finally, in the third (and perhaps the last), Machlinia did something altogether new in the English print trade: he put that shorter title on its own page, producing the first printed title page in an English book.[7]

The story of Machlinia's efforts to reformat a translation of a French copy of a centuries-old plague treatise can be read as its own portent of things to come within the English printing industry. Over the next century, English printers of practical books followed in Machlinia's footsteps, tinkering with the presentation of very old medical recipes, agricultural treatises, herbals, and other pragmatic texts drawn either from manuscript sources or from continental editions in order to capture the interest of English readers. In this chapter and the two that follow, we will turn away from the compilers, collectors, and readers of manuscripts and look instead at the printers whose labor made vernacular natural knowledge more widely accessible following the arrival of the printing press in England in 1476. Because the English print trade was wholly centered in London, and after 1520 or so, almost entirely confined to shops that lined Fleet Street or ringed St. Paul's churchyard, that's where we'll remain, too. When we glimpse English printers jockeying for position or building on the innovations of their rivals, we should remember that most lived and worked just a stone's throw from one another, passed each other in the streets every day, and formed alliances and rivalries that shaped the trajectory of information exchange in early modern England.[8] We don't know much about these printers' personal lives, but we do know a good bit about what they published, thanks to decades of bibliographical research collected in the *English Short Title Catalogue*.[9]

Building on the publication data available from the *ESTC*, in this chapter I compare multiple editions of the same practical texts to reconstruct how printers tweaked the presentation of natural knowledge to catch and keep readers' interest. This was no easy task. Printers working in early sixteenth-century London only had a few dozen practical texts to work with, nearly all of them drawn from fifteenth-century manuscript sources. With so much competition in the market for natural knowledge, but so little innovation within the genre itself, early printers had to devise ways to convince consumers that each new edition of the same old text was worth their hard-earned coin. Printers had to imagine what readers wanted, based only on evidence of what they'd bought before, and then shape their publications to meet those desires. These marketing decisions in turn shaped what readers came to expect from printed practical books.[10] Over several decades, incremental changes to the presentation of practical books resulted in real transformations to the way English people assessed the value of natural knowledge. Indeed, by the middle of the sixteenth century, when the establishment of the Stationers' Register finally put some constraints on the free-for-all competition among early printers, readers' tastes had already been conditioned by the commercial pressures of a speculative print market. Years of shopping for the latest recipe collection or herbal had taught readers to value novel, original, and experiential knowledge—and to see themselves as participants within England's bustling economy of information exchange.

What's Old Is New Again

In 1476, William Caxton, a prosperous English merchant, set up the first printing press in England at the sign of the "Red Pale" in Westminster.[11] Shortly thereafter, rival printers established presses in Oxford in 1478, St. Albans in 1479, and finally, London in 1480. Five years after its establishment, William Machlinia's press in London published the first medical book printed in English, discussed at the start of this chapter. But surprisingly, the success of Machlinia's plague treatise did not inspire other printers to follow suit. Only two other practical books were published over the next five years in England: a collection of directions on hawking, hunting, and heraldry known as the "Book of Saint Albans" after the location of its publication, and a dietary manual and health regimen published by William Caxton, titled the *Governayle of helthe*.[12] This was Caxton's first practical book, and it would also be his last. In 1492, Caxton died, and his printing operation passed to his foreman, Wynkyn de Worde, a native of the Duchy of Lorraine.[13] From the moment that De Worde assumed control of Caxton's press, his shop—which he moved to the "sign of the Sun" on Fleet Street in London around 1500—dominated the market for practical books in England. Over the next quarter century, in addition to issuing his own editions of earlier publications like the "Book of Saint Albans" (1496 and 1518), *The governall of helthe* (1506), and *A treatyse agaynst pestelence* (1509 and 1511), De Worde also brought out a number of new practical books: the *Proprytees & medicynes of hors* (1497 and 1502?), the *Boke of husbandry* (1508), and *The crafte of graffynge & plantynge of trees* (1518).[14]

Many of these practical books were printed editions of texts widely available in the manuscript collections examined in the previous chapter. The "Book of Saint Albans" featured veterinary medical recipes and instructions for hawking and hunting drawn from an early fourteenth-century Anglo-Norman treatise called *L'Art de Venerie*, which was translated into Middle English and redacted at least three times in the fifteenth century.[15] *The governall of helthe* was an English translation of a health regimen attributed to the fourteenth-century physician John of Burgundy, but it also included a dietary in verse by John Lydgate. Both texts circulated widely in fifteenth-century manuscripts, Lydgate's alone in over sixty witnesses.[16] *A treatyse agaynst pestelence* was a reedition of Machlinia's plague tract, which had circulated in at least one Middle English translation prior to its publication.[17] The veterinary recipes offered in De Worde's *Proprytees & medicynes of hors* were part of the same popular collection excerpted by Peter Cantele at the opening of his medical manuscript, discussed in chapter 1.[18] The *Boke of husbandry* was a Middle English translation of a thirteenth-century Anglo-Norman estate management treatise by Walter de Henley.[19] And finally, *The crafte of graffynge & plantynge of trees* was a combination of two popular Middle English treatises on viticulture and arboriculture, one by Nicholas Bollard and the other a medieval commentary on the fourth-century author Palladius.[20]

That every one of the earliest printed English practical books was a direct adaptation of a text that circulated in fifteenth-century manuscripts should not surprise us.[21] As we saw in chapter 1, those manuscripts were themselves vehicles for the transmission of much older medical and scientific knowledge. Printers simply carried forward that legacy within a new medium. The dual technologies of moveable type and the handpress allowed for many more copies of these texts to be produced than ever before, but they did not immediately inspire a wave of new content—for good reason. The up-front costs of producing a printed edition were enormous: paper, typesets, woodcuts, and ink had to be procured first; next, multiple laborers had to be paid to set the type, print hundreds of pages of text, and then fold, cut, and collate those pages into as many as 500 unbound editions. None of these costs could be recuperated as profits from sales until every last sheet had been printed.[22] In the early sixteenth century, printers shouldered these production expenses, but later they would be fronted by a separate publisher who hired out the manual labor of printing.[23] In either case, the responsible party had to be sure that enough books would sell to recover their expenses. Selecting from among a corpus of already popular texts from manuscript sources would probably have seemed like a safe bet. And yet, for all that medical recipes, plague tracts, and guides to animal husbandry look in hindsight like obvious choices for cash-strapped printers, De Worde was the first to really exploit that market.[24] What did he know that other early printer-publishers didn't?

De Worde's success as a pioneer in printed practical books came from his ability to imagine the speculative printing industry as more than a mere extension of manuscript culture. Caxton and other early printers had mostly focused their energies on publishing expensive copies of poetry or devotional texts—exactly the kinds of works that wealthy patrons commissioned in beautifully illustrated fifteenth-century manuscripts, too.[25] But De Worde saw that there was a much broader market for print so long as his books were produced and sold inexpensively. Of the nine editions of practical books printed by De Worde between 1496 and 1520, only one was longer than twelve pages and printed in a format larger than quarto.[26] These small booklets could be sold at very low prices. According to the records of the Oxford bookseller John Dorne, unbound copies of De Worde's editions of the *Proprytees & medicynes of hors* and the *Boke of husbandry* sold for between one and two pence in the year 1520—about the same price point as two printed ballads, or somewhere between a quarter and a fifth of a day's wages for a laborer in Oxford or London.[27] These pamphlets on gardening, estate management, or husbandry could be purchased in simple paper wrappers, or, alternatively, customers of greater means could bind them together into larger, composite volumes, reminiscent of the manuscript compendia discussed in the previous chapter. That may be exactly what Lady Margaret Beaufort, mother to King Henry VII, did with the "certain small books" she "bought of Wyknyn de Word" for sixteen pence in 1507.[28]

Whether sold individually as pamphlets or grouped together in larger books, De Worde's ability to turn a profit in the market for practical books depended on selling quite a lot of these small publications, and to do that, he needed to entice his readers to come back for more. Like William Machlinia, whose 1485 edition of a plague treatise contained the first stand-alone title page printed in England, De Worde also recognized the marketing power of an eye-catching title page. Caxton and other English printers had continued to use title pages after Machlinia introduced them, but De Worde was the first to think of using woodcut images alongside text on those pages.[29] Today, we think of a book's cover as prime real estate for visual marketing, but sixteenth-century shoppers were accustomed to buying their books unbound, so the woodcut illustrations on title pages would have been readily visible to would-be consumers.[30] When De Worde put an image of a horse and its master on the title page of the 1502 edition of the *Proprytees & medicynes of hors*, or when he used an image of an astronomer on the title page of his 1509 edition of Machlinia's plague tract (a tract that described the signs of plague as falling stars), he was signaling to readers what they could expect within that book's pages.

The practice of clearly delineating the contents of printed practical books with title page imagery certainly must have helped sell more editions, but it also represented a significant shift in the presentation of natural knowledge. Whereas, as we saw in the previous chapter, manuscript collections often featured a variety of medical, astrological, craft-related, or agricultural texts within one volume with little sense of organization, printed practical books isolated genres of natural knowledge in individual imprints. From a brief glance at their title pages, readers could quickly differentiate between information about uroscopy or herbal lore. Because printers very reasonably selected relatively short and self-contained practical texts for publication in stand-alone pamphlets, and because these pamphlets were marketed such that readers could quickly evaluate what made them different from one another, the threads of natural knowledge that had once been woven together within manuscript compendia began slowly to unravel in print.

Yet, for all that print may have altered how readers thought about the interconnectedness of natural knowledge, the very earliest printed practical books did not prompt readers to reassess the origins of that knowledge. Most early practical books had no authorial attribution, though a few did cite prominent ancient and medieval figures. For example, Caxton's *Governayle of helthe* cited Galen as its authority; Machlinia's plague treatise mentioned Avicenna; and De Worde's *Boke of husbandry* prominently (and incorrectly) attributed the treatise to "master Grosehede" (Robert Grosseteste).[31] But not a single English printer working in the first four decades after the arrival of the press in 1476 marketed any of these printed practical books as *theirs*, either in the sense that the texts were their compositions, or in the sense that these texts were unique to their press. Instead, the short titles they chose for their books were informative rather than catchy. In many cases, they simply echoed the incipits of their manuscript sources: *Here begynneth the Proprytees and medycynes for hors* or *Here begynneth a lytell treatyse called the gouernall of helthe with the medecyne of the stomacke.* In those early years, when manuscripts were just as prevalent among readers as printed books, no printer had yet thought to claim ownership over such a vast and diffuse body of natural knowledge.[32]

But as competition grew, the financial pressures of speculative publishing forced a change in how printers thought about their rights to sell and profit from their publications. For early sixteenth-century printers, who might outlay considerable capital to produce a run of up to 500 printed books, no legal mechanism existed to protect their investment. A rival printer could easily undercut the competition by bringing out another edition of the same old text. For nearly all of his career, De Worde had to contend with this kind of competition from Richard Pynson, a Norman-born printer who became active in London right about the same time that De Worde took over Caxton's operation. Whereas De Worde specialized in less expensive works for a more popular audience, Pynson cornered the market for legal books, which eventually led him to become the official printer to King Henry VIII, a role that granted him exclusive rights to print Parliamentary statutes.[33] Despite their differing specialties, however, Pynson and De Worde still went head-to-head in the publication of practical almanacs and prognostications (genres discussed in the following chapter) throughout the first two decades of the sixteenth century. In the end, however, Pynson's close ties to the Crown gave him an outsized advantage. In 1518, Pynson petitioned the king for "royal privilege" to publish a work that had nothing to do with Crown business. He became the first English printer to receive a royal patent protecting his exclusive rights to publish new texts—in his case, for a period of two years.[34]

In the words of bibliographer Peter Blayney, the establishment of royal privilege gave Richard Pynson the means "to protect books that were both 'new' and 'his' from competition."[35] Prior to Pynson's receipt of royal privilege, neither of those concepts had been associated with practical books. As we have seen, most of the earliest printed practical books were sourced from continental publications or older manuscript copies. They were neither "new" in the absolute sense, nor were they claimed as the property of a single printer. But with royal privilege, Pynson had incentive to produce a practical book that he *could* protect from his rivals. In 1523, he commissioned John Fitzherbert, brother of the famous Tudor legal scholar (and prolific author) Anthony Fitzherbert, to compose a new treatise on estate management, published that year as *Here begynneth a newe tracte or treatyse moost profitable for all husbandmen*.[36] Lest anyone doubt Pynson's exclusive rights to this text, he made sure to advertise the work as "new" right in its title and added a phrase to his colophon at the back: "With privilege to him granted by our said sovereign lord the King."[37]

For royal privilege to apply to a text, it had to be novel, and John

Fitzherbert's treatise on husbandry (mostly) qualified as such. Though it was not a reprint of De Worde's 1508 edition of the *Boke of husbandry*, there are echoes of that earlier work throughout Fitzherbert's treatise. For example, chapter 4 of the *Boke of husbandry* (written by De Henley in 1280) outlines the merits of using a team of oxen over a team of horses to plow a field, and—because neither horses, oxen, nor the expense involved in caring for one animal over the other had changed substantially in three hundred years—John Fitzherbert repeated nearly identical advice.[38] Herein lay the problem with attempting to generate novelty within editions of practical, natural knowledge: the authority of that knowledge rested on its fidelity to ancient principles or tried-and-true methods. This dynamic had long structured the circulation of natural knowledge in manuscript, which is one reason why practical manuscripts contain so much redundancy and repetition. Even "new" recipes, like those of Nicholas Neesbett discussed in the previous chapter, were very often adaptations or riffs on established treatments. In other words, though royal privilege made novelty a critical factor for printers to consider when publishing practical books, we should be very careful lest we associate the novelty they claimed for *their editions* with the discovery of truly new knowledge.

Moreover, as royal privilege was granted to more of London's printers, claims to novelty became commonplace in the titles of practical books, as illustrated in table 4.1, below. But since these printers could only rarely claim to have published "new" natural knowledge, they instead foregrounded their editorial work, highlighting "newly corrected" or "newly augmented" editions. In essence, that is what Richard Banckes, only the second of London's printers to receive royal privilege, did with the three medical books he published in 1525 and 1526: an herbal, *Here begynnyth a newe mater, the whiche sheweth and treateth of the vertues & proprytes of herbes,* [hereafter *The vertues & proprytes of herbes*], published in 1525; a guide to uroscopy, *Here begynneth the seynge of urynes, of all the colours that uryns be of* [hereafter *The seynge of urynes*], published in 1525; and finally, a medical recipe collection, *Here begynneth a newe boke of medecynes intytulyd or callyd the treasure of pore men* [hereafter *The treasure of pore men*], published in 1526.[39] Banckes hoped readers would purchase these books as a set, using the final page of *The seynge of urynes* to advertise the other two volumes, urging "all they that desire to have knowledge of Medicines for all such Urines as be before in this book go ye to the Herbal in English or to the book of medicines."[40] At least one early modern reader took him up on the suggestion: a single volume in Cambridge University Library contains all three of Banckes's medical treatises, and the same sixteenth-century handwriting dots the margins of each one (fig. 4.1).[41]

Table 4.1. Editions of practical books printed in England from 1520 to 1559 whose titles advertise novelty

Book titles	1520–1529	1530–1539	1540–1549	1550–1559
Here begynneth a newe tracte or treatyse moost profytable for all husbandmen	1	1	3	4
Here begynnyth a new mater, the whiche sheweth & treateth of the virtues & proprytes of herbes	2			
Here begynneth a newe boke of medecynes intytulyd or callyd the treasure of pore men	1		1	1
The vertuose boke of distyllacyon . . . now newly translate [sic] *out of duyche into englysshe*	2	1		
A declamation in the prayse and commendation of the most hygh and excellent science of physicke . . . newly translated out of Latyn into Englyshe		1		
The great herball newly corrected		1		
The byrth of mankynde, newly translated out of Laten into Englysshe			2	1
The questyonary of cyrurgyens . . . with the spectacles of cyrurgyens newly added			1	
A new booke entyteled the regiment of lyfe: with a syngular Treatise of the pestilence			3	1
A compendyous regyment or a dyetary of healthe . . . newly corrected and imprynted with dyuers addycyons			1	1
A newe herball of Macer, translated out of Laten in to Englysshe			1	
A lytel herball of the properties of herbes, newly amended & corrected . . . by Anthony Askha[m] physycyon				1
A new herbal, wherin are conteyned the names of herbes in Greke, Latin, Englysh, Duch, Frenche				1
A newe boke of phisicke called ye gouernment of health				3
The kalender of shepardes. Newely augmented and corrected				2
Totals:	6	4	12	15

Source: English Short Title Catalogue, accessed at http://estc.bl.uk.

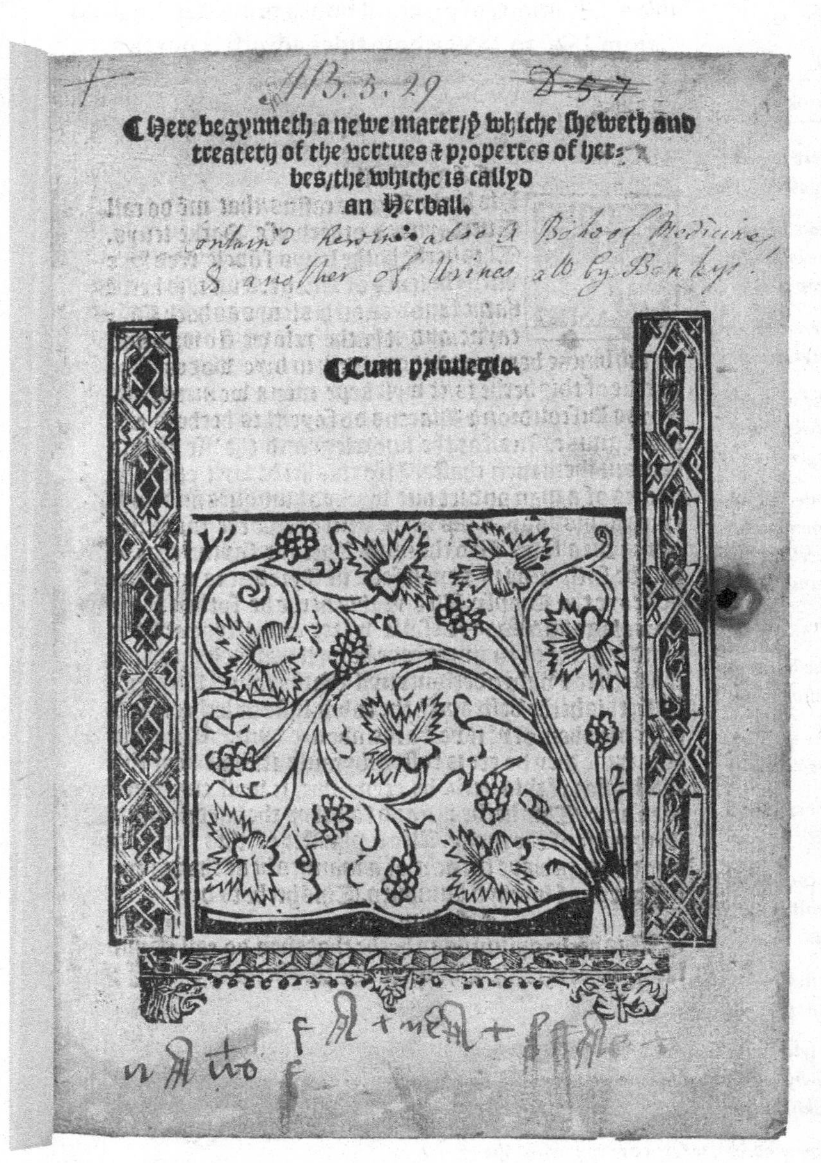

¶ Here begynneth a newe marer/ŷ whiche shewethe and treateth of the vertues & propertes of her=bes/the whiche is callyd an Herball.

¶ Cum priuilegio.

FIGURE 4.1. The title page of the 1526 edition of *The vertues & propertes of herbes*, STC 13175.2, bound into Cambridge University Library Sel.5.175 with Richard Banckes's two other vernacular medical publications, *The seynge of urynes*, STC 22153a (1526), and *The treasure of pore men*, STC 24199 (1526). The reader who left marks at the bottom of this title page left other marginal notes throughout all three treatises in the book. By kind permission of the syndics of Cambridge University Library.

All three of Banckes's medical editions were the first of their kind printed in English, which certainly gave him the right to claim royal privilege for each one. Like Pynson before him, he made sure to stress their novelty in each of their titles, lest a rival doubt his claim to privilege. *The seynge of urynes* was subtitled "not imprinted in English before this time," which was entirely true.[42] The other two publications were described as "new" in their titles, yet both were not so much "new" as newly arranged versions of older texts. A brief survey of the contents of *The treasure of pore men*, Banckes's remedy collection, shows that he sourced his recipes from Middle English recipe manuscripts. For example, the first recipe in the book's chapter on diseases of the head, directing a reader to "take lye of Vervain, Betony, Chamomille, Southernwood & wash thy head therewith three times in a week," is close to identical to one that opens a section of recipes "for the head" in British Library Sloane MS 372.[43] The fifth recipe in that same chapter of *The treasure of pore men*, calling for similar herbs to be chopped, boiled, and mixed with wheat bran to form a plaster, is nearly identical to a recipe that opens the remedy collection in British Library Arundel MS 272.[44] Still another recipe for headache in *The treasure of pore men*, this one requiring "Galingale," "Ginger," "Nutmegs," and "Sugar," is the same recipe that opens a different Middle English remedy collection in British Library Additional MS 12056.[45] Finally, a recipe to determine "if the brain pan be broken" is the same one found in the recipe collection of Nicholas Neesbett, Bodleian MS Ashmole 1438, discussed in the previous chapter.[46]

Yet while the contents of *The treasure of pore men* were not "new" in any absolute sense, the book itself was, in many respects, novel. Banckes clearly had access to a number of fifteenth-century manuscripts whose contents he arranged by ailment, from head to foot, adding a table of contents so the reader could consult his volume with ease. These organizational strategies weren't new to print, of course.[47] A number of fifteenth-century remedy collections were organized from top to toe, and many contain tables of contents added by scribes or compilers after the fact. The difference between those and Banckes's printed version is that in manuscript, it was difficult for compilers to maintain an organizational structure in circumstances of haphazard and unpredictable knowledge collection. British Library Sloane MS 372, for example—one of the manuscripts with a headache remedy similar to that in Banckes's collection—contains a twenty-one-page table of contents at its opening. Its compiler attempted to follow an orderly progression of recipes from head to foot, just like Banckes. But, where Banckes had the luxury of accruing all his many recipes, collating them, and arranging them before setting the type for the printed edition, the compiler of British Library Sloane MS 372 had no such opportunity. The collection starts with recipes to cure ailments of the eyes and ears, headaches, and toothache, but it ends with recipes for preventing moths from eating clothes, for curing palsy, making aquavit, healing burns and, at the last, curing migraines.[48] Though the compiler set out to organize knowledge efficiently, that desire was in tension with an equal desire to collect as much knowledge as possible. At the very least, Banckes's recipe book gave the impression that those two aims could finally be reconciled in print.

Banckes was able to perform a similar organizational feat in *The vertues & proprytes of herbes*, a printed version of the Middle English *Agnus castus* herbal. In manuscript copies, the *Agnus castus* was loosely arranged alphabetically from *A* for *Agnus castus* to *S* for *Solatrum nigrum*, but the difficulties of hand-copying left most manuscript witnesses incomplete in some way. For example, the *Agnus castus* herbal in British Library MS Sloane 1315 ends with *Q*; another in British Library MS Royal 18 A.vi ends with *P*; and still another copy in Wellcome Library MS 409 ends with *L*.[49] Banckes's printed edition is complete, however, and in fact, it incorporates several herbs not original to the Middle English version so that the printed edition ends with *W* for *Wormwood*. Again, Banckes made these efforts at organization and extension legible to readers through the inclusion of an index on the book's final pages. Thus, though the contents of his herbal, like the contents of *The treasure of pore men*, were not really "new" as their titles promised, Banckes's organization of those contents was novel. Both printed collections suggested to readers that something like comprehensiveness had finally been achieved, albeit within a single category of practical knowledge like medical recipes or herbal lore.

In *The treasure of pore men* and *The vertues & proprytes of herbes*, Banckes had done something "new" in collecting and arranging knowledge that had once been scattered. Yet he could not claim to have discovered new knowledge, or even to have selected only the best or most effective knowledge for these books. *The treasure of pore men* retains all the same redundancy and repetition common to manuscript remedy collections. For example, the recipe for headache requiring ginger and sugar, found in British Library Additional MS 12056 and in *The treasure of pore men*, repeats in Banckes's printed edition just three pages after the first iteration, with only the most minor differences between the two versions: one gives ingredients in ounces while the other gives measurements in drams; one attributes the medicine to "Galen the good philosopher" while the other does not.[50] This is far from the only example of such repetition within *The treasure of pore men*. Yet Banckes included no prologue, dedication, or introduction to help readers understand why so many recipes repeated, or why he included certain recipes over others. He didn't see himself as an author of a new medical text whose logic he needed to explain. He saw himself, first and foremost, as an editor of medical manuscript sources.

In that respect, Banckes was not so very different from humanist printers like the renowned Aldus Manutius. Coincidentally, Manutius produced his famous, first-of-their-kind Greek editions of the Galenic and Hippocratic corpus over the same two-year period in which Banckes published his suite of first-of-their-kind vernacular medical books in England.[51] The two printers' aims were similar. When Manutius edited the Hippocratic corpus for publication, his job was to publish Hippocrates's words with as much fidelity to the original Greek as possible. Any questions of interpretation that arose would center on the nature of the Hippocratic writers' language, and not on the nature or theory of Hippocratic medicine.[52] Though Banckes may not have wrestled with thorny questions about the syntax of Middle English as Manutius must have done with ancient Greek, he, too, recognized that his job as editor and publisher was to produce editions of Middle English medical recipes, uroscopy texts, and herbal lore that reflected the original intentions of anonymous medieval medical authorities. When Banckes did make editorial decisions—such as when he chose to combine medical recipes for the same ailment from different manuscripts to create chapters in his recipe collection, or when he amended the *Agnus castus* herbal to extend all the way to *W*—he did so following the organizational principles he found in his manuscript sources (alphabetical order in the case of the herbal, and top-to-toe arrangement for the medical recipes). In the same way that Manutius sought to correct authoritative texts that had been degraded in manuscript copies, so, too, did Banckes seek to correct and complete what he believed to be a deficient textual tradition in Middle English manuscripts.

And, in the same way that Manutius saw no reason to evaluate the precepts of Hippocratic medicine before publishing his editions, Banckes saw no reason to evaluate the methods and ingredients of the Middle English medicine he published in his suite of books. He reproduced the recipes he found in his manuscript sources, redundancies and all. Even though his sources were vernacular compilations unattributed to any major medical authority, and even though they were practical rather than theoretical in nature, Banckes—like other early sixteenth-century editor-printers—had no notion that his textual sources could or should be verified through experimentation. Neither did English readers. The advent of royal privilege may have taught English readers to value novelty in their recipe collections or herbals, but in the 1520s and 1530s, these readers did not yet expect that practical books would contain knowledge that had been practiced and proved.

Marketing Medicine to the Masses

The year is 1540. We are in London shopping for a practical guide to medical care—perhaps an herbal or a recipe collection. The bookshops on Fleet Street and around St. Paul's Cathedral are well stocked with pamphlets, unbound books, and expensive volumes produced by the ten printers working in London at that time. Thomas Berthelet and William Middleton both have their shops on Fleet Street where a few decades earlier Wynkyn de Worde and Richard Pynson had sold their wares from shops at the "sign of the Sun" and the "sign of the George," respectively. Berthelet's was close to the stream (or "conduit") that ran under Fleet Bridge and out into the Thames to the south, only about a five-minute walk east on Fleet Street from Middleton's shop next to St. Dunstan's church. Past Berthelet's shop, another brisk walk five minutes east would take us into the old walled city through Ludgate and immediately into the courtyard around St. Paul's Cathedral. There the printer Thomas Petyt had his shop at the "sign of the Maiden's Head," but there were other bookstalls around St. Paul's, too, managed not by printers but by booksellers. William Tilletson was not a printer himself, but he sold editions printed by Edward Whitchurch and Richard Grafton at his shop by the west door of the great cathedral. No doubt there were other vendors of secondhand books and even a few who sold older manuscripts, too, all of them set up in single-room stalls between the buttresses that held up the church's massive walls.[53]

If we have the good fortune to know a bookseller with connections, theoretically his shop might have as many as eleven different editions of herbals for us to choose from in 1540—all but three of them editions of Banckes's *The vertues & proprytes of herbes*. Banckes himself had very little to do with later editions of his herbal, however. He left the printing business in 1526 and did not return until 1539.[54] By the time he came back to the trade, his royal privilege for *The vertues & proprytes of herbes* had expired, and there was nothing to prevent other printers from publishing their own editions. John Scot was the first to print a new edition of Banckes's herbal in 1537, after which Robert Redman issued a reprint in 1539. Redman's widow, Elizabeth, issued another reprint in 1541, as did Thomas Petyt at his shop in Paul's Churchyard in that same year.[55] But Banckes's herbal had competition, too. By 1540, Peter Treveris had issued three editions of *The grete herbal*, published for the first time in 1526 at Treveris's shop just across the Thames in Southwark. Treveris's herbal was an expensive folio edition of yet another popular text from medieval manuscripts—in this case, an English translation of a French edition of the *Circa instans* herbal, discussed in chapter 1.[56]

Even more competition emerged over the course of the 1530s within the market for medical recipes like those found in Banckes's *The treasure of pore men*. In 1531, Robert Wyer published *This is the myrour or glasse of helthe*, a medical compendium that featured a short treatise on the plague, as well as a remedy collection featuring some of the same recipes in Banckes's book.[57] It was an immediate hit with English readers. By 1540, eight editions had been printed in London, along with another three editions of *The antidotharius, in the whiche thou mayst lerne howe thou shalte make plasters, salves, and oyntment*.[58] *The antidotharius* contained not remedies for ailments, but recipes for the surgical care of wounds, broken bones, cuts, and bruises: recipes for plasters, ointments, salves, balms, and "wound drinks," ultimately derived from Latin surgical manuals.[59] And, by 1540, Robert Redman and Thomas Petyt had brought out four new editions of Banckes's *The treasure of pore men*.[60] Besides these recipe collections and herbals, there were also recently published works on distillation, dietary regimens, plague tracts, and surgical treatises, as well as new editions on animal husbandry, horse medicine, and agriculture. The business in practical books was booming.[61]

Even so, practical books made up just a small percentage of the books published over the 1520s and 1530s, though it is difficult to estimate exactly what percentage. Before the creation of the Stationers' Register in 1557, we can only account for those editions that have survived the intervening centuries. Peter Blayney has estimated that anywhere between forty and fifty percent of books published before 1560 in England have been lost—and inexpensive practical books were probably lost more often than weighty legal tomes.[62] Based on surviving editions recorded in the *English Short Title Catalogue*, Wynkyn de Worde was far and away the most prolific printer of practical books in England until his death in 1534. Even so, the twenty-two editions he issued over the course of his career represent less than three percent of his total output. Far more frequent were his publications of grammar textbooks, which numbered around 230 editions, as well as editions of English verse (eighty-five in total).[63] Practical books make up a significantly larger percentage of the vernacular books recorded in the *ESTC* for Thomas Berthelet, who took over the role of official printer to the king following Richard Pynson's death in 1530. Between the start of his printing career in 1525 and 1540 (the date of our imagined tour through London's bookshops), Berthelet published fifteen practical books. But, for comparison, in just the three years between 1531 and 1534, Berthelet (in his role as printer to the king) published twenty-two books justifying Henry VIII's divorce and explaining the royal supremacy.[64]

De Worde and Berthelet both cornered niches within the sixteenth-century print market thanks to their prominent status within the trade. But vernacular medicine was no one's specialty, really. Sickness and injury were commonalities that united all Londoners, and it seems every printer thought they could make money on a book of herbal cures or health regimens. Some were certainly more active as printers of medical books than others, but nearly every printer working in London before 1550 produced a recipe book or herbal at some point or another. Indeed, between 1485 and 1550, as many as thirty different printers housed in or around London produced at least one edition of vernacular medicine. Yet, most of these editions were reprints of the same old texts.[65] Though in total there were 114 editions of vernacular medicine printed in England over these sixty-five years, those 114 editions represent just thirty-seven unique texts, most of which were derived from popular manuscript sources.[66]

In her study of printed herbals, Sarah Neville has speculated that London's printers reproduced the same texts over and over again because they were constrained by censorship policies, first laid out by London's Bishop Cuthbert Tunstall in 1526, and then subsequently strengthened by Henry VIII himself in November 1538.[67] The conditions were these: between 1526 and 1538, any new book printed in English needed to pass muster with the Bishop of London's agents, and after 1538, every new book published in English had to be examined "by some of his Grace's Privy Council, or other such as his highness shall appoint."[68] In practical terms, these policies encouraged publishers to reprint older texts that had once been granted "royal privilege"—and thus had presumably been approved (though not recommended) by the Crown—rather than appeal to the Crown for clearance to publish something new. It was therefore all the more critical for printers to distinguish each new edition of an old text from those of their rivals.

One printer had a considerable leg up on the competition in this regard. As official printer to King Henry VIII, Thomas Berthelet, whose shop abutted Fleet Bridge on Fleet Street, had a much easier time getting approval for his vernacular publications. Moreover, the might and authority of the king served as guarantor that none of Berthelet's rivals would attempt to undercut him. As a result, Berthelet could afford to commission a number of practical books that were truly "new" to the English reading public. Moreover, because he was well connected, he could put the best minds of England to work on these new commissions. In the late 1520s, Berthelet was able to convince Thomas Paynell, at one time canon of Merton Priory and chaplain to Henry VIII, to translate the *Regimen of health of Salerno* (*Regimen sanitatis Salerni*) into English for publication from his press.[69] This dietary regimen in verse (falsely attributed to Arnaud of Villanova) was an enormous success and was issued a total of four times by Berthelet between 1528 and 1541.[70] Following the success of the *Regimen*, Berthelet again asked Paynell for his help with another medical translation that he believed would be "beneficial to the commonwealth," this one having to do with treatments for syphilis, commonly known as the French pox.[71] Paynell again set to work, this time rendering the German reformer Ullrich von Hutten's *De morbo gallico* into English. It was the first book published in England to describe using a New World botanical, the Caribbean shrub known as guaiacum, to cure a disease that had become epidemic in Europe beginning in the last decade of the fifteenth century. Berthelet published it at least five times.[72] But Berthelet's most successful medical publication by far was *The castell of helth*, composed by the well-known humanist scholar Sir Thomas Elyot, who likely trained in medicine with Thomas Linacre, translator of Galen's Greek texts and founder of the College of Physicians. *The castell of helth*, issued a total of eight times by Berthelet, offered English readers a synthesis of the Galenic medical theory that Linacre knew so well, accompanied by medical recipes. It was the first printed book to present this relatively learned material in English.[73]

Berthelet, official printer to the king, could invest in publishing medical works that were truly "new" in the late 1520s and 1530s, secure in the knowledge that he had the backing of the Crown and that few would dare undercut him. He was able to publish thirty-four practical books over the course of his career, and nearly all of them remained exclusive to his press. But no one else had the security of his position as King's Printer. The rest of London's printers and publishers would need to appeal directly to the tastes of English readers, using whatever strategies they could to sell their edition of the same old herbal or recipe collection to increasingly discerning and selective consumers. Whichever printer could master the art of marketing natural knowledge would rise above the rest.

Robert Wyer was that printer. In a career that spanned from 1529 to 1556, Wyer printed fifty-nine editions of practical books from his press "at the Sygne of saint John Evangelyst."[74] But Wyer's shop was not in the churchyard at St. Paul's or even set up alongside Fleet Street near his rivals. He did business on the outskirts of London, just next to Charing Cross, around a mile to the west of the next closest print shop—Robert Redman's at St. Dunston's church. Though Wyer was physically closer to the court at Westminster, he had none of the royal connections of Berthelet. To succeed in London's commercial print market, he had to find some way to give his publications the same air of novelty that Berthelet could count on. And so, over the course of a twenty-seven-year career, Robert Wyer steadily tweaked and altered the contents and titles of already popular practical books to market them to English readers, in the process tweaking and altering what those readers thought they wanted from a practical book.

For example, after issuing two successful editions of the remedy collection and plague treatise *The myrour or glasse of helthe*, in 1540, Wyer tweaked the text for a new edition, which he titled *This is the glasse of helth, a great Treasure for pore men*. He removed part of the plague treatise that comprised the first entry in the first edition, choosing only to leave in the section that gave remedies for the plague. In so doing, he turned his edition into more of a remedy collection than a compendium of medical knowledge—much more in line with Banckes's *Treasure of pore men*, whose title he also poached.[75] Wyer pursued the same strategy with Banckes's herbal. By 1541, there had been just six editions of the work published in London: two by Banckes's press in the 1520s, and four published between 1537 and 1541. These four were all nearly identical to one another and had the same new title: *A boke of the propertyes of herbes the whiche is called an Herbal*.[76] Again, Robert Wyer wanted to stand apart from this crowd. In 1540, he issued his own edition of the herbal with a slightly amended text and a new title: *Hereafter foloweth the knowledge, properties, and the vertues of herbes*.[77]

Though Wyer's 1540 edition of Banckes's herbal was once criticized as a work of plagiarism, Sarah Neville has since rehabilitated Wyer's reputation and demonstrated that his efforts to revise and augment Banckes's original text were typical in sixteenth-century England.[78] Even if Wyer did reset and reissue Banckes's original text for his own edition, that would hardly distinguish him from other sixteenth-century printers. And in fact, there is reason to believe that Wyer was even less of a plagiarist (to use the term anachronistically) than his rivals. Whereas the four printers who reissued Banckes's herbal between 1537 and 1541 copied his text exactly, Wyer seems to have followed in Banckes's footsteps and revised the herbal using other fifteenth-century manuscript sources. In Banckes's original edition, for example, the herb *Agnus castus* is described as follows: "it hath yellow flowers and beareth black berries and it groweth in dry woods."[79] In Wyer's edition, the same entry is expanded: "he hath yellow flowers as great as a penny, and this herb beareth above at the crop black berries when they been ripe, & if they be not ripe they been yellow as the flower," a nearly word-for-word copy of the description of *Agnus castus* in British Library MS Sloane 2460.[80] Like Banckes before him, Wyer saw an opportunity to expand a deficient text based on a version he found in manuscript. He gave that newly corrected text a new title and, in so doing, set his edition apart from those published contemporaneously by Petyt and Redman.

In 1543, Wyer issued yet another updated edition of Banckes's herbal under yet another new title: *A newe herball of Macer, translated out of Laten into Englysshe.*[81] This time he cut some of the material from the 1540 edition while adding finding aids to the book's margins. Other than that, there is little else that is "new" in the 1543 edition, and the only portions of the text recently "translated out of Latin into English" are a few of the titles for the herbs. It would be easy, then, to dismiss this new title, and especially the authorial attribution to "Macer," as nothing more than a marketing ploy. The name Macer would have been well known to English readers as the author of the popular Latin herbal *On the powers of herbs* (*De viribus herbarum*), which, in Middle English translation, was often prominently attributed to the "cunning and sage clerk Macer" or sometimes simply given the title "Macer."[82] To be sure, Macer had no relationship to the *Agnus castus* herbal, and it is not altogether clear why Wyer chose to attribute it to him. Perhaps, as some have suggested, he misread a typo in the title of Banckes's second edition (the word "matter" misprinted as "marer," which in blackletter type could easily be mistaken for "macer").[83] Or, alternatively, Wyer may have believed that the manuscript he used as a source for his 1540 edition was, in fact, a translation of Macer's Latin text. Several manuscript copies of the *Agnus castus* herbal—including that in British Library Sloane 2460, cited above, with a version of the *Agnus castus* nearly identical to the one in Wyer's edition—circulated under the Latin title *De virtutes herbarum* (*On the virtues of herbs*), which is awfully close to the Latin title of Macer's herbal, *De viribus herbarum*. Attaching Macer's name to Banckes's herbal was a total misattribution, but one we can perhaps excuse again as the result of a closer attention to manuscript sources.

Whether it was an honest mistake or a clear-eyed marketing strategy, Wyer's decision to market Bancke's old herbal as "new" and attach a famous name to its title inspired other printers to do the same. In 1550 William Powell printed yet another edition of Banckes's herbal, this time with the title *A lytel herball of the properties of herbes newly amended and corrected, with certain addicions at the ende of the boke.*[84] This version was, indeed, amended and corrected: several of the original errors of transcription in Banckes's herbal were fixed, and Powell made sure to put the herbal entries in correct alphabetical order by their Latin titles, beginning his edition with *Abrotonum* instead of *Agnus castus*. But, like Wyer, he also put a famous name in the title: *made in M.D.L. the xii. day of February by A. Askham.* This A. Askham was Anthony Askham, a physician-astrologer from Yorkshire and prolific producer of almanacs throughout the 1540s and 1550s, who was also brother to Roger Askham, tutor to the princess Elizabeth. Though Askham probably was the author of the "certain additions at the end of the book" mentioned in the title (probably a short treatise on astrology and herbal lore), he had nothing to do with the "lytel herball." William Powell certainly knew as much, but he may have positioned Askham's name prominently in the title in hopes that readers might believe that Anthony Askham, famous astrologer, had produced the "newly amended and corrected" herbal, too.[85]

If Powell's use of Askham's name only stretched the truth slightly, Wyer's 1552 edition of Banckes's herbal sidestepped it entirely. This edition, identical to the one from 1543, was titled *Macers herbal: Practysyd by Doctor Lynacro: Translated out of laten, in to Englysshe.*[86] Doctor Lynacro, or Thomas Linacre, was England's most famous physician in the sixteenth century, having founded the College of Physicians in 1518. Wyer's addition of Linacre's name to this edition would seem to be an undisputed case of spurious authorial attribution—though such attributions were awfully common in the era of early print.[87] In fact, however, Linacre isn't credited in Wyer's title as *author* of the herbal. Wyer advertised his edition as one that had been *practiced* by the famous doctor. Instead of touting Linacre's considerable intellect or mastery of ancient languages, Wyer invoked Linacre's skill as a healer. He sought to entice readers to his edition by implying that it had been used by the era's most important physician, someone whose experience could be trusted.

Three years later, in 1555, Wyer tried the same tactic on another of Richard Banckes's original publications from the 1520s. Wyer published a slightly amended version of *The seynge of urynes* with a new title: *Hereafter foloweth the judgement of all urynes: and for to knowe the mannes from the womannes, and beastes both from the mannes & womans, with the coloure of euerye uryne.* Again, he used the book's subtitle to demonstrate that its contents were valuable because they had been tried and tested by knowledgeable experts: *Exercysed, & practysed by Doctor Smyth, with dyerse other. and other at Mountpyller.*[88] Wyer's pivot away from claims of novelty and authorial attribution toward a new emphasis on experience marks an important shift in the marketing strategies adopted by English printers. Prior to Wyer's departure from the printing trade sometime in 1556, there were just ten editions of medical, agricultural, or household-related practical books printed with titles that advertised their contents as "practiced" or "proved" by some expert authority, six of which were from Wyer's press. By contrast, between 1556 and 1600, sixty editions of practical books made such a claim.[89] Readers who purchased these editions grew to recognize that novelty and originality weren't enough: efficacy (or, at least, the promise of it) was what really mattered.

Rights and Registers

The emergence of claims to experiential authority in the titles of practical books around the mid-sixteenth century corresponds with a real shift in the quality of natural knowledge for sale to English readers around the same time. By the 1550s, a whole host of new practical books, many of them based on the experiential knowledge of their author (and not, as it were, on very old manuscript sources), appeared on the English market. After 1551, English readers interested in English pharmacopeia no longer needed to rely on Banckes's printed edition of the medieval *Agnus castus*. If they had the (considerable) means, they could purchase William Turner's *A new herbal*, which very much lived up to its title. Turner's massive work, undertaken to be of use to "physicians of the mean sort, many surgeons and apothecaries, and many of the common people," contained descriptions of more than 170 plants from across Europe, illustrated in hundreds of finely carved woodcuts, nearly all of them directly copied from Leonhart Fuchs's *De historia stirpium*, published in Basel in 1542.[90] In 1553, English readers without much Latin could study the skeletal form and musculature of the human body thanks to Thomas Gemini's *Compendiosa totius anatomie delineatio*, a vernacular treatise on human anatomy based on a thirteenth-century surgical treatise, featuring woodcuts copied directly from Andreas Vesalius's *De humani corporis fabrica*, also published in Basel the year following Fuchs's text.[91]

Given what we have already seen from publishers like Robert Wyer, it makes sense that Turner and Gemini saw no problem repackaging revolutionary knowledge from continental publications in vernacular practical books. Turner and Gemini felt free to pilfer from Fuchs and Vesalius, tweaking and amending their texts to make something new that would appeal to English readers, just as generations of printers had done before them. Yet, it must be said that this sort of piracy worried men like Vesalius and Fuchs, who had labored for years to produce books that were carefully conceived visual depictions of their scholarly arguments.[92] After seeing Gemini's copies of his anatomical drawings, Vesalius lamented that he would be "ashamed to have any one think that I had published these illustrations in such a form."[93] That even a renowned figure like Vesalius would have difficulty maintaining control over the fruits of his intellectual labor is precisely why historian Adrian Johns has argued that authors and printers felt they had to shore up the "credit" and "trustworthiness" of early printed texts. Scientific authority could be easily undermined by piracy and plagiarism.[94]

The partial solution to the problem of piracy, and one examined at great length by Johns, was the incorporation of the Company and Mystery of Stationers in May 1557.[95] That year the very first entries were recorded in the Stationers' Register, a manuscript record of licenses granted to members of the guild for the printing of a text. The incorporation of the Stationers gave that guild total control over the regulation of the print trade in England. Instead of petitioning the king for privilege after producing an edition, the Register became the means by which a printer could secure his rights to publish a text before he had set even a single row of type. A printer need only visit the old castle that had become the Stationers' Hall just to the west of St. Paul's Cathedral to pay a small fee, after which the printer's rights to the title would be entered into the Register. Once registered, not only could that printer bring suit against anyone who published a competing edition of that text, he could also bring suit against anyone who printed a text similar to those he had registered. If a suit was brought against another printer, two senior members of the Stationers' guild would investigate the claim. They might examine the two editions, visit both printers' shops, and compare what they found to the Register itself. If the offending printer was truly afoul of the Register, he would pay a small fine, and his editions might be confiscated.[96]

According to Johns, the system of self-regulation that printers developed over the sixteenth and seventeenth centuries in England made scientific authority possible, widespread recognition of which enabled the scientific revolution as we know it.[97] While that very well may be true, consumers of natural knowledge in mid-sixteenth-century London might have felt differently. They would have had no idea that the regulation of the book trade would give rise to a culture that recognized intellectual advancements as the product and property of individuals—one wherein Turner's and Gemini's pirated copies of Fuchs's and Vesalius's images would be self-evidently problematic. Rather, shoppers looking to buy a recipe collection or an herbal in the later 1550s or 1560s would have felt the effects of the Stationers' Register not as a win for authorial integrity, but rather as a loss of the texts on which they had come to rely over decades.

Figure 4.2, below, shows the publication dates for the most popular vernacular practical books published in England over the course of the sixteenth century, defined as those with five or more editions. On the left side of the graph, titles have been arranged chronologically by date of the first edition, beginning with Machlinia's plague treatise. Looking toward

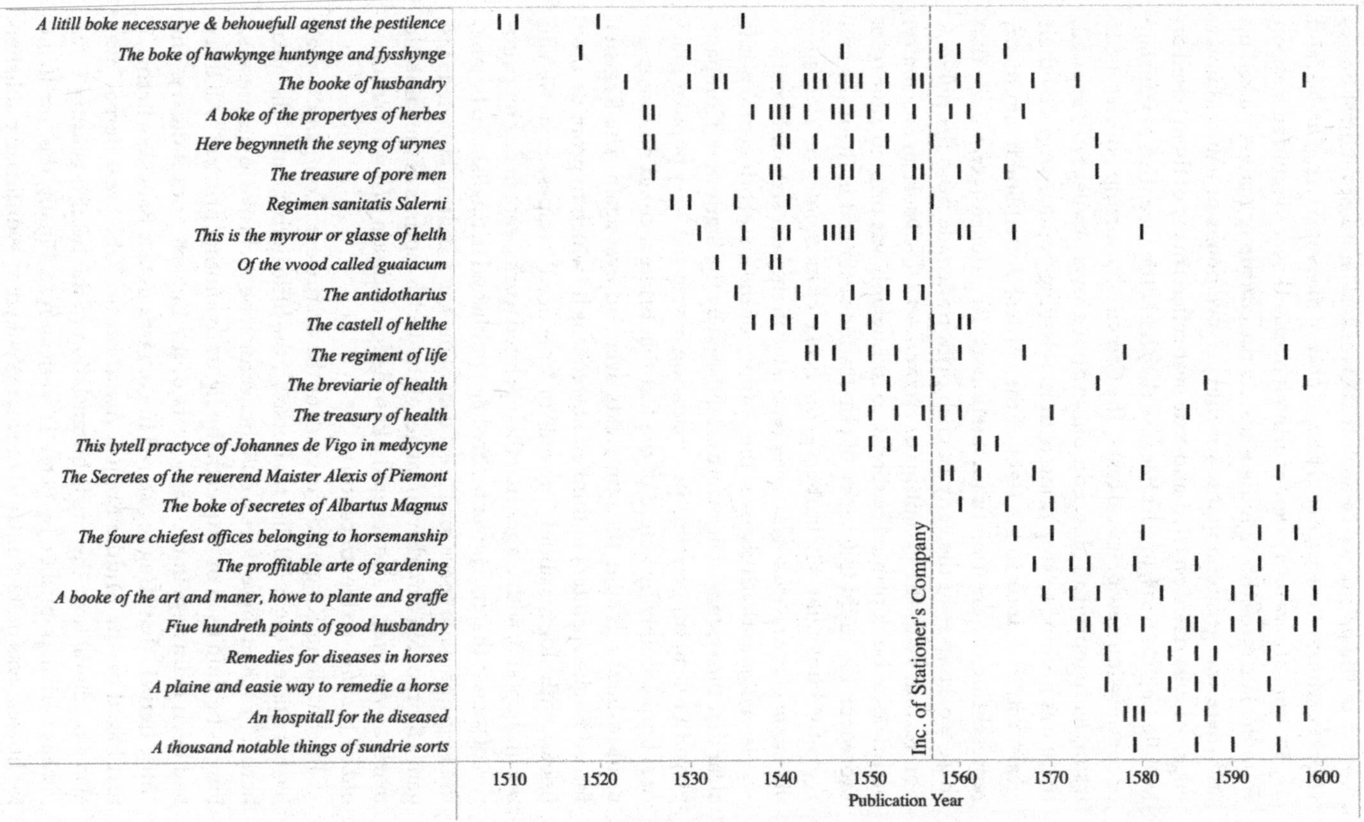

FIGURE 4.2. Gantt plot showing the publication dates for books on medicine, animal husbandry, or agriculture with five or more editions published between 1500 and 1600. Titles are arranged in chronological order by date of first edition. The incorporation of the Stationers' Company in 1557 is represented by the vertical dotted line. Publication date sourced from the *English Short Title Catalogue*, http://estc.bl.uk.

the top of the graph, we can see that Banckes's suite of medical books (*Here begynneth the seynge of urynes, A boke of the propertyes of herbes*, and *The treasure of pore men*) were reprinted over and over in rapid succession from the late 1530s through the 1550s. As the consumer market picked up and there was greater appetite for inexpensive books of natural knowledge, printers relied on tried-and-true best-sellers, many of them based on Middle English texts, in part because they had little incentive to translate or commission new practical texts. The Crown's censorship policies made it simpler to reprint an older edition, and once royal privilege had expired, there was nothing to stop printers from reissuing popular works by their rivals. Yet, if we look to the right of the vertical dotted line in figure 4.2, representing the creation of the Stationers' Register in 1557, we see that these workhorses of the early English print trade fade from the publication record. With the exception of Fitzherbert's *The booke of husbandry*, no practical book printed before 1540 in England was printed after 1580. Moreover, the rapid frequency of publication sustained in the 1540s and 1550s for books like Fitzherbert's *The booke of husbandry* or *This is the myrour or glasse of helthe* was never replicated in the later sixteenth century.

Viewed quantitatively over the entirety of the sixteenth century, it is clear that the system of licensure established by the Stationers' Company played a role in suppressing the circulation of very old, very popular practical books simply by nature of the fact that printers could no longer go head-to-head reissuing the same texts over and over again. The Register forced English printers to think of new means by which to promote their books: with slogans about "proved" or "practiced" recipes, or, as we will see in chapter 6, with claims that their printed works were in fact repositories of secrets that had never before been published in English. To be sure, the Register made space for a great deal of new and exciting knowledge to gain a foothold in a market no longer glutted with reprints of the same old material. And yet, it also severed a knowledge tradition that was centuries old by the mid-sixteenth century.

In 1560, John King paid the clerk of the Stationers' Company a small fee to register "the medicine for horses," "the little herbal," and "the great herbal." Each of these titles would henceforth be exclusive to his press, so long as he didn't sell or bequeath the rights to another printer.[98] All three had been mainstays in the early years of the English print industry. The "little herbal" that King registered in 1560 was in fact Banckes's herbal, a text based on the Middle English *Agnus castus*. "The great herbal" was also a medieval text: the English translation of the twelfth-century *Circa instans* herbal, published by Peter Treveris in 1526. Finally, the "medicine for horses" was quite similar to the text Wynkyn de Worde had published in 1497 and 1502 under the title *Proprytees & medicynes of hors*, which itself was drawn from manuscript sources.[99] After paying his fee to the Company and Mystery of Stationers, John King possessed the rights to a body of practical knowledge that had been circulating, first in manuscript and later in print, for as many as three centuries. Within two years of registering these texts, King had published editions of all of them. In 1560, *A treatyse: contaynynge the orygynall causes, and occasions of the diseases, growynge on horses* was published with a subtitle that emphasized its value as a text that had been practiced, whose authority rested on experience rather than on the antiquity of its manuscript sources: *Collected, and gathered together, by a cunynge horse mayster, very longe usyng, and practysynge the experience therof.*[100] In 1561, King issued a new edition of Banckes's herbal, as well as an edition of Peter Treveris's *The grete herbal.*[101] And then, King died. The medieval collection of "medicine for horses" was reissued just once more, in 1565.[102] Banckes's herbal was printed again for the last time in 1567.[103] *The grete herbal* was never published in English again.

Conclusion

In 1485, when "the sweat" first appeared in England and Londoners watched as friends and family members perished in a matter of hours, one man thought to assuage the anxieties of the English with a pamphlet on the plague. William Machlinia's publication of a fourteenth-century French physician's treatise met the moment, though it had nothing whatsoever to do with "the sweat." The sweat would appear again in England in 1506 and 1511, when the humanist philosopher Erasmus caught it after a summer in London. It would come again in 1517, when it raged so ferociously that Sir Thomas More wrote that it was "safer on the battlefield than in the city" of London, and again in 1528, when Henry VIII's paramour but not-yet-wife Anne Boleyn fell ill.[104] But it wasn't until after the outbreak of 1551 that someone in England finally saw fit to write a medical treatise specifically dealing with a disease that by all accounts was not the bubonic plague. In 1552, John Caius, one of England's most respected medical authors, a man who had once lived with Andreas Vesalius while both were completing their medical training in Italy, composed *A boke, or counseill against the disease commonly called the sweate, or sweating sicknesse.* In it, he described the "begynning, nature, accidents, signs, causes, preservations, and cures natural of this disease the sweating sickness . . . so shortly & plainly as I could for the commune safety of my good countrymen."[105]

In the sixty-seven years between the first outbreak of the sweat and what would be the last, the market for English natural knowledge had changed significantly. Whereas Machlinia was a pioneer, tweaking the presentation of an older text to catch the eye of readers he hoped might purchase a plague tract, Caius didn't need to worry about capturing an English reading public's interest. In Caius's estimation, he had rather the opposite problem. Though he took pains to write his treatise "plainly and in English," he lamented that he would be subjected to "the judgement of the multitude, from whom in matters of learning a man shall be forced to dissent . . . for that the common setting forth and printing of every foolish thing in English, both of physic imperfectly and other matters indiscreetly, diminished the grace of things learned set forth in the same."[106] The proliferation of practical books over the first half of the sixteenth century had made all manner of Englishmen and -women judges and critics of natural knowledge. As Caius knew very well, these readers' tastes and opinions could be swayed by "every foolish thing" they read in print. A learned physician could only throw up his hands at the free-for-all market for medical knowledge in mid-sixteenth-century London. Who would discipline these readers' minds, instruct them properly in what they ought to believe?

No one, in fact. Printers would continue to do their best to attract readers with spurious authorial attributions or advertisements of novelty or even assurances that their texts had, in fact, been "practiced" by notable authorities. Readers would signal their receptiveness to these marketing strategies with their hard-earned coin. No doubt, many of these readers made the "wrong" choices, by Caius's (and our) estimation. Perhaps they shunned William Turner's *A new herbal*, published in 1551, in favor of Wyer's edition of Banckes's herbal, supposedly "practiced by Doctor Lynacro," published the following year. They may have done so because they had an allegiance to a bookshop that carried Wyer's wares, or because his herbal was much cheaper, or simply because their mother or grandmother had relied on that same text, and they wanted their own copy, too. No matter what motivated their choice, it matters that they had a choice to make. And, as this chapter has shown, even when ordinary English readers were duped by printers' marketing strategies—even when they chose the "wrong" texts—they were doing so for what historians of science have long recognized as the "right" reasons. They valued novelty over ancient authority and experiential knowledge over received wisdom. Though early printed practical books had very little that was new or original within them, though they were not the product of experiments in the workshop or tests in the stillroom, printers' efforts to sell those books gradually pushed readers to value these criteria above others, laying the foundations for their wholesale embrace of the truly new practical books that would replace the old in the later sixteenth century.

5

Prognostications Past and Future

On the thirtieth of September 1530, the twenty-second year of Henry VIII's reign, Sir Edmund Walsingham and Sir John Daunce set to work interrogating a prisoner in the Tower of London, accused of the crime of prophecy. The prisoner in question was not Elizabeth Barton, the famous Holy Maid of Kent, who would go to the gallows for the same offense in 1534.[1] The prophet in question was William Harlokke, and in this case, it was not Harlokke's words that brought him to the attention of Henry's Lieutenant of the Tower, but rather the pictures he possessed. Harlokke stood accused of showing some of his countrymen "a calendar of prophecy wherein there were pictures of kings and lords arms," which, according to Harlokke, he had received from a "doctor of physic and astronomer" with whom he had lived a decade or so before. After the doctor's death in 1521, Harlokke had shown this "calendar of prophecy" to a man in Somerset, who interpreted the pictures as evidence that there "should be a great battle of priests." At some point, word of Harlokke's calendar reached Sir Nicholas Wadham, onetime sheriff of Somerset, who told Harlokke to burn it. He did not. Instead, he showed the pictures to Richard Loweth, a goldsmith, who interpreted an image of a dragon in the calendar as a sign that the King of Scotland would arrive imminently on English shores.[2]

What was this "calendar of prophecy"? It certainly sounds a great deal like the manuscripts discussed in chapter 2, which contained pictorial versions of a number of cyclical prognostications. With their combination of prognostic imagery, religious iconography, and medical illustration, the folding calendars and codex-style almanacs discussed in that chapter were exactly the sort of practical book that would appeal to the "doctor of physic and astronomer" whom Harlokke claimed as the calendar's original owner. Moreover, the pictorial prognostication most commonly found in those fifteenth-century manuscripts was one that contained an image of a

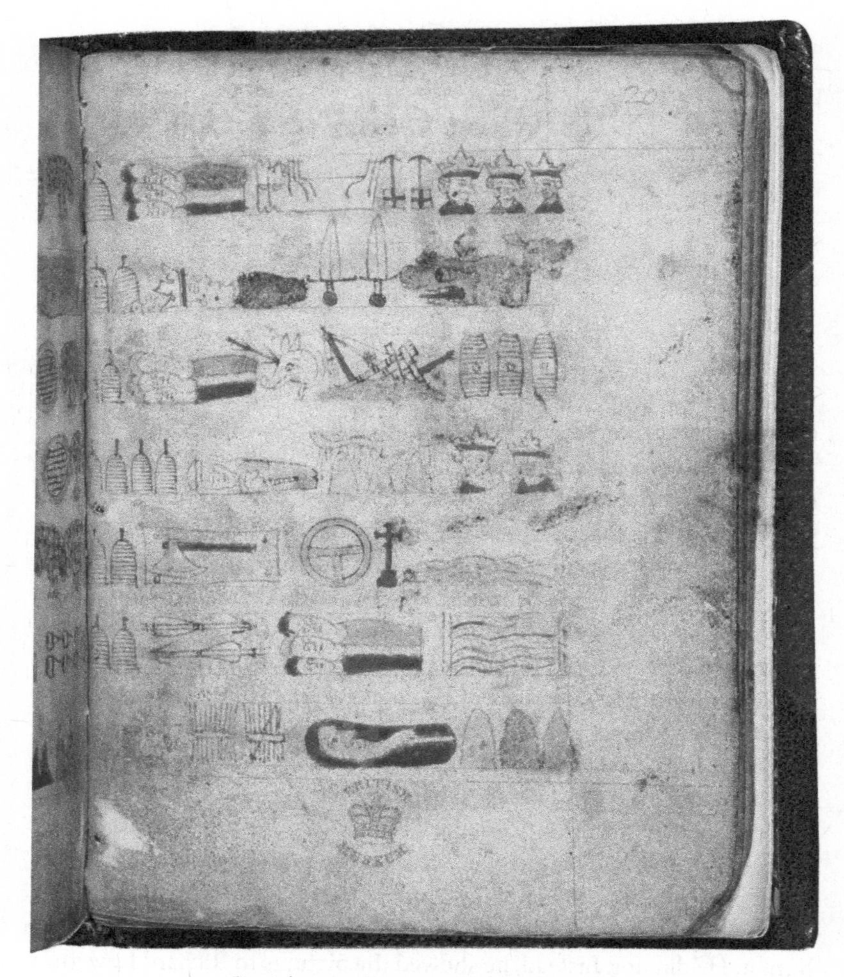

FIGURE 5.1. Partial view of iconographic prognostications according to dominical letter from an early fifteenth-century almanac. The second row of icons (for dominical letter *B*) concludes with an image of a dragon. © The British Library Board, MS Harley 2332, f. 20r.

dragon (second row, far right in fig. 5.1 above) and two swords or two arms engaged in combat (to the left of the dragon), the same images cited by Richard Loweth as portents of the Tudor dynasty's demise.

Of course, in the end, these icons did not correctly predict the coming of the King of Scotland or a "great battle of priests." Yet Harlokke's ordeal in the Tower of London nonetheless illustrates the seriousness with which Henry VIII responded to prophecies in 1530, even when they were pronounced by men of little consequence. In 1530, Henry was in open warfare with his own clergy, and thus a "great battle of priests" was perhaps very much on his mind. He needed an heir, and when the Pope refused to annul his marriage to Catherine of Aragon so he could get one by his beloved Anne Boleyn, he turned to his trusted advisor and the most important man in England, Cardinal Thomas Wolsey, to find a precedent for ridding himself of Catherine without papal sanction. When even his own churchmen wouldn't bend to his will, he had no choice but to break them. By September 1530, Henry and his advisors had begun to formulate a plan for declaring him head of the church in England, which meant entering the religious fray that had engulfed Europe in the thirteen years since Martin Luther nailed those ninety-five theses to the door of the Wittenberg cathedral.[3] Nothing about Henry's future, nor the future of England, was certain. It was a dangerous time to own a picture book of prognostications.

And yet, if the "calendar of prophecy" owned by Harlokke was one of the pictorial manuscripts discussed in chapter 2, containing pictures intended to predict weather, crop yields, and calamities for a seven-year cycle, it is notable that it attracted such attention from the agents of the Crown. That same seven-year prognostication had been widely read in England in its textual format (first in Latin, then in Old English, and finally in Middle English) since at least the ninth century, and a number of practical manuscripts with this exact set of icons circulated in fifteenth-century England.[4] Harlokke's calendar might have been dismissed as a run-of-the-mill physician's handbook, but it wasn't. In 1530 a "calendar of prophecy" was an unusual object. In the fifty years or so since the printing press's arrival in England, the perpetual almanacs and folding calendars that had been such a staple of fifteenth-century medical practice had been replaced by printed ephemera. Though almanacs and prognostications were just as popular among English readers as they had been a century before—perhaps even more so—their format had changed entirely thanks to the technology of moveable type and the pressures of a commercial market.

Printed almanacs weren't meant to convey the rhythms of the church year and the interplay between biblical time and individual experiences as their manuscript predecessors had. They were, instead, the product of a single astrologer's expertise, and they offered English readers a concise prediction for a single year or a short set of tables of lunar and solar alignments for just a few years. These throwaway publications would have been inconceivable within the manuscript culture of the fifteenth century, but they were perfectly made for the world of commercial print. Every year, English consumers would need to purchase another almanac or prognostication. Every year, they would get a brief glimpse into the immediate future. Whether that glimpse proved accurate depended not on their own observations of the natural world, but on the acumen of some remote reader of the stars. William Harlokke's crime was his possession of a book that would encourage his own, inexpert reading of the future—a book of pictures that made the patterns of divine time legible. And though Harlokke couldn't have guessed as much when he was sitting in the Tower, the question of who could read what books to access knowledge of the divine would determine not just Harlokke's fate, but the future of the English church in the decades to come.

As William Harlokke's no doubt harrowing ordeal reminds us, there was much else besides the format of almanacs that was changing in sixteenth-century England. This chapter is as much about the political and religious turmoil of the English Reformation as it is about the emergence of printed almanacs and prognostications in later fifteenth- and sixteenth-century England. For sixteenth-century readers, printed almanacs and prognostications provided a new framework for making sense of a newly tumultuous world. Because almanacs, *kalendaria*, and prognostications had always reflected the rhythms of the church year and the order of ecclesiastical ritual, when those rituals changed, so did the format of almanacs and prognostications. In learning to read these new printed pamphlets and broadsides, English readers also learned to adapt to the new rhythms of the reformed church. If the icons and illustrations of pictorial manuscripts taught English readers to assess their position within the eternal cycle of divine order, then the printed ephemera that rolled off sixteenth-century presses helped English readers account for rupture and look ever toward the future.

Preparing the Heavens for Print

Well before the printing press arrived on English shores, almanacs and *ephemerides*—or tables of solar and lunar positions for a given set of years—were staples in continental print shops. In 1457, just five years after Johannes Gutenberg printed his famous Bible, he issued the very first printed almanac.[5] Then, in 1474, the famed mathematician and printer Johann Müller (known to most as Regiomontanus) produced the first printed *ephemerides* with precise locations for the sun, moon, and five planets for every single day from January 1, 1475, to December 31, 1506. At nearly 450 pages of exquisitely produced tables, the *Ephemerides* was a monumental achievement, both as a feat of mathematical genius and as a work that pushed the limits of the new technology of moveable type.[6] That same year Regiomontanus also produced a *Kalendarium* to meet the needs of less affluent or educated readers, with dates and times for the new moon and full moon for the three Metonic cycles from 1475 to 1531, as well as explanations in German for how to calculate the dates of moveable feasts using the Golden Number and the dominical letter—exactly the same contents found in the many manuscript "physician's almanacs" popular in fifteenth-century England.[7]

While Regiomontanus was busy printing astronomical tables in Königsberg, William Caxton was learning the printing trade from a master printer in Bruges, and yet, when Caxton brought his printing press to England in 1476, there was no subsequent surge in the publication of almanacs or prognostications in Westminster. Caxton was no Regiomontanus. When the first astrological publications did finally appear in English bookshops in 1498, they looked nothing like the manuscripts of fifteenth-century physicians, nor did they resemble the printed *Kalendarium* from Regiomontanus's press. Instead of tables of astronomical figures, English readers were treated to narrative predictions for the coming year, produced by William Parron, a native of the Duchy of Milan who served as official astrologer to King Henry VII.[8] These were not perpetual or cyclical prognostications based on the weather or the lunar cycle, but instead (supposedly) precise predictions for one year calculated according to the position of the sun, moon, and planets.

As court astrologer, Parron was a practitioner of judicial astrology, which fifteenth-century contemporaries recognized as a different sort of astrological practice from that of the local physician, whose interest in astrology might only extend to determining auspicious days for bloodletting, administering purgatives, or taking a journey. A physician could practice astrology using a book like the *Kalendarium* printed by Regiomontanus, or using similar versions in manuscript, like those produced by the Friars John Somer and Nicholas Lynn for fifteenth-century England.[9] Judicial astrology, by contrast, was the science of predicting the future through attention to the precise positions of the planets in relation to one another and within the astrological houses at a single moment in time: at the moment of birth, for example, or at the onset of illness, or—thanks to the invention of the printing press, which made the practice lucrative—at the instant the year turned from old to new. The annual prognostications that resulted from this practice were first popularized in the 1470s in Italy, where astrologers were abuzz with excitement over the rediscovery of classical texts on astronomy and astrology, the most prominent of which were Ptolemy's *Tetrabiblos* in the original Greek and the body of texts attributed in the fifteenth century to Hermes Trismegistus.[10] The newly reinvigorated judicial astrology was not without its critics, and yet, even though debates raged over the authority and orthodoxy of astrological prediction, the powerful in Europe were more than happy to benefit from what they believed to be astrologers' expertise.[11] In the Duchy of Milan, birthplace of William Parron, the ruling Sforza family employed astrologers throughout the fifteenth century to prognosticate on their political fortunes.[12]

When William Parron arrived in England from Milan in the late 1480s, it seems he brought with him some set of ideas about what a court astrologer should provide to his sovereign: namely, lavish gifts and a good deal of flattery to soften what were always dire predictions about the future.[13] For example, though his prognostication for 1498 warned the king to "beware of perils & sicknesses & other such hurts," he followed this warning with immediate assurances that the king "shall overcome his enemies, shall be exalted & fortunate with his children."[14] The role of court astrologer was never an easy one, however. In 1503, Parron fell out of favor with the king, probably because he had failed to foresee the death of Henry VII's queen, Elizabeth of York.[15] When the king's support evaporated, so too did Parron's career as the first prognosticator for the English press. It wasn't until twelve years later, in 1516, that the Dutch astrologer Jasper Laet filled the gap left by Parron's fall from grace. From 1516 to 1548, the Laet family exported annual prognostications composed in their native Antwerp to England, thereby providing the English reading public with descriptions of upcoming eclipses for the year, weather forecasts, and political prognostications. Even so, English readers must have been sorely disappointed with what amounted to just a few sentences about England in compositions otherwise wholly devoted to their Dutch readers. In Laet's prognostication for 1520, for example, England only gets three sentences, sandwiched between predictions for the Duchy of Brabant, Louvain, Antwerp, Brussels, Flanders, Ghent, and Bruges.[16]

Just as English authors were slow to embrace roles as prognosticators for the nation, so too were English printers slow to produce printed tables of solar and lunar positions like those pioneered by Regiomontanus in the 1470s. It wasn't until 1507 that Richard Pynson finally printed a table of lunar and solar conjunctions calculated for the meridian of Oxford, based on the astronomical tables of the fourteenth-century scholar William Rede. Unfortunately, the work was very poorly done. Pynson and his typesetters did not have the technical skill to replicate Regiomontanus's use of spacers set between rows of metal type to produce printed tables, nor does it seem that he had a large enough stock of type to print the dates, hours, and minutes of conjunctions and oppositions in Arabic numerals. Instead, the numerical data indicating the day, hour, and minute for every new moon and full moon for the years 1507 to 1519 are printed in his *ephemerides* as lists of lengthy Roman numerals.[17] Wynkyn de Worde did no better than Pynson in his arrangement of a rival *ephemerides* published in 1508.[18] The poor quality of these printed tables meant that truly committed astrologers must have continued to use manuscripts to guide their practice, perhaps like Bodleian MS Ashmole 340 or Morgan Library MS M.1117, both of which contain the same contents as fifteenth-century manuscript almanacs, with solar and lunar positions updated for the Metonic cycles that began in 1520 and 1539.[19]

And yet, despite what seem like significant problems with both early printed English prognostications and *ephemerides,* they seem to have sold well. The Oxford bookseller John Dorne's register for the year 1520 records the sale of thirty-five "*ciclus vel almanac,*" twenty-three "*prognosticata,*" fifty-five "prognosticon in english," and two books specifically described as "prognostication jasper [Laet]."[20] According to Dorne's records, all of these publications sold for about a penny, and therein lies the most likely explanation for their success: these were very inexpensive publications, attainable by many in England, which offered regularity and predictability in a world racked by uncertainty and upheaval. Fifteenth-century readers had sought exactly the same kind of certainty in the pages of manuscript almanacs, too, but the technology of moveable type made almanacs and prognostications much more accessible. At the same time, however, print also dramatically altered those genres' form and function. Instead of a perpetual prognostication based on the lunar cycle or on the dominical letter—texts that asked the reader to participate in the identification of patterns or signs indicative of an ordered universe—readers of printed prognostications had to trust that men like Jasper Laet had the expertise to read the heavens correctly. None of these early publications gave readers the tools to replicate the predictions themselves.

Moreover, printed almanacs and prognostications offered a very different vision of the passage of time than their manuscript antecedents. Astrologers and printers quickly realized that one-year predictions were far more lucrative than perpetual prognostications: if just one year's prediction was on offer, then readers would have to come back again and again for more. Cheaply printed, disposable prognostications made sense in a competitive print industry, wherein novelty sold books. From a reader's perspective, however, the pivot away from perpetual prognostications to annual ones represented a break with a sense of time that was cyclical, repetitive. The pictorial prognostications discussed in chapter 2 were legible precisely because their patterns repeated: the moon waxed and waned every month and the seasons changed every year just as they had always done. In images and in text, these prognostications offered a means of reading the future as an extension of the past. What was to come was knowable only because of what had been. Printed prognostications described a future that was foreordained according to the exact position of the heavenly bodies at a single moment in time. The future was an outgrowth of the present, not a repetition of the past.[21]

The pressures of commercial print thus incentivized English publishers to eschew perpetual prognostications in favor of annual ones, but even if they had wanted to replicate manuscript almanacs and pictorial prognostications with all their illustrations, they would have had a great deal of difficulty. Illustrations for printed books required a totally different set of skills than those required for manuscript illustration. Elizabeth Eisenstein may have believed that "in duplicating crude woodcuts, publishers were simply carrying on where fifteenth-century copyists left off," but the scarcity of illustrations in early English almanacs and prognostications would suggest otherwise.[22] Even the "crude" replication of manuscript imagery in woodcut required specialized skill, and English artisans simply didn't possess it. Where manuscripts had been illustrated by artisans who used pen and ink, paints, and washes applied directly to a manuscript's page, the black-and-white figural drawings common to early printed books were printed from carved wood blocks that were set into the printers' plates, often surrounded by metal type. A woodcut image began its life like a manuscript illustration, with a draftsman who sketched a figure by hand either on paper or onto the woodblock, typically made of strong pear wood. But the quality of a woodcut image depended on another artisan, the cutter, who carefully chiseled away the negative space of the image so that only the lines of the drawing remained raised to receive ink for impression.[23]

European woodblock printing preceded the introduction of moveable type by a few decades, and in the Low Countries and German lands where the art form was already established, artisans quickly adapted the technique for incorporation with moveable type.[24] Such was German artisans' skill that in 1474, when a rival printer in Nuremberg wished to publish his own edition of Regiomontanus's German *Kalendarium,* he chose to reproduce the whole thing as a block book, carving all those many tables and numbers in relief and in reverse into wood, rather than attempting to replicate Regiomontanus's use of metal type.[25] In England, such a task would have been impossible. Not only did English artisans lack the experience of their continental counterparts, but early English printers seem also to have lacked the vision to even consider images in their early publications. Caxton did not produce a single illustrated book for the first five years of his operation in England. When he finally did, in 1481, the book he produced has been described by Edward Hodnett, cataloguer of early

English woodcuts, as containing "some of the poorest cuts ever inserted between covers."[26]

It wasn't until 1506 that an English printer (or, at least, a printer who worked in England) produced a book that in many ways replicated the astrological and calendrical compendia of the fifteenth century, complete with instructions for remembering the order of the feast days in perpetuity, descriptions of the relationship between the astrological signs and the complexion of the human body, horoscopes for persons born under the influence of each of the seven planets, and—most importantly—tables, diagrams, and images, including a beautifully rendered wheel of time depicting the zodiac signs and labors of the month.[27] *Here begynneth the kalender of shepherdes*, published by Richard Pynson, was a translation of an earlier French publication: *Le compost et kalendrier des bergiers*, popular among French readers since the 1490s.[28] In other words, it wasn't an English printer or artisan who conceived of incorporating images within this practical guide to astrology and medicine. To the contrary, Pynson made quite clear in the preface to his edition that he had only chosen to publish the book after seeing a Frenchman do it first. The Parisian printer Antoine Vérard had published the first English edition of the text in 1503, but he had chosen a Scotsman to do the translation from French.[29] Pynson, a native French speaker himself, saw an opening for a new translation, given that—as he put it in the prologue to his own edition—"no man could understand" the "corrupt English" of Vérard's version.[30] But, though the English of Vérard's edition left something to be desired, his woodcuts were clearly superior to anything Pynson could hope to produce. Rather than make his own illustrations for the 1506 edition, Pynson somehow or another got his hands on Vérard's woodcuts and used those.[31] Continental woodcut artistry was just that much better.

Then, in 1516, Wynkyn de Worde published his own quarto-sized edition of *The kalender of shepeherdes*, at which point readers finally had the chance to purchase an English book that more closely resembled a medieval almanac in its format.[32] The opening pages of De Worde's edition feature a liturgical calendar with small woodcut labor-of-the-month illustrations, as well as the dates and times of the new moon for two Metonic cycles. There are tables of upcoming solar and lunar eclipses from 1480 to 1552, with woodcut illustrations showing the extent and position of each event. Finally, though it doesn't feature the same beautifully rendered wheel of time created by Vérard's French masters, De Worde's edition does include a zodiac man, or *homo signorum*, and two circular renderings of astrological symbols.[33] Pynson and De Worde would each reprint their editions of the shepherd's calendar two more times before 1530, suggesting that the illustrated calendar with motifs drawn from manuscript culture continued to appeal to English readers, even though print shops were stocked with annual prognostications.[34]

The conceit of *The kalender of shepeherdes,* whether in French or English, was that it contained the wisdom of a shepherd, one who "understood no manner of scripture nor writing" but whose simple life tending sheep and sleeping in the fields under the stars gave him precious insight into the best way to live a holy life in sync with the rhythms of nature.[35] We might think of the figure of the shepherd as performing the role of "Harry" or "Peris," the two manorial laborers depicted on the covers of Bodleian MS Rawlinson D.939, the earliest pictorial calendar to include pictorial prognostications, discussed in chapter 2. In both the printed shepherd's calendar and the late fourteenth-century pictorial calendar, agrarian laborers were representatives of a kind of natural knowledge born not from books but from experience. Yet in neither case did *real* agrarian laborers have access to the books their fictional counterparts inspired. Pynson's folio edition of *The kalender of shepherdes* would have cost a small fortune in the early sixteenth century, and De Worde's quarto edition, at nearly 150 pages, was worth more than any shepherd could expend.[36] That was precisely the point, of course. The shepherd in his fields or the hayward with his dog were supposed to exist outside the world of books and learned science. They were English folk who possessed valuable knowledge about the world and the stars, but who "understood no manner of scripture nor writing." These were the sort of people who might only ever read a simple calendar of pictures like the lewdecalendars discussed in chapter 2.

But printed pictorial calendars tracking the rhythms of agrarian life, the cycle of the zodiac, and the pattern of the church year did not exist in England for the first several decades following the introduction of the printing press. It wasn't until the early 1520s that inexpensive printed lewdecalendars finally did begin to circulate among sixteenth-century readers. These updated versions were made on a single sheet of parchment that could be folded into a small square, just like their fifteenth-century antecedents, but instead of hand-drawn labors of the month or saints' icons, these calendars, like the one in figure 5.2 below, were printed from carved woodblocks.[37] Once again, English readers could use images as guides to the passage of different scales of time: the ritual year, the agricultural year, the astrological year, and even the deep time of biblical ages, represented by icons (on the far right of fig. 5.2, below) indicating the years since the world's creation, since the great flood, and since the birth of Christ.

The English artisans who made these woodblock calendars couldn't rival the German woodcut masters, or even the French artisans working

FIGURE 5.2. A xylographic or woodblock printed pictorial calendar. Icons of saints for the months of September through December are visible, as is a brief pictorial history giving the years elapsed since the creation of the world, since the birth of Christ (1523), since the martyrdom of Thomas Becket, and since the coronation of the king. The Bodleian Libraries, The University of Oxford, Douce A 632, recto. CC-BY-NC 4.0.

for Antoine Vérard in Paris, but by the third decade of the sixteenth century, they had learned how to produce inexpensive woodcut illustrations for the masses.[38] Not surprisingly, these woodcut illustrators were driven to improve their craft by exactly the same forces that inspired Benedictine artisans to create pictorial calendars and almanacs in the fifteenth century: the church's insistence on the merits of "reading" religious images in order to contemplate the divine. Well before there were woodcut lewdecalendars in England, these artisans had begun to produce single-sheet woodcut indulgences featuring the ubiquitous "Image of Pity," a depiction of the crucified Christ surrounded by a border of the icons of the Passion. William Caxton first published one of these simple devotional images in 1487, and over the next few decades, hundreds of these woodcuts were published in at least twenty-seven different editions (and probably many more than that, given the survival rate of cheaply printed broadsides).[39] For around a penny or perhaps even less, the English faithful could purchase a woodcut "Image of Pity" as a small remittance for their sins, tack it on their wall, and then "read" this illustration of the crucifixion exactly as described in the fourteenth-century devotional guide *Dives & pauper*: taking "heed by the image how his head was crowned with a garland of thorns" and "how his hands were nailed to the cross" and "on this manner I pray thee read thy book."[40]

When printed pictorial calendars did finally appear on the English market in the 1520s, they were part of a growing momentum to translate the visual rhetoric of medieval devotion into the medium of woodcut illustration for a broad readership. They built on pictorial and devotional traditions first developed in the fourteenth and fifteenth centuries and carried them forward into the era of print. But then, everything changed. Henry VIII demanded an end to his first marriage and precipitated a wholesale break with the Catholic church. His rejection of papal authority opened the door for members of his council to initiate a sweeping program of reform within the newly formed Church of England. By the mid-1530s, the visual rhetoric of the "Image of Pity" and the litany of feast days depicted in printed lewdecalendars had come under direct attack.

A Reformed Astrology

The Act of Supremacy of 1534 made Henry VIII the official head of the church in England, completely severing all its ties to Rome. It did not, however, establish what sort of church Henry would lead. Henry was not particularly inclined to Lutheranism—he had, after all, once been the Pope's lauded "defender of the faith." Yet the two men who might have steered the church away from what they believed to be Lutheran heresy—Cardinal Thomas Wolsey, Lord Chancellor of England, and Sir Thomas More—did not survive Henry's impatience for a divorce from his queen, nor his unsparing demands that the English recognize his authority over their immortal souls. When these two men fell from favor, there were others in Henry's orbit like Thomas Cranmer, Archbishop of Canterbury, and Thomas Cromwell, chief advisor to Henry, who seized the opportunity to rid the church of what they saw as medieval excess and superstition.

Among Henry's councilors, debates raged over whether the English church would preserve or reject a range of medieval beliefs and practices: transubstantiation, clerical celibacy, monasticism, and episcopal hierarchy, to name just a few. Yet for the ordinary English parishioner, the issue that stood to have the most impact on day-to-day worship was this: would the English continue to venerate the saints as they had done for centuries, and would they do so through images and icons? The English church had confronted the question before, in the later fourteenth century, when John Wyclif's followers, the Lollards, condemned the veneration of religious images as idolatry. Drawing on that legacy, Protestant sympathizers in England quickly embraced a similar set of iconoclastic beliefs.[41] Calls to abandon the veneration of the saints in images had grown so loud by 1529 (well before Henry's break with Rome) that Sir Thomas More felt compelled to publish a defense of the practice, recognizing that attacks on icons threatened to bring down the whole of traditional religion.[42] That was precisely the point, of course. Reformers like Hugh Latimer and William Tyndale believed that the accumulated precedents of medieval Popes—like Gregory the Great, who had once praised the veneration of religious images—had turned the church away from its roots and toward idolatry and heresy.[43]

In 1534, Cromwell sanctioned a new English *Primer*, or Book of Hours, that did away entirely with the liturgy of the saints.[44] Then in 1535, while More, foremost defender of Catholic ritual, sat awaiting his death in the Tower of London, William Marshall published a translation of the German reformer Martin Bucer's treatise condemning religious images.[45] In 1536, when the English church issued the Ten Articles, its first official doctrinal document, religious images were sanctioned under the vaguely Gregorian principle that they might be "kindlers and firers of men's minds." Yet, just a few days after the Articles were confirmed, the "Act for the abrogation of certain holydays" did away with much of the traditional Church calendar, removing both local saints' feasts and numerous major feast days from the ritual year.[46] *The Institution of a Christian Man* (more commonly known as the "Bishop's Book"), published in 1537 as a means of disseminating the new church doctrine to the English people, again followed a moderate line and reaffirmed the presence of images in holy spaces.[47] But then, the very next year, a new set of Injunctions to the clergy outlawed acts of devotion before images, including the burning of candles or tapers before depictions of the saints.[48] What looked to be a move toward outright iconoclasm was halted in 1539, however, when Thomas Cromwell was executed and Henry's innate conservativism propelled the English church back toward a more traditional use of religious images.[49]

These bait-and-switch policy changes throughout the 1530s must have contributed to a general feeling of uncertainty among English worshippers, and yet, for all that, it appears to have had little effect on the production of xylographic pictorial calendars like those in figure 5.2, with icons of the saints marking the passage of the church year. Of the four printed pictorial calendars that survive from sixteenth-century England, three were produced in the last decade of Henry's reign: in 1537, 1538, and 1542.[50] Around the same time, an anonymous manuscript artisan produced British Library MS Additional 17367, the only surviving English pictorial manuscript calendar dating from the sixteenth century, created in exactly the same format as those discussed in chapter 2 from the first half of the fifteenth century.[51] With its perpetual prognostications (written in Middle English, not in icons) and endlessly repeating cycle of saints' days copied onto durable parchment, Additional 17367 is a manuscript premised on the stability of English religious practice, on the repetition of meaningful patterns. Saints' days would follow saints' days just as summer followed spring, and Additional 17367 would serve as a guide to these comforting patterns for as long as there was someone to read it.

But of course, that's not what happened. The English church would not follow the same patterns of worship, year after year. Nor would the reader of Additional 17367 continue to reference its calendar. Reader marks on the manuscript show that it was used for just about a decade—no longer than a cheaply printed pamphlet almanac would have been. Though the owner of the calendar made updates to its table of "names & reigns of all the kings," adding regnal years for Henry VIII (who died in 1547), no subsequent regnal years were added for Edward VI (who died in 1553).[52] We cannot know why its owner set the manuscript aside sometime after 1547, but a series of policy changes enacted by Henry's Protestant-educated son may provide a partial answer. A new set of Injunctions drawn up after Henry's death in 1547 reaffirmed the reformist positions of the Injunctions of 1538, and in fact doubled down on the condemnation of images, declaring that church authorities should "take away, utterly extinct and destroy . . . pictures, paintings, and all other monuments of feigned miracles, pilgrimages, idolatry, and superstition" from places of worship.[53] Removing offending images was just the first step in Edward's counselors' ambitious plans for reform. The next was to expunge any traces of the old beliefs. Parish priests would need to swear that they had "taught the people the true use of images; which is only to put them in remembrance of the godly and virtuous lives of them that they do represent." If English worshippers believed otherwise, they were guilty of "idolatry to the great danger of their souls."[54]

The 1549–1550 "Acte for the abolishing and puttinge away of diverse Bookes and Images" went even further in its purge of the material legacy of the medieval church, enjoining the English faithful to destroy Primers and other old religious books.[55] In 1551, Edward put the final nail in the coffin of the Catholic ritual year with the "Acte for the keeping of Hollie daies and Fastinge dayes," which purged the calendar of local saints and reduced the number of festival days in the church year to twenty-three.[56] By 1551, there was no reason to use a pictorial liturgical calendar to celebrate feast days that no longer existed, and there was certainly no reason to hold on to a book that might draw the attention of church or Crown authorities.[57] Pictorial calendars—and by extension, the pictorial prognostications that often came with them—made no sense in a world without the litany of the saints.

The Edwardian Reformation's particular hostility toward images thus put an end to the genre of pictorial calendars, almanacs, and prognostications, which had always relied on the rhythms of medieval worship. Yet these reforms did not diminish English readers' appetite for astrology, nor did they entirely disrupt the widely held belief that close readings of the immutable stars and the changeable earthly sphere could reveal elements of divine order. One of Cromwell's associates, the physician and former Carthusian monk Andrew Boorde, reaffirmed exactly those practices in his 1547 book, *The pryncyples of astronamye*, which explained that the "heavens doth show the glory of god & the firmament doth show the works of god."[58] Boorde believed mankind ought to study the stars in order "to laud & to praise god in his works," and he shared that belief with a number of other important Protestant reformers.[59] Philip Melancthon, one of Martin Luther's closest allies, went so far as to argue that almanacs and *ephemerides* were actually tools of reform because they offered more direct access to God's time in their printed tables with precise figures for the positions of the sun, moon, and planets.[60] The principal aim of Melancthon and many of England's most fervent reformers was to strip away the detritus of medieval superstition and ritual, to lay bare the essence of God's truth in his creation and in his Word. In that regard, spare tables of lunar and solar conjunctions were an obvious improvement over calendars of saints' days, marked by irregular divisions of time that were human-made. The regular and predictable movements of the sun and moon were proof of God's omnipotence and order. Moreover, almanacs and *ephemerides* were a means by which reformers could acquaint ordinary folk with the stripped-down quality of reformed theology, which insisted on individual worshippers' direct relationship to the divine, without the intercession of superstitious medieval priests.[61]

Astronomical tables purported to reflect the bare truth of God's creation, without the imposition of man's artificial time. Yet they were also intended to inspire man's intellection and form the basis from which he offered an interpretation of divine will. A judicial astrologer would use an *ephemerides* to calculate the position of the stars and planets precisely because these positions were supposed to be meaningful predictors of God's intentions. And therein lay the problem for many reformers. Though many favored almanacs over liturgical calendars, they were starkly divided on the question of whether those publications should be used to read the future. Though Melancthon wholeheartedly embraced judicial astrology, Luther was not so sanguine about the practice of reading the stars. He drew a distinction between the study of the movements of the stars (what we would call astronomy) and the formulation of predictions based on those movements (or astrology): "God recognizes certain signs, such as solar and lunar eclipses, and these are not uncertain. However . . . we are not to divine from them." Prognostication, according to Luther, was nothing more than "popecraft [*Papsthum*]" with its "ceremonies."[62] Luther suggested that the esotericism and ritual of prognostication linked that practice with Catholicism, while proponents of judicial astrology like Melancthon saw it as a means of directly accessing God's truth in nature.

Yet if Melancthon's church comprised a universal priesthood of believers empowered to interpret the divine—as Luther suggested in various writings that it should—then the practices of judicial astrology should have been accessible to everyone.[63] That was hardly the case in the printed prognostications that proliferated in Germany and England in the 1530s and 1540s. Instead, in most of those printed pamphlets, the author-astrologer obliquely proclaimed the meaning writ in the stars, and in so doing, performed the role of intercessor to the divine. For precisely that reason, Church officials in England warned that astrologers made themselves false prophets when they intoned about future events.[64] If printed almanacs presented the truth of God's time more directly than manuscript calendars, printed prognostications did the opposite, erecting barriers between reader and the intricacies of the book of nature. Debates about the validity and authority of astrology thus often centered on questions about *how* to read God's truth and *who* should be doing it. And because both proponents and opponents of the practice were so concerned with *reading*, both groups naturally carried out their arguments in the press, in books like Andrew Boorde's *The pryncyples of astronamye*, or in polemical pseudo-prognostications that riffed on the conventions of the genre to satirize the astrologer's craft.

The pseudo-prognostications published by reformers in England were, not surprisingly, first authored in Germany, the epicenter of astrological print in the first half of the sixteenth century and the heartland of the Lutheran Reformation. Perhaps because the English were already accustomed to reading imported prognostications in the 1530s and 1540s, reformers had no problem adapting these German texts for an English readership, too. In 1536, John Ryckes (who called himself a "priest" in his book's introduction) published his English translation of a prognostication by Otto Brunfels, a German naturalist and onetime student of Martin Luther.[65] In a sign of those contentious times, Ryckes dedicated his translation to the "king's noble council: to whom his Grace hath assigned to examine English books."[66] He claimed to have found the original text "cast in a corner among other pamphlets," upon which discovery he chose to make it available to English readers as a correction to those "common prognostications, that yearly goeth about" but are merely "trifles."[67] However, though the book claimed to be an "almanack most true, & ever for to endure," it was not calculated according to the positions of the planets, but rather according to "the will of Him, which is most mighty maker & ruler (not only of stars) but of all things in heaven & earth."[68] The book was filled with excerpts from scripture, as was the pseudo-prognostication published ten years later in 1547 by Miles Coverdale. Just like Brunfels's earlier publication, Coverdale's translation was a riff on a traditional prognostication, with the usual predictions regarding political leaders, crop yields, illnesses, and warfare. But again, *A faythfull and true pronostication* was based not on readings of the stars but on a close reading of scripture. As Coverdale's translation insisted, the stars "are tokens only" because "all things are in Gods hand, and governed by Jesus Christ."[69] Both Coverdale and Ryckes saw in the judicial astrologer a figure not unlike a Catholic priest, offering vain and corrupt interpretations of a set of truths that might be better accessed directly in scripture.

Yet Coverdale's and Ryckes's animosity toward judicial astrology was not broadly representative of the reformist view in England. No such coherence existed. Whereas Coverdale's prognostication insisted on the stars' subservience to divine power, Andrew Boorde's manual on astrology described to readers how the "heavens doth show the glory of god."[70] Both Coverdale and Boorde were closely allied with Cromwell and the reformist cause. And for that reason, even though Coverdale and Boorde sat on either side of the debate over astrology, both men were united in their desire to give English readers direct access to the truths by which they could direct their lives. For Coverdale and Ryckes, that meant turning to scripture. Boorde's astrological manual, on the other hand, imparted to readers the principles they would need to do their own reading of the stars. Published in the same year as Coverdale's pseudo-prognostication, Boorde's book offered his readers a primer on the qualities and influences of the zodiac and the seven planets on the human body. No matter which side of the debate over astrology they stood on, all of these men agreed that readers should be trusted to make judgments about the manifestations of God's power in the world and in his Word. Each of their books prodded English readers to consider: would they rather have the truth dictated to them by self-appointed authorities or would they prefer to use their own faculties to ascertain it for themselves? In 1555, Leonard Digges, a Protestant rebel, mathematician, and astrologer, hoped fervently that English readers would prefer the latter.

Prognostication Everlasting

The year 1554 had not been a good one for Leonard Digges. It started, in January, with his very poor choice to participate in Sir Thomas Wyatt's rebellion against Queen Mary. The rebellion collapsed before it had even really begun, and Wyatt and his noble co-conspirators, including the Duke of Suffolk, were found guilty of hatching a plot to dethrone Mary and put her half sister Elizabeth in her place. Wyatt was tortured and then publicly executed, along with around ninety other rebels. Though Digges was spared public torture, he had his lands and goods confiscated by the Crown, and until he was pardoned on April 1, 1554, he too was destined for a hangman's noose.[71] Even after he felt the reprieve of a royal pardon, however, Digges's ordeal was far from over. By December 1554, the salacious details of the rebellion had been published in John Proctor's *The historie of Wyates rebellion with the order and maner of resisting the same*, wherein readers were treated to exposés of the "degenerate and sedicious" rebels.[72] In March 1555, Digges was ordered to pay the enormous sum of 400 marks to the English Crown in restitution for his participation in the rebellion.[73] In short, in 1555, Digges found himself landless, in debt, and stinking of treason.

Digges needed money and he needed to rehabilitate his image—but how? Just prior to Wyatt's failed rebellion of early 1554, Digges had authored a (now-lost) prognostication, published to no great acclaim. With circumstances as dire as they were, Digges decided to revisit his past work and once again offer his astrological expertise to the English reading public. His revised and extended prognostication was published in 1555 by Thomas Gemini as *A prognostication of right good effect*.[74] The book was dedicated to Sir Edward Fines, Lord Clinton and Saye, one of the nobility who had a pivotal role in putting down Wyatt's rebellion. As something of a *mea culpa*, Digges wrote that he hoped the book might "declare me thankfully minded toward your lordship, among other honorable, to whom I owe myself."[75] It seems Fines could solve the problem of Digges's poor reputation, but would Digges's prognostication help improve his financial situation? Would it sell well among English readers who were already inundated with annual almanacs and prognostications?

By 1555, an English reader in the market for astrology was spoiled for choices. After several decades in which the Laet family of Antwerp were the only prognosticators in the game, over the course of the 1540s, several native English astrologers had entered the market.[76] Andrew Boorde, discussed in the previous section, produced annual prognostications for the years 1545, 1546, and 1547.[77] Anthony Askham, a practicing physician and Cambridge-educated priest, produced his own almanacs and prognostications calculated for the meridian of York for every year from 1548 to 1557, as well as a guidebook to medical astrology, published in 1550.[78] But perhaps no one did more to make the basic tenets of astrology accessible to the average reader than the printer Robert Wyer, whose marketing savvy was documented in the last chapter. In 1540, Wyer issued *The pronostycacyon for ever of Erra Pater: A Jewe borne in Jewery, a Doctour in Astronomye, and Physycke*, a short pamphlet of sixteen octavo-sized pages wherein a reader could find short passages on the humoral complexion of the body, the influences of the planets and astrological signs and, notably, two sets of prognostications: a table of perilous days for the lunar month and annual prognostications calculated according to the dominical letter. Just like the prognostication in icons shown in figure 5.1 at the start of this chapter, Wyer's book described for readers how, in "the year that January shall enter upon the Sunday, the winter shall be cold, and moist" with "abundance of corn," though "great wars and robberies shall be made, & many young people shall die."[79] In its contents and inexpensive octavo format, *The pronostycacyon for ever of Erra Pater* was the first book published in England to closely approximate the sort of material found in medieval vernacular almanacs. The result was an enormously popular practical book, published another eight times in the sixteenth century and in various other formats right up to the early eighteenth century.[80]

Digges had his work cut out for him if he was going to find a way to distinguish his publication from both the inexpensive perpetual prognostications of the fictitious Erra Pater and the annual almanacs and prognostications produced by his real-life competitors. First, to distance his book from long-form works of popular astrology like the *Erra Pater,* Digges peppered his text with references to respected authorities both old and new. In a preface to the book directed "Against the reprovers of Astronomy and sciences Mathematical," Digges positioned his work within an ancient lineage established in Aristotle's *Posterior Analytics* and reinvigorated by "that excellent Guido Bonatus," the thirteenth-century Italian whose textbook, *Liber astronomicus,* was the preeminent resource for European astrologers prior to the fifteenth century. For those interested in more recent authorities, Digges cited the letters of Philip Melancthon.[81] Finally, to those readers who might think to compare his calculations with those of other astrologers, Digges had a ready rejoinder: "be assured, they are false, or at the least for other Elevations."[82] Digges wished to present his perpetual prognostication as a serious work, not derived from the folk wisdom of a wandering Jew, but rather based on the principles of the "sciences Mathematical."

Mathematically inclined consumers would thus readily choose Digges's work over a book like the *Erra Pater* from Wyer's press, but the case for choosing *A prognostication of right good effect* over competing annual calculations from Anthony Askham was much more difficult to make. Askham also presented his publications as cutting-edge works of scholarship. Not only did his almanac for 1555 feature precise times for the rising and setting of the sun and the constellations in England, it also featured the latest in geospatial data from the New World. At the start of every month's entry, Askham gave a rough set of coordinates to track the sun's apogee across the globe: in April it rose "over a corner of Bedrosia in Inde, and at noon he goeth over the coasts of Libya interior in Africa, and at one of the clock at after noon he goeth over Cap de Wyer in Africa, and he setteth over certain isles of America called Rio de Gracias."[83] If that weren't enough, Askham was an outspoken Catholic who used his almanac to praise Queen Mary, "Our noble Judith," who "through power of the deity, Hath vanquished Holofernes."[84] For Digges, who had only recently been in open rebellion against that "noble Judith," Askham's annual publications were a particular challenge.

Digges's solution was to impugn his rivals' profit motives. According to Digges, men like Askham calculated their prognostications "for one year's profit only, compelled thereby of necessity to make a yearly renewing of them: whereupon errors many increased." Their work was filled with "manifest imperfections, and manifold errors" committed "partly by negligence, & often through ignorance."[85] Digges framed the failings of his competitors as an outgrowth of the commercial market: their desire for profits drove them to calculate new prognostications year after year for sale to English readers, and, according to Digges, those annual recalculations meant considerable errors. Not only would his prognostication be more accurate, it might also serve as an antidote to the rank commercialism of the print market. Others might publish with personal profit in mind, but *A prognostication of right good effect* would be "profitable to a common wealth."[86]

In a number of respects, Digges's perpetual prognostication really was different from everything else on the English market in 1555. Where most practical guides to do-it-yourself prognosticating centered on astrological medicine, Digges's book foregrounded astrometeorology, offering rules by which to predict the weather according to the movements of the stars or the cycle of the calendar. Notably, *A prognostication of right good effect* also contained tables for calculating the tides on English coastlines and detailed instructions for constructing a sundial, with accompanying tables to ensure accurate time-telling throughout the year. The 1556 edition, which was a total resetting of the previous year's edition, retitled *A prognostication everlasting of ryght good effecte*, contained a diagram of the relative sizes of the planets, as well as detailed illustrations of both a quadrant and a sundial, with instructions on how to manufacture them in metal.[87] The presence of these instructions may explain why the printer-engraver Thomas Gemini was willing to take a chance on Digges even after Wyatt's rebellion and his attainder for treason. Gemini was an experienced printer and instrument maker, who later produced an astrolabe for Queen Elizabeth.[88] Indeed, it was Gemini's fine Italic type and eye for pleasing spatial arrangement of text that really set Digges's perpetual prognostication apart from the competition.

But while Digges's book was markedly different from most other astrological publications on the English market in 1555, it still drew on earlier ancient and medieval traditions. After listing a number of tokens from the natural world indicative of poor weather—cattle lowing continuously, for example, or dogs wallowing—Digges was quick to cite Virgil's *Georgics* as his source.[89] When describing thunder as "the quenching of fire, in a cloud" or, alternatively, as "an exhalation hot and dry, mixed with moisture," he was careful to cite both competing authorities: Pliny, in book two of the *Natural History*, and Aristotle in his *Meteorology*.[90] Like most sixteenth-century authors, Digges eagerly referenced the ancients. However, there was at least one set of very old sources in his book that Digges did not reference. Digges included the popular medieval prognostication calculated according to dominical letter in *A prognostication of right good effect*, but he made no mention of its origin in medieval manuscripts (or, indeed, of its presence in Wyer's edition of *The Pronostycacyon for ever, of Mayster Erra Pater Astronomyer*, reissued the same year as Digges's prognostication).[91]

Even so, there was at least one person in England who recognized that Digges's prognostication by dominical letter had once circulated in pictures. This individual, who styled himself J.A., composed his own perpetual prognostication in 1556: a little volume of a dozen octavo-sized pages, titled *A perfyte pronostycacion perpetuall*. It was published by Robert Wyer, the same printer responsible for the *Erra Pater* prognostication and so many other best-selling printed practical books, but it was unlike any of Wyer's other publications.[92] J.A.'s *A perfyte pronostycacion perpetuall* was a printed edition of the pictorial prognostications from dominical letter, discussed at length in chapter 2. It was a book filled with woodcut icons and scattered English captions, and the title of the book made clear its intentions: "Very easy to understand . . . and also for them which knoweth not a letter on the Book" to "perceive and have some understanding how the year doth go about, [just] as well as the learned men."[93] Many of the icons in this little printed book—like the man's head in profile to indicate the seasons or the three bodies wrapped in shrouds to represent death, shown in figure 5.3—replicate those in fifteenth-century manuscripts almost exactly.

It seems both Leonard Digges and the anonymous J.A. had the same idea concerning printed prognostications around the middle of the sixteenth century: instead of providing prognostications for a single year, they would offer their readers the tools to perform their own readings of the natural world. Ironically, this was exactly what Henry VIII had feared when he imprisoned William Harlokke in the Tower in 1530 for the crime of owning a "calendar of prophecy wherein there were pictures." If readers of generic pictorial prognostications were encouraged to draw their own conclusions, they might determine any number of dangerous things, like that "great battle of priests," predicted by one of Harlokke's acquaintances.[94]

But in the end, the English authorities needn't have worried about prognostications in pictures, whether in Harlokke's manuscript or in J.A.'s printed book. By the time *A Perfyte pronostycacion perpetuall* was

When Newyeres daye fal-
leth on the Tuesday. There doth
belonge a.xi. Sygnes. As here after
foloweth, that is to saye.

Wynter wyndye. Sōmer rayny.

A dyuers har-
uest.

Corne meane-
lye.

Fruyte indyffe-
rente.

Syckenes of wō-
men.

FIGURE 5.3. Woodcut icons prognosticating a windy winter, a rainy summer, a diverse harvest, little corn, "fruit indifferent," and sickness of women for years in which New Year's Day falls on a Tuesday. © The British Library Board, J.A., *A perfyte pronostycacion perpetuall*, British Library 717.a.46, STC 406.3 (Robert Wyer, 1556), fol. vii *v*.

published, it seems the English were no longer interested in reading pictures to predict the future. J.A.'s little book was not a great success. Whereas Wyer's other perpetual prognostication, the *Erra Pater*, was reprinted over and over again in the 1540s and 1550s, *A perfyte pronostycacion perpetuall* survives in just one copy, from just one edition. There are no records of other editions or reprints in the Stationers' Register. A number of factors may have contributed to the failure of this, the last book of pictorial prognostications produced in England. For one, Robert Wyer left the printing business the same year that this little book of icons was published. Perhaps if he had stayed he would have reissued it again, as he did with so many of his other practical books. And yet, there is reason to suspect that the book would not have found a readership even if Wyer had published it another few times. The basic premise of the book, that its icons might be read by "them which knoweth not a letter on the Book," had been undermined considerably by the iconoclasm of Edward VI's Reformation. Though Queen Mary was on the throne in 1556, and thus religious images were no longer illicit, the iconoclasm of Edward's reign from 1547 to 1553 had destroyed the legible pictures painted on the walls of English churches.[95] Reformers had railed against the premise that images might be "laymen's books," as Pope Gregory had long ago asserted. Though the icons in *A Perfyte pronostycacion perpetuall* are not religious in nature, the tradition they depended on was deeply interwoven with later medieval devotional practices.

Parishioners in sixteenth-century England were living in a world replete with text, and not just because reformers had done so much to emphasize the Word as a conduit to the divine. Print was, first and foremost, a technology of text, and not one especially well suited to reproducing the richness of manuscript illumination. Indeed, technological limitations may have spelled the end of pictorial prognostications in England. In order for the icons in those prognostications to be legible, they needed to be consistent: the same forms, lines, or colors repeated such that a slight variation between images carried meaning. But the woodcuts in *A perfyte pronostycacion perpetuall* are so crude that a reader could not expect to make meaning from minor variations to line or form. Ironically, though print has been credited with making possible the "exactly repeatable pictorial statement," in the case of pictorial prognostications, it was the coming of the press that disrupted what were already repeatable (and oft repeated) pictorial statements in manuscript.[96] The legibility of *A Perfyte pronostycacion perpetuall* was totally undermined by the poor quality of the woodcuts within it.

Even more problematic is the inconsistent positioning of those woodcuts within the text. In figure 5.3, the prognostication "Winter windy" is represented as a man's head in profile, with lines indicating breath blowing from his mouth. This single icon is easy enough to interpret. The difficulty arises, however, if we compare that icon to the same prognostication for Friday—"Winter windy"—which is represented by a different head in profile with no lines indicating breath. The prognostication for Sunday has that same head in profile (without breath lines) accompanied by the caption "Winter temperate, inclined to heat," and so does the prognostication for Monday, with the caption "Winter black." The icons no longer have stable meaning. The woodcut artisan who made these images may have known that it was important to keep them consistent across the seven days of the prognostication, but it wasn't his job to place his woodcuts in the frame with type for printing. That was the job of the typesetter, who may simply have put the blocks in the wrong place. Perhaps a laborer whose job it was to live in a world of words had no idea what it would mean to really read a series of pictures.

Pictorial prognostications did not flourish in print, perhaps because the practice of reading pictures to comprehend the order of nature failed to translate across the chasm created by both the coming of the press and the Protestant Reformation. Patrick Collinson has argued that the iconophobia of the Edwardian Reformation actually intensified in the later sixteenth century, making it even less likely that pictorial prognostications would have found a readership in Elizabethan England.[97] Even so, the basic premise of medieval perpetual prognostications—that a reader could be trusted to perform his or her own acts of interpretation following observation of the natural world—lived on. Indeed, that is exactly what Digges promised his readers in his *A prognostication of right good effect,* though he certainly didn't put it in those terms. Though medieval readers had also thought of prognostication as a set of practices which were ongoing and subject to myriad individual acts of observation, and though those practices were as old as the prognostication by dominical letter that he included in his book (which is to say, centuries old), Digges made no mention of this deep tradition. Instead, he defined the novelty of his publication against the annual prognostications and almanacs that glutted the English market in the 1550s, a genre of book that was only as old as the printing press itself. He claimed that his methods were entirely new, that his was the first book to offer "infallible rules taught forever, a truth of all such things as heretofore have been put forth of others for one year's profit only."[98] Digges offered permanence—a perpetual prognostication—as a revelation in a media environment dominated by ephemerality.

Conclusion

Digges's claim that his prognostication was unlike anything else available to English readers was a marketing technique, similar to those outlined in the previous chapter. Novelty sold books, and Digges certainly needed a best-seller. In the end, that's what he got: Digges's prognostication was published a total of twelve times between 1555 and 1605. Eventually, Digges's book did become the unprecedented work of astronomy he had once promised his readers. In 1576, Digges's son Thomas extended his father's original work to include the first English translation of a portion of Nicolaus Copernicus's *On the revolutions of the heavenly spheres* (*De revolutionibus orbium caelestium*), accompanied by the first diagram of the heliocentric universe published in England.[99]

And yet, like his father, Thomas Digges saw no reason that the old couldn't coexist with the new. A reader of the 1576 edition of *A prognostication everlastinge of right good effecte* would find Leonard Digges's original diagram of the universe with Earth at its center on the verso of folio 4, and then, stretching across the entirety of folio 43, that same reader would encounter a very different universe, regular with concentric circles, but oriented around the sun. It was no oversight on Digges's part that the book contained two fundamentally incompatible diagrams of the universe, either. Thomas Digges chose to leave both in and to explain away their inconsistencies as "sundry faults that by negligence in printing have crept into my father's General Prognostication: Among other things I found a description or Model of the world and situation of Sphere Celestial and Elementary according to the doctrine of Ptolemy."[100] Rather than rebut this model of the universe or excise it entirely from his father's popular prognostication, Digges softened the blow of upending an entire worldview by presenting Copernicus's revolutionary text and his model of the heliocentric universe as a mere *corrigenda*, only necessary because of the frenetic pace of an overly commercialized print market.

Both father and son gave English readers of prognostications something new in their editions of *A prognostication of right good effect*, but both were also more than happy to keep the old intact, too: to keep prognostications by dominical letter alongside precise timetables for English tides or to keep a model of the universe with Earth at its center right alongside one in which the planets orbited the sun. Both men recognized that English readers would make the final judgment about what to believe, and that they could win over these discerning readers, earn their trust, if they framed their publications as corrections to an overcommercialized information economy, characterized by careless printers and greedy astrologers. It was readers who would determine the success or failure of their publications, and readers who would adopt (or not) the science that altered their place in both space and time.

6

Printing Women's Knowledge, Censoring Secrets

Sometime around 1510, Wynkyn de Worde published a little book called *The gospelles of dystaves.*[1] The book was a translation of a French text that had been composed within the Burgundian court, where it circulated in manuscript copies until it reached the printing shop of Colard Mansion. Mansion, who had taught William Caxton the printing trade in Bruges, produced a sumptuous folio edition of the text in 1479, based on a manuscript owned by Marie de Luxembourg, wife of Louis de Luxembourg, Count of Saint-Pol.[2] But though it had courtly origins, soon the "distaff gospels" reached a more popular readership. Between 1482 and 1498, at least five quarto editions and one tiny (and presumably cheap) sedecimo edition were published for French readers in Lyon.[3] By 1510, when De Worde published the first translation of the "distaff gospels" from the original French, he had already established himself as the leading printer of practical books in England, with editions of the *Proprytees & medicynes of hors* (1497 and 1502?), *The governall of helthe* (1506), the *Boke of husbandry* (1508), and *A treatyse agaynst pestelence* (1509). In certain respects, *The gospelles of dystaves* was very much like these earlier publications. It was a printed version of a text that had once circulated in fifteenth-century manuscripts (though in Burgundy, not in England), and it offered advice on how to cure fevers, treat smallpox, or heal a lame horse, among other things. But in one important respect *The gospelles of dystaves* was unlike De Worde's other editions of useful, natural knowledge: the book claimed to record the particular knowledge of women, and it was the first publication in England to advertise itself as such.[4]

For all that De Worde's edition claimed to foreground "women's knowledge," however, it was probably not published with women readers in mind. Put simply: most sixteenth-century English women wouldn't have been able to read *The gospelles of dystaves.*[5] The few who could—like Margaret Beaufort, mother of Henry VII and an active literary patron in her own right—might not have liked what they found in its pages. The book was a satire in the tradition of the *querelle des femmes*. Written from the perspective of an unnamed male scribe, it records the wisdom of six women, shared over the course of the six evenings between Christmas and New Year's Eve. A sampling of the women's advice finds Dame Isengryne, a former prostitute turned midwife, describing how a pregnant woman may "know if it be a son or a daughter" on Monday evening. On Tuesday, Dame Transeline, a priest's concubine and an expert on divination, describes how to remove a wart by rubbing it with a leaf from an alder tree. On Wednesday, Dame Abunde, also a former prostitute, expounds on the protective powers of holy water. Dame Sebylle advises that a lady should give a "feeble cock" garlic and "anoint his crest" until he becomes "stronger and more vigorous" on Thursday, and Friday's teacher, Dame Gambarde, instructs on how to conceive a son. Finally, on Saturday, Dame Berthe, introduced as the daughter of a great physician trained at Montpellier, teaches various medical remedies even though she knows that men "set but little by us, for they hold their parliament . . . in the reproach of our sex."[6]

The point of the "distaff gospels" was to reveal women's knowledge as silly, superstitious, and potentially dangerous, all the while lambasting the female sex as lusty and promiscuous. With its frame narrative, *The gospelles of dystaves* draws a contrast between the rationality of the literate male narrator and the superstition and gossip shared among the illiterate women. This satirical framing is especially effective as a literary device because it picks up the widely held early modern belief that there were categorical differences between "women's knowledge" and "men's knowledge"—the latter the rational stuff of writing and books, the former the superstition of secret coteries and gossipy, oral exchange. The overt misogyny of the "distaff gospels" was meant to shape how readers thought of the simple rituals and recipes recounted in the book. And this reshaping was important, because much of the advice that was meant to be read as superstitious when it came from the mouths of Dame Abunde or Dame Berthe in *The gospelle of dystaves* was widely available in fifteenth-century English manuscripts and in other sixteenth-century printed books. For example, numerous practical manuscripts contain recipes to know the sex of an unborn child similar to that which Dame Isengryne shares on the first night of the women's spinning.[7] One of Dame Sebyll's tips on how to tell if rain was coming was reproduced almost word for word in Leonard Digges's *A prognostication of right good effect*, discussed in the previous chapter.[8] And one of Dame Berthe's cures, a charm to "hele fevers," is nearly identical to the "sage leaf" charm found in seventeen fifteenth-century practical collections, discussed at length in chapter 3. What was superstitious "women's knowledge" in one context was run-of-the-mill practical knowledge in another.

This chapter explores how readers came to associate certain categories of natural knowledge with femininity, and how those associations were amplified or minimized by sixteenth-century publishers, depending on the contents and contexts of the books they were selling. In the case of *The gospelles of dystaves* and, later, popular "books of secrets," associating practical knowledge with women was a marketing tactic intended to make that knowledge *more* appealing, not less so. Like so many other practical books published in England, De Worde's "distaff gospels" and the first editions of "books of secrets" to appear in London bookshops were quite literally translated from European vernaculars into English. But, as we will see, the narrative framing that structured these books performed its own kind of translation, turning inoffensive categories of natural knowledge into "women's secrets." These acts of translation had real consequences for English women, because in the decades between the publication of De Worde's 1510 translation of the "distaff gospels" and the 1573 publication of John Partridge's *The treasurie of commodious conceits, & hidden secrets and may be called, the huswives closet, of healthfull prouision* (the first "book of secrets" advertised for women readers), the English Reformation hardened readers' attitudes toward "women's secrets" considerably.[9] Authorities within the new Church of England were particularly suspicious of the kind of knowledge that De Worde had once published in *The gospelles of dystaves*. Church officials demanded to know if England's parishioners had witnessed the use of charms, incantations, or Latin prayers, especially "in the time of women's travail," even though, at the very same time, these parishioners might read charms and Latin incantations in printed books like *The boke of secretes of Albertus Magnus*, attributed to a venerated philosopher.[10] In effect, these questions reinforced what readers of books of secrets already knew: there was a difference between knowledge that had been authorized and published by men and knowledge that hadn't—even if that knowledge was for all intents and purposes identical.[11] For later sixteenth-century readers of practical books, defining "women's knowledge" was a question of context rather than content.

The Problems and Possibilities of "Women's Secrets"

Throughout the fifteenth century, male medical practitioners and owners of practical manuscripts were deeply invested in gathering knowledge that, within *The gospelles of dystaves*, was gendered feminine. As we know from chapter 3, practical manuscripts very often featured charms, including many to aid women in childbirth, a few to help a woman conceive, and even one or two to make a woman love her husband.[12] These charms were exactly the sort of "superstition" that medieval church officials condemned as the sins of "foolish women" and yet, Middle English manuscripts featuring these charms were almost certainly compiled and read by men.[13] The church's official condemnation of simple ritual magic did little to dissuade would-be collectors of natural knowledge, probably because common or simple magic was rarely prosecuted in fifteenth-century England.[14] So, for instance, while the male scribe who copied a charm for childbirth in Huntington MS HM 58 did leave a comment in the margins noting the church's disapproval—"These writings are prohibited by the Catholic church"—he nevertheless copied it in full.[15]

The scribe who copied Huntington MS HM 58 probably completed his work sometime in the last two decades of the fifteenth century, perhaps a few years after Innocent VII issued his 1484 papal bull authorizing Heinrich Kramer, a German Dominican and zealous inquisitor, to prosecute witches in the Holy Roman Empire. This English scribe may even have put the finishing touches on MS HM 58 in the same year that Kramer composed his infamous manual on witch hunting, the *Malleus Maleficarum*, or *Hammer against Witches*, published in 1486.[16] In that book, Kramer elaborated a theory of witchcraft that linked women's susceptibility to temptation, epitomized in the figure of Eve, to a proclivity for *maleficium*, or demonic magic. As an educated Dominican, Kramer knew that *maleficium* resulted from submission to demonic forces, and in his mind, women's exceptional lust made them especially prone to succumbing to the devil's wiles.[17] Moreover, Kramer shared with fellow fifteenth-century intellectuals an essentially binary view of the world: God was opposed with the devil, good with evil, and virtue with vice. These categories mapped easily on to the primary binary opposition that structured early modern thought: man and woman.[18]

And yet, Kramer's anxieties about nefarious women and diabolical magic weren't shared by the scribe of MS HM 58, nor were they adopted by English clerical authorities. In the thirteenth and fourteenth centuries, English authorities either accepted charms, folkloric prognostications, and healing rituals as "natural magic" that tapped into the hidden, or occult, properties of the divinely ordained natural world, or dismissed them as superstition—not ideal, but certainly not a serious threat.[19] Those accused and tried for witchcraft in fourteenth-century England were predominantly male, and in 1406, when Henry VI issued a writ to the Bishop of Lincoln to seek out "sorcerors, magicians, and necromancers," he used the plural masculine forms of those Latin nouns.[20] As late as 1480, when London's commissary court oversaw a case against a healer for using "sorcery" to cure a fever, the healer on trial was a man—John Stokys—not a woman.[21] The first ripples of the witch craze that emerged in the fifteenth-century Swiss and Italian Alps had not yet reached England by the time MS HM 58 was composed in the 1480s or 1490s. In practical manuscripts, charms, love magic, and prognostications circulated widely among (mostly male) readers.[22]

If charms and prognostications were not yet gendered as "women's knowledge" in fifteenth-century manuscripts, other categories of useful knowledge common to practical manuscripts did have a more overt association with women. Of the total 182 manuscripts surveyed for this book, fifty-six contain recipes related to menstruation, lactation, or childbirth, most of them integrated within larger head-to-toe collections of medical recipes that address all manner of ailments. In British Library MS Additional 34210, for example, recipes to bring on and cease menstruation are followed very shortly thereafter by recipes for avoiding stinking breath.[23] A recipe collection in British Library MS Sloane 382 that begins with instructions for treating gout also contains recipes for a woman to "have great plenty of milk" after childbirth.[24] And, in the popular herbal known by its incipit, "Here men may see the virtues of herbs," mugwort—often known as "moderwort" in Middle English—is suggested as a treatment to aid in the delivery of a deceased fetus.[25] In fifteenth-century practical manuscripts, these recipes relating to women's concerns were rarely separated from generic directives for treating a headache or curing a fever.

Yet, even so, learned medical authorities did understand women's ailments to be distinctive from those suffered by men, in large part because they believed that a woman's general health was directly related to processes we associate only with reproduction. For example, both menstruation and lactation were believed to be critical for maintaining the balance of women's humors, and therefore both were subject to the theories of male physicians.[26] Moreover, according to premodern medical theory, a woman's reproductive organs made her body vulnerable to particular ailments and illnesses, like hysteria or green sickness. Since the time of the Hippocratics, learned male practitioners had composed and collected medical knowledge about these special "diseases of women."[27] Indeed, both Avicenna and Gilbert the Englishman included chapters on "women's diseases" in their general-purpose medical guidebooks. These learned Latin collections provided the source material for Middle English recipes to bring on menstruation or encourage lactation, just as they were the sources for recipes to cure headache or gout. In fact, Gilbert's chapters on women's ailments in the *Compendium of medicine* circulated on their own in Middle English translation under the title "The Sickness of Women."[28]

But even though recipes to aid lactation or bring on menstruation were very often authored by men, reproductive knowledge was still very often associated with women, because women were in charge of routine medical care surrounding pregnancy and childbirth. Though male physicians were understood to have superior expertise, and though they were sometimes called in to assist in complicated deliveries, propriety dictated that female midwives should oversee nearly all births in premodern Europe.[29] Therein lay the problem: there was no male supervision of the birthing room, and because women were so rarely literate in premodern Europe, midwives' experiential knowledge couldn't be communicated in writing where it might be assessed by men. Indeed, the only female practitioner to have authored a Latin collection of reproductive recipes in the Middle Ages was Trota of Salerno, whose legacy has been exhaustively studied by Monica H. Green. Trota gained notoriety for her healing in twelfth-century Italy, but as Green argues, she would never have achieved the same reputation if she had lived just a century later, after the establishment of medical faculties at medieval universities. At the universities, medical expertise became firmly associated with Latin literacy, which made medieval intellectuals even more suspicious of the sort of empirical knowledge developed and shared among women.[30] The kind of experiential knowledge that made Trota so renowned became dangerous precisely because it was totally inaccessible to learned male authorities. There was no telling what kind of knowledge about generation was transmitted orally within the all-female space of the birthing chamber. These women's "secrets" needed revealing.

The first text to claim to do so was the late thirteenth-century treatise *On the Secrets of Women* (*De secretis mulierum*), falsely attributed in the Middle Ages to the twelfth-century Dominican philosopher Albertus Magnus.[31] The work claimed to reveal the secrets of "prostitutes and women . . . learned in the art of abortion" who had access to knowledge related to conception and generation.[32] Even though the treatise was coded in the language of natural philosophy and written in Latin, it was astonishingly popular: eighty-three manuscript witnesses survive in addition to the fifty or more fifteenth-century printed editions of the text.[33] According to Green, the popularity of the pseudo-Albertine *Secrets of Women* inspired a wholesale reassessment of earlier medical treatises relating to women's reproductive health. By the later Middle Ages, all sorts of reproductive medical texts that had once been understood to relate to the "diseases of women" were repackaged as texts on the "secrets of women."[34] Moreover, as Katharine Park has argued, the female body, supposedly plagued by diseases caused by the wandering womb or an accumulation of too much menstrual blood, became a "secret" that needed decoding, too.[35]

Park and Green have both documented how the emergence of language around "secrecy" marks a significant shift in attitudes toward women's bodies and reproductive knowledge, particularly among learned fifteenth-century physicians and philosophers. Yet, it is not altogether obvious that the same shift was felt within the realm of vernacular medicine, and still less in Middle English manuscripts. None of the reproductive recipes in the fifteenth-century manuscripts I've examined are couched as the secrets *of* women. Rather, if reproductive recipes are framed with any commentary at all, it is to describe them as knowledge that women *lack*. For example, the prologue to one Middle English adaptation of the *Trotula*—the reproductive recipe collection that was at least partially authored by the female practitioner Trota of Salerno—explained that its author had translated "the treatises of diverse masters" into English so that "every woman lettered [may] read it to other unlettered and help them and counsel them in their maladies without showing her disease to men."[36] In this translator's opinion, women had an obligation to care for other women's bodies in order to preserve their modesty, but they could only do so if their ignorance was ameliorated through study of the "masters," whom this translator presumed to be male. The point of this translation was to share expert male knowledge *with* women, not to reveal the secrets *of* women—though ironically, the expertise this translator hoped to share was at least partially that of a woman, Trota of Salerno.[37]

The presentation of reproductive knowledge in this Middle English copy of the *Trotula* is clearly at odds with that in *The gospelles of the dystaves*, published only around fifty years after that manuscript was created. Instead of offering reproductive knowledge *to* women who lacked expertise, the scribe of the "distaff gospels" claimed that his role was to record knowledge produced *by* women. In a frame narrative akin to that in the *Decameron* or *The Canterbury Tales*, the male narrator describes how the six wise women invited him to "bring paper and ink enough" with him over six evenings so that he could "put in scripture a little volume" of their wisdom. Though the ladies of the "distaff gospels" express dismay about the prospect of entrusting their precious knowledge to a man, they know that without a literate interlocutor, it would remain confined to their all-female spinning circle. For his part, the narrator explains that he accepted the ladies' request because he hoped the "words and authorities of the ancient women . . . should not be lost nor in such wise vanished but that the memory should remain fresh among the women of this present time."[38]

And then, over the course of these six evenings, the male scribe experiences a change of heart. Although the scribe commences his narrative with showers of praise for the female sex—"naturally noble, honest, sweet, fair, and courteous and full of sapience"—by the end of the book, his view of the women has transformed from respect to derision.[39] By the final page of the book, that same scribe is busy making excuses for his earlier praise, explaining that "the words that [the women] had spoken was without reason and without any good sentence as I thought well it should be at the first beginning."[40] The male scribe of *The gospelles of dystaves* thus follows the model established in Latin "secrets" literature, claiming to reveal knowledge that was "secret" and somewhat dangerous by virtue of the fact that it was possessed by women, and so outside the sphere of literate, male discourse. It doesn't matter that the reproductive lore or charms and prognostications attributed to women in the "distaff gospels" were also widely read and recopied by men and for men in contemporary fifteenth-century manuscripts. *The gospelles of dystaves* performs a feat of translation: everything in the volume becomes "women's knowledge" because the narrative framing makes it so. If readers weren't already mistrustful of those "worthy female doctoresses" when they opened *The gospelles of dystaves*, the narrator's gradual change of heart guided them toward that revelation. By the close of the book, as the scribe pleads for readers to "pardon me & repute the said fault" to the women whose wisdom he has recorded, readers understand that the men responsible for writing and publishing this "little volume" were victims of the women's cunning, too.[41]

And that may, in fact, be precisely what Wynkyn de Worde wished them to believe. After all, he had published a book of superstition and folklore, attributed to nefarious and bawdy women, in an easily accessible, organized format, not at all unlike the other practical books rolling off the English presses in the first few decades of the sixteenth century. In *The gospelles of dystaves*, each evening forms a chapter in the book, and each piece of advice offered by the women is numbered, in reference to the literal gospels. Though this structure is part of the satire, it also had the added benefit of making the book into an exceptionally useful reference. Perhaps De Worde, a savvy pioneer of printed practical books, understood all too well that a printed edition of charms, rituals, and recipes hitherto only accessible in manuscript would appeal to English readers.

Crucially, however, *The gospelles of dystaves* could only fulfill both functions—serve as both satire of women's superstition and useful practical book—thanks to the male narrator/scribe, the only figure with the power to put women's knowledge into writing and thus to act as mediator between those dangerous "worthy doctoresses" and his readers. The narrator's conversion arc is a particularly powerful rhetorical strategy, as it directed readers' attention where it belonged: away from the men who recorded and published charms and reproductive lore and toward the women whose superstitious knowledge they would profit from. It is perhaps worth mentioning that the earliest manuscript of the "distaff gospels" created around 1470 has a very limited prologue and conclusion. The frame narrative with the male scribe only developed in subsequent copies, as the text began to circulate more widely in court circles—perhaps as a workaround to the problem of sharing "women's secrets."[42] Whatever its origins in the "distaff gospels," the figure of the male narrator/scribe capable of revealing "women's secrets" in print would reappear in practical books long after De Worde's little publication had lost its appeal among English readers. He would emerge again in the "books of secrets" that proliferated on the English market in the latter half of the sixteenth century.

Discovering Women's Knowledge

In 1555, readers browsing the bookstalls in Venice were introduced to another male interlocutor capable of revealing hidden knowledge: the Reverend Master Alexis of Piedmont, pseudonymous author of the *Secreti del reverendo donno Alessio piemontese*. This book, the first of many "books of secrets" published in Italy, was an instant best-seller. Like *The gospelles of dystaves*, it was quickly translated into English, French, Dutch, and German so that by 1600, seventy different editions and translations of Master Alexis's *Secreti* were circulating among early modern readers.[43] The 1558 English edition of *The secretes of the reverende Maister Alexis of Piemount*, translated by William Ward, was just as popular as the Italian original. Its publication sparked a frenzy for "books of secrets" among English readers.[44]

The contents of *The secretes of the reverende Master Alexis* are quite different from *The gospelles of dystaves*. The book doesn't contain much in the way of healing charms, domestic superstitions and rituals, reproductive lore, or prognostications. Over 1,700 individual recipes appear in the four volumes of secrets attributed to Master Alexis published in the 1550s and 1560s, and of those, only five are explicitly magical, and only thirty-one relate to menstruation, pregnancy, childbirth, or fertility.[45] The genres of knowledge so closely associated with superstitious women in *The gospelles of dystaves* are mostly absent. Instead, Master Alexis's book offered a wealth of knowledge to appeal to all genders: medical recipes, instructions for making household goods, culinary instruction, and chapters on cosmetics, metallurgy, and alchemy. With the exception of alchemy and metallurgy, these were exactly the kinds of recipes that would remain standard in later books of secrets specifically targeted at English women, like *The treasurie of commodious conceits & hidden secrets and may be called the huswives closet of healthful provision*, published in 1573.[46]

In one important respect, however, *The secretes of the reverende Maister Alexis* and *The gospelles of dystaves* were alike. Like the anonymous scribe of the "distaff gospels," Master Alexis presented himself to readers as a mediator able to use the written word to reveal hidden knowledge. But, where the narrator of the "distaff gospels" recounted his invitation to record knowledge as it was dictated to him, the pseudonymous Master Alexis described his act of mediation in a narrative of discovery. In the preface to the book, Alexis described how his "natural inclination" led him to take a solitary, twenty-seven-year journey across the world in search of practical knowledge.[47] In many respects, the discovery narrative of *The secretes of the reverende Maister Alexis* can be understood as the culmination of the marketing techniques catalogued in chapter 4. As we know from that chapter, medical recipe books and herbals printed in England from the 1520s onward often suggested through their titles that knowledge might be discovered in the pages of "new" printed books that were anything but novel. Books of secrets took this strategy a step further. Their authors, like Master Alexis, offered to reveal new knowledge gained through one single man's "diligence and curiosity."[48] While we can be sure that fifteenth-century compilers had exercised diligence and curiosity in the collection of recipes for manuscript collections, those earlier implicit practices of knowledge collection were never spelled out explicitly for readers. In making explicit the search for new knowledge, books of secrets encouraged readers to do something similar.

William Eamon first drew attention to this quality of books of secrets, and the role they played in the development of observation and experimentation as epistemic practices, several decades ago. Drawing a direct line from medieval natural philosophers' search for "secrets"—operations that supposedly unlocked the hidden powers of nature—to early modern "books of secrets," he argued that the *Secreti* and other rival publications encouraged readers to go on a "hunt" for the "secrets of nature" via experimentation.[49] Eamon is certainly correct that books of secrets were notably different from earlier printed recipe collections, like Richard Banckes's *Treasure of pore men*, and they may indeed have encouraged readers to seek experiential knowledge in ways that those earlier collections or manuscript miscellanies did not. But Eamon also overlooked a simpler, perhaps more obvious, definition of "secrecy" at work in the *Secreti*—one spelled out by Master Alexis himself in the book's opening preface. Alexis writes that he collected "goodly secretes, not alonely of men of great knowledge and profound learning" but also from those without the ability to put their knowledge into writing. When enumerating these illiterates with useful knowledge, the first group Alexis explicitly identifies are "poore women."[50] Indeed, in both the original Italian and the English translation, the structure of this particular sentence places those "poor women" in direct opposition to "men of great learning"—"not alonely of men . . . but also of poore women"—reinforcing the two categories as diametrically opposed to each other.[51] Whatever secrets Alexis collected from the books of learned men may very well have been the "secrets of nature" sought by medieval natural philosophers, but the "secrecy" of women's knowledge was an entirely subjective determination. Knowledge that might be characterized as "secret" by a man like Master Alexis was perhaps entirely mundane to the "poor women" from whom it was collected.

On closer analysis, then, the discovery narrative of Master Alexis begins to look a great deal less like experimentation and more like editorial practice: collecting and organizing knowledge already extant in the world, held by those with no access to a press. If that dynamic was only hinted at in the preface to the original Italian edition of the *Secreti*, it was hammered home by William Ward, translator of the English edition. In the "Epistle" that precedes Alexis's preface, Ward explained that "this book hath been published and communicated to the world by the said Alexis" explicitly so "that men of all countries might have the knowledge of that with ease, sitting at home in their studies."[52] *The secretes of the reverende Maister Alexis of Piemount* invoked a model of "secrecy" predicated on a distinction between published and unpublished knowledge, which, just as in *The gospelles of dystaves*, was often indistinguishable from a distinction between the masculine and the feminine. Like the anonymous scribe of the "distaff gospels," Master Alexis laid particular claims to knowledge-making that simultaneously advertised and obscured the contributions of women. Both male writers presented themselves as the only figures with the ability to render women's voices in text, bridging a perceived divide between a preliterate world of unprinted knowledge characterized as feminine and an ordered, literate, and masculine world of printed knowledge.[53]

The parallels between the frame narrative of the "distaff gospels" and the discovery narrative of Master Alexis are worth our attention because they speak to the emergence of an epistemological distinction between published and unpublished knowledge, a distinction that had particular consequences for the circulation and reception of charms, reproductive lore, and natural magic among sixteenth-century readers. Because unpublished knowledge increasingly came to be associated with women, reproductive medical texts or natural magic that circulated in manuscript or without an established authorial or publishing pedigree could be suspect as "women's secrets." By contrast, magic or reproductive lore that would seem much more dangerous than the supposedly nefarious content in *The gospelles of dystaves* could circulate without opposition, so long as it was associated with a male authority or couched with disclaimers about its long history in print. A perceived distinction between published and unpublished knowledge thus became an important criterion for English readers who wished to determine the legitimacy and authority of natural knowledge associated with women.

A couple of examples should illustrate this distinction. Fifteen years before Master Alexis published his *Secreti*, Richard Jonas had translated another continental book of women's knowledge for English readers. *The byrthe of mankynde*, published in 1540, was a translation of a popular German midwifery manual, first published in Strasbourg in 1513.[54] It provided instructions and medical recipes relating to fertility, conception, prenatal care, labor and delivery, and postpartum conditions. A few of the recipes in Jonas's book, like rituals to aid conception and instructions to determine the sex of an unborn child, even resemble passages from *The gospelles of dystaves*.[55] Unlike that earlier edition of women's knowledge, however, *The byrthe of mankynde* was explicitly intended for women readers. Jonas dedicated his work to Katherine Howard, Henry VIII's fifth wife, and he wrote that he hoped "all honorable & other honest matrons" would read it.[56] Like the compiler of the Middle English version of the *Trotula* quoted in the previous section, Jonas believed that women should care for other women and, like that earlier translator, he believed that access to learned texts would mitigate feminine ignorance.[57]

Unlike the narrator of *The gospelles of dystaves* or Master Alexis, Jonas didn't claim to have collected knowledge *from* women to bring it to literate men at "ease, sitting at home in their studies." He wasn't an interlocutor between the feminine sphere of oral knowledge and the masculine sphere of literate knowledge. Rather, his book made literate, masculine knowledge *about* women available *for* women's benefit. And, lest a reader suspect him of trading in "women's secrets," he made sure to emphasize his learned sources. If by chance a reader missed the book's credentialing in the subtitle (*Newly translated out of Laten and into Englysshe*), Jonas devoted considerable space in the dedicatory epistle to praising that "famous doctor in Physicke called Eucharius" and discussing the work's subsequent publication in French, Dutch, Latin and, finally, English.[58] In fact, the man responsible for publishing *The byrthe of mankynde*, Thomas Raynald, was a physician himself. Not only had Raynald fronted the money for John Herford to print Jonas's translation, he also made sure that the manual included copper-plate engravings of fetal positions—the first engraved images printed in an English book.[59] In 1545, Raynald took it upon himself to completely revise Jonas's original translation in order to "correct and amend such faults in it . . . and to advise the readers what things were good or tolerable to be used, which were dangerous, & which were utterly to be eschewed." But again, Raynald's text was intended to share expert knowledge *with* women rather than the other way around. The "salutary & effectual medicines" Raynald added to this edition were only those that "I myself, or other physicians being yet alive at this day, have experimented & practiced."[60] Raynald's revised edition was printed a total of thirteen times, every one with a new subtitle: *The womans booke*.[61]

Then, in 1560, only two years after Master Alexis's book of secrets burst onto the scene in England, John King printed another book of secrets: *The boke of secretes of Albertus Magnus of the vertues of herbes, stones, and certayne beasts*.[62] King's edition wasn't a new work out of Italy, but was rather a reedition of the *Liber secretorum Alberti magni de virtutibus herbarum lapidum et animalium*, printed only once before in England by William Machlinia in 1483, possibly at the same time that he printed that other book of secrets falsely attributed to Albertus Magnus, *On the Secrets of Women* (*De secretis mulierum*).[63] In a sense, then, *The boke of secretes of Albertus Magnus* was yet another reprint of a much earlier practical book, much like the other practical books King printed in 1560 and 1561 (discussed in chapter 4), which were all reeditions of earlier best-sellers. There was one key difference between *The boke of secretes of Albertus Magnus* and King's 1561 publication of Banckes's herbal, however. Whereas *A little herbal of the properties of herbes* was a reedition of an original vernacular text, *The boke of secretes of Albertus Magnus* made an old Latin text newly available in the vernacular. And, given that the book was filled with charms and magical rituals, publishing this book of secrets in English certainly might have posed a problem for King. Readers flipping through the book's pages might learn, for example, that pennyroyal, when mixed with a stone found in the nest of a black plover, could bring the dead back to life and engender children; that a lodestone (or magnet) placed under the head of a woman when she slept could reveal her chastity (or lack thereof); or that the foot and the heart of an owl placed on a man would induce him to tell all his secrets.[64]

But where magic of this kind in *The gospelles of dystaves* would have been characterized as superstitious or even dangerous, *The boke of secretes of Albertus Magnus* presents these rituals as the secret wisdom of a learned male philosopher. They are unobjectionable because they are attributable to a literate, male expert. Even so, King seems to have been at least a bit concerned about authorities' potential objections to the book's contents. To bolster the authority of the work even further, he penned a dedication "To the reader" at the opening of the English edition not-so-subtly reminding those readers of the work's pedigree: "since it is manifestly known that this book of Albertus Magnus is in the Italian, Spanish, French, and Dutch tongues, it was thought if it were translated into English tongue it would be received with like goodwill and friendship."[65] And it was. *The boke of secretes of Albertus Magnus* was printed another thirteen times in England, the final edition appearing in 1684.[66]

Plenty of reproductive lore, some of it very much like that of Dame Isengryne, was available in Thomas Raynald's *The byrthe of mankynde*. All sorts of magical rituals were accessible in *The boke of secretes of Albertus Magnus*. But neither John King nor Thomas Raynald claimed to act as mediators between the world of feminine, unpublished knowledge and that of masculine, published knowledge. Nowhere did either book claim women as their sources. Rather, these books—which really did contain knowledge that in other contexts would certainly be associated with dangerous femininity—founded their claims to legitimacy and permissibility on the premise that they were reproducing content available in other printed editions. *See*, they seem to say, *if it exists in print elsewhere, it can't come from women, so it can't be that bad!* By contrast, popular editions like the *Secreti* of Master Alexis or the English book of secrets known as the *Treasurie of commodious conceits* did claim women as sources. For these books, filled with run-of-the-mill medical and culinary recipes or domestic instruction, the label "women's secrets" made the mundane enticing by suggesting that it had never before been made available in print. In sixteenth-century England, neither secrecy nor women's knowledge were particularly stable as categories, leaving readers to determine what was what in their own encounters with natural knowledge in print and in manuscript.

Censoring Secrets

The publication of the four volumes of the *Secreti* in England between 1558 and 1578 sparked a frenzy among English readers for all kinds of "secrets," and authors and printers were quick to capitalize on it. Master Alexis had provided a useful model that was easy enough to follow: one could be the discoverer of secrets even if those acts of discovery simply meant publishing knowledge already circulating among "poor women, artificers, peasants, and all sorts." When native English authors began publishing their own books of secrets on everything from gardening to cookery to horsemanship, they followed Alexis's lead. For example, in 1579, just one year after the fourth and final volume of Master Alexis's secrets was published in English, Thomas Lupton composed his own book of "rare, strange & excellent things, as many could have been content to hide and keep secret."[67] Lupton dedicated his work to a powerful woman, Lady Margaret Stanley, cousin of Queen Elizabeth I—though he may have regretted it when Lady Margaret was detained and placed under house arrest that same year. Like Master Alexis before him, Lupton opened his book with a preface to the reader, in which he described the contents of his book as "so many notable, rare, pleasant, profitable and precious things . . . as never were yet set forth in any volume in our vulgar or English tongue, nay diverse of them were never hitherto printed, nor written, that ever I knew: but only that I writ them at such time as I heard them credibly reported."[68] Once again, English readers were presented with knowledge that was "secret" by virtue of the fact that it hadn't been published, and a discovery narrative that entailed little more than setting pen to paper to record what others simply couldn't, or perhaps wouldn't, write down.

If secrets were nothing more than knowledge that had never before been published, then marketing a work as a secret meant signaling that it was only now, for the first time, available in print. This was an especially promising strategy following the incorporation of the Stationers' Company in 1557. Once a printer had registered a title in the Stationers' Register, that title (and the text that went with it) was off limits to other members of the Stationers' guild. But, if a printer were to claim that a book was full of secrets that "were never hitherto printed," the Company might have reason to believe that its author had not run afoul of the Stationers' regulations.[69] Conversely, if secrets were defined as such because they had never before been published, then knowledge that *had* been published could not really be secret. It might be little noticed or underappreciated, but it wasn't secret. Thomas Lupton made exactly this distinction in the preface to *A thousand notable things, of sundry sortes*. In addition to recording secret knowledge that was "never yet set forth in any volume," Lupton explained that he also selected excerpts from books that were "not long since printed," for readers who "never could or would have bought, or looked on the books."[70] In the case of printed knowledge, Lupton could only be an anthologizer, not a discoverer.

So, it seems, if readers did want to follow Master Alexis on a hunt for secrets, as William Eamon suggests they did, those readers couldn't very well turn to printed books, even if those printed books made claims to secrecy. Yet, most didn't have the means to travel the world for twenty-seven years to gather knowledge from "poor women, artificers, and all sorts," like the famous Italian Master. What they could do—and what Thomas Lupton also claimed to have done in *A thousand notable things*—was gather "notable and precious things" from "old English written books."[71] In other words, they could look to decades- or even century-old manuscripts, those "old English [hand]written books" whose unpublished contents were by definition "secret." In the hands of sixteenth-century readers, practical manuscripts that had circulated widely in their grandparents' or great-grandparents' generation without any semblance of secrecy attached to them became objects prized for the hidden knowledge they contained. Early modern readers sought them out, collected them, and—in some cases—censored the secrets they found within them. Just shy of two-thirds of the fifteenth-century manuscripts examined for this book bear the hallmarks of these early modern readers' search for hidden natural knowledge: of the total 182 fifteenth-century practical manuscripts studied for this book, 112 contain reader marks written in later sixteenth- or seventeenth-century hands.[72]

When these early modern readers approached fifteenth-century manuscripts, however, they brought a whole new set of expectations with them, thanks not only to the proliferation of printed books but also to the English Reformation, which had precipitated the most dramatic transformation in English readers' attitudes toward the charms, magic, and women's superstition they found in fifteenth-century manuscripts. As chapter 5 illustrated, English reformers were intent on snuffing out the rituals of the Catholic church that had no precedent in scripture. In this category, they included both the prayers and rituals that were licit in the medieval church—like reciting the *Ave maria* or praying before images of the saints—and practices that the medieval church characterized as the "superstition" of "foolish women": charms, healing rituals, and the like. Because charms and healing rituals were very often evocative of authorized medieval religious practice, it had been quite difficult for medieval church authorities to draw a line between licit practices (like praying the *Ave maria* over a sick patient) and illicit ones (like writing the *Ave maria* on a sage leaf for the sick to eat). For zealous sixteenth-century reformers, however, the faint line between ritual prayer and healing charms was immaterial. The "Bishop's Book" of 1537 lumped all of these practices together in one condemnatory passage. All of it was superstition, or—depending on who you asked—blasphemy on par with witchcraft or sorcery.

As the first official publication of the newly reformed English church, the Bishop's Book was meant to familiarize the English with the tenets of the new faith. As such, it banned "not only all such as use charms, witchcrafts, and conjurations . . . but also all those, that seek and resort unto them for any counsel or remedy, according to the saying of god."[73] Lest these superstitious practices persist in the far-flung parishes of England, the newly formed Church of England implemented what it believed to be an effective bureaucratic mechanism for rooting them out. Every year, agents of the church would visit the many parishes of England, making inquiries at each stop to determine whether there were priests or parishioners who still held on to the old ways.[74] The 1547 Articles of Visitation issued by Lord Protector Somerset, chief advisor to Edward VI, enjoined church authorities to demand of parishioners "whether you know any that use charms, sorcery and enchantments, witchcraft, soothsaying, or any other wicked craft invented by the devil."[75] On their surface, these questions had nothing directly to do with women. In theory, anyone could have been prosecuted for using charms. And yet, just as the medieval church characterized these activities as the sins of "foolish women," so too did the reformed Church of England present superstition and magic as feminine.

In 1548, John Bale, a former Carmelite friar turned Protestant firebrand, composed *A comedy concernynge thre lawes, of nature Moses, & Christ, corrupted by the Sodomytes, Pharysees and Papystes*, one of his famous morality plays lambasting the corruption of the Roman church. The play offered a dialogue between characters who were personifications of popish heresies, including Sodomismus, Infidelitas, and the figure of Idolatria, described by Sodomismus as a magical healer. In the play, when Infidelitas remarks that Idolatria was once a man ("What, sometime thou were a he"), Idolatria countermands him, stating "now I am a she / And a good midwife by day / young children can I charm." What follows is a veritable checklist of accusations lobbied against supposed witches: "I can make corn and cattle, / That they shall never thrive . . . Their wells I can up dry, / Cause trees and herbs to die, / And slay all poultry."[76] Idolatry may once have taken the solemn masculine form of the priest venerating images of the saints in the Mass, but in the reformed church, idolatry took the feminine form of the midwife-witch.[77]

When Catholicism was restored in England under Queen Mary in 1553, the solemn male priest venerating the saints once again took up a position of authority, but the figure of the idolatrous midwife-witch remained. In 1554, Bishop Edmund Bonner, restored to episcopal authority by Queen Mary, added a new section of questions to the Visitation Articles he drafted for the diocese of London. This new section pertained specifically to the fraught time of childbirth, and demanded of parishioners whether they knew "any midwife or other woman, coming to the travail of any woman with child, [who] do use or exercise any witchcraft, charms, sorcery, invocations or prayers, other than such as be allowable"?[78] When Mary's sister, Elizabeth, reversed course and returned the English church to Protestantism, the precedent set by Bonner's Articles of 1554 stood. Elizabethan parishioners were similarly questioned as to whether they had witnessed the use of charms, incantations, or Latin prayers, especially "in the time of women's travail."[79] On at least one belief Protestants and Catholics could agree: the all-female space of the birthing chamber was especially vulnerable to the infiltration of magical and superstitious practices, not least because of the women who served as authorities there. What those superstitions entailed depended upon which side of the confessional divide one stood. As Mary Fissell has noted, Catholic rituals related to childbirth were often the focus of reformers' ire precisely because they spoke to the central issues of the Reformation: the role of Mary in Christ's incarnation, the nature of the Eucharist, and the intercessory power of the saints.[80]

Not surprisingly, then, when allegations of witchcraft increased in the later sixteenth century, the themes of pregnancy, maternity, and childbirth were often invoked in descriptions of witches' supposed crimes.[81] As elsewhere in Europe, the English women accused of witchcraft were often the most vulnerable: those who were old and indigent, whose precarious position in society was felt more keenly after the Reformation destroyed traditional institutions of charity.[82] And, because rituals surrounding reproduction, childbirth, and postnatal care were still strongly correlated with women's secrets, it was relatively easy for rumors involving those rites to coalesce into accusations against less socially secure women.[83] The case of Elizabeth Frauncis, accused as a witch in 1566 and tried with Agnes Waterhouse, the first woman hanged for witchcraft in England, is illustrative of this dynamic. The drama of Frauncis's story was spelled out for English readers in a splashy, tabloid-like pamphlet, *The Examination and Confession of Certaine Wytches at Chensforde in the countie of Essex*, published in 1566 by William Powell, a well-established London printer with at least eighteen practical books to his name by the time of this pamphlet's publication. According to this pamphlet, Frauncis had found herself pregnant and abandoned by her lover. She then made a pact with her "familiar"—a white cat by the name of Sathan—who "bade her take a certain herb and drink it, which she did, and destroyed the child forthwith."[84]

The story of Elizabeth Frauncis is an extreme example of where heightened fears about the power of women's reproductive rituals could lead in sixteenth-century England, but it is nonetheless useful context for understanding the climate of hostility and suspicion toward "women's secrets" that swirled around the early modern readers who browsed fifteenth-century manuscripts searching for hidden knowledge. With Elizabeth Frauncis's story in mind, we should not be surprised that many were troubled by what they found. These readers had been prompted to treat unpublished knowledge in "old English written books" as "secret." So, when they encountered charms, love spells, and reproductive recipes in their manuscripts, they had reason to believe they had found "women's secrets." At least a few of them responded to these secrets with censorship. They got out their inkpots or their penknives and blotted out or scraped off charms and reproductive recipes from the pages of their fifteenth-century manuscripts.

In British Library MS Additional 34210, for example, a late sixteenth-century reader (perhaps the Simon Barrsdall who left his signature on the last folio of the manuscript) took care to annotate scores of medical recipes with marginal notes updating odd Middle English spellings with more contemporary language. On the first leaf of the manuscript, this reader corrected "drowe" to "dry," "peleter" to "pellitory," "sepe" to "shepe," "werwot & celodyne" to "wormwood & celondine," and "wex & encense" to "wax & franckincense."[85] At the very least, this early modern reader wished to make the knowledge on this manuscript's pages legible, and perhaps useable, too. And yet, midway through the manuscript, this same reader took care to totally obliterate two childbirth-related recipes. A recipe for "A medicine for woman [illegible] of child" is completely blacked out with ink, and another following it "for the same" has been partially scraped from the parchment with a penknife.[86] Both appear to have been recipes for expelling a fetus, perhaps using herbs somewhat like those Elizabeth Frauncis admitted to ingesting in her trial for witchcraft.

Another early modern reader combined both of Barrsdall's censorship techniques in British Library MS Sloane 962, first scraping over and then saturating with ink the last words of the title of a fertility recipe "For a woman that may not have a child," so that "a child" is only barely legible.[87] In British Library Sloane MS 393, nearly a full page of successive reproductive recipes "For to deliver a woman of a quick child," the first in Latin and the following two in English, have been partially scraped from the page with a penknife. The first is difficult to read now, though it appears to have been a recipe for an abortifacient, and the other two are charms to encourage a woman to deliver, "if that it be so that her time is come."[88] In Wellcome Library MS 409, shown in figure 6.1 below, five different charms are now only barely legible from underneath thick strokes of black ink, perhaps added by John Starys, who left a late sixteenth-century *ex libris* ownership mark on the final page of the manuscript.[89]

By and large, evidence of censorship in practical manuscripts is rare. Early modern readers were far more likely to make marginal additions than they were to undertake deletions in fifteenth-century recipe collections.[90] That makes it all the more remarkable that when censorship does appear

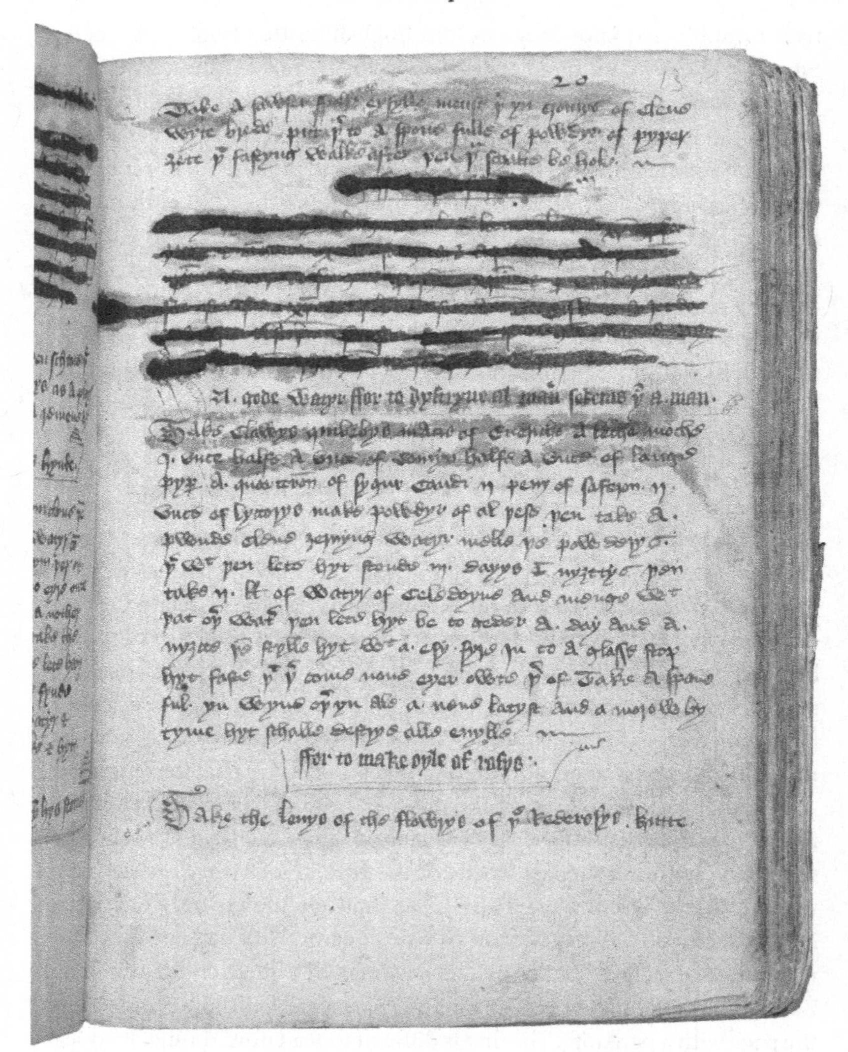

ffor to make oyle of rosys.

FIGURE 6.1. One of four charms in a Middle English medical miscellany (Leech-Books, VI) that have been entirely blacked out by a later reader. London, Wellcome Library, MS 409, f. 13r. Public Domain Mark.

within practical manuscripts (as it does in twenty-seven of the 182 in my corpus), erasures and cancellations nearly always disguise or attempt to discredit knowledge that was—according to longstanding associations discussed in this chapter—perceived as "women's knowledge."[91] The professional scribes who copied these charms and recipes in the fifteenth century left no sign or comment that they aroused suspicion. But by the later sixteenth century, attitudes toward this sort of "women's knowledge" had changed considerably. With the story of Elizabeth Frauncis and her feline familiar in mind, we can perhaps understand why the anonymous censor of British Library Sloane MS 393 chose to scrape away the details of a charm "to deliver a woman of a quick child."

But secrets—even dangerous ones—are hard to give up. For many early modern readers, the thrill of powerful and secret knowledge was enough to risk its preservation. A good number of early modern readers used marks or notes to signal their disapproval of certain charms or recipes while nonetheless leaving them intact. One commentator wrote the phrase "evil this" in a late sixteenth- or early seventeenth-century hand alongside charms in British Library Sloane MS 1315.[92] Whoever systematically canceled the dozens of charms within Bodleian Library MS Ashmole 1477 left a few entirely legible, only lightly crossing over them with a single X.[93] A similar technique was used by a later reader of Cambridge University Library MS Ee.1.15, who crossed out a version of the sage leaf charm (the same one offered as a good medicine for fevers by Dame Berthe in *The gospelles of dystaves*), as well as another charm for stanching bleeding and one for toothache.[94] Sometimes readers merely blacked out the Latin words within a charm and left the rest—a savvy move given that Elizabethan church authorities were especially on the lookout for "Latin incantations."[95] Henry Dyngley, whose inveterate manuscript collecting will be discussed in the following chapter, used different techniques in different books. Though he only lightly crossed out a charm to stanch blood in Wellcome Library MS 5262 and added what looks like a finding aid in the margin for "a charm" (see fig. 6.2) alongside it, Dyngley totally rubbed over with ink a recipe "For to charm a woman" in Bodleian MS Rawlinson C.506.[96]

For every one of these early modern censors, however, we can probably assume that there were an equal or greater number of early modern readers who saw no reason to rid their manuscripts of valuable knowledge, even if it was the sort associated with women. The sixteenth-century owner of Wellcome Library MS 404, for example, clearly valued the natural knowledge in that manuscript. He took the time to add marginal finding aids to its many recipes, adding brief captions like "for the headache" or "for the gout & falling evil" alongside blocks of Middle English text.[97] One page full of that reader's annotations contains a charm for fevers utilizing the *abracadabra*, and just underneath that, a recipe "for the flowers [menstrual flow] of a woman that may not hold."[98] Neither are blotted out or canceled in any way.

As these examples make clear, the individual motivations of do-it-

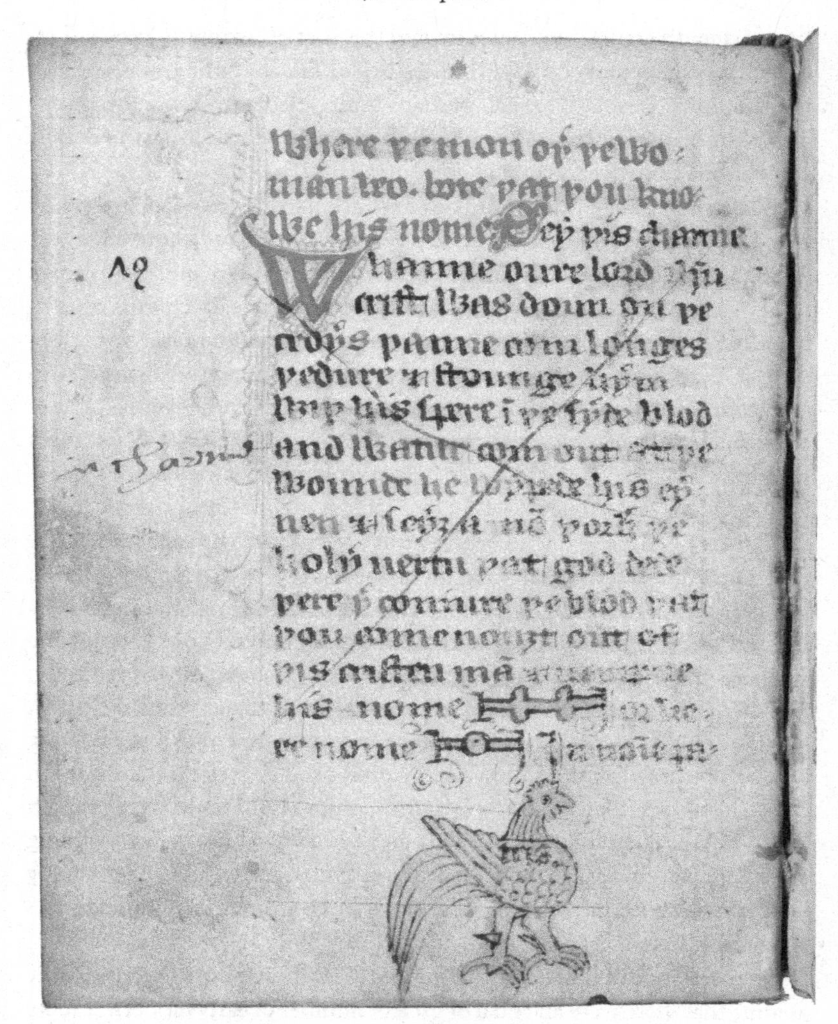

FIGURE 6.2. Canceled charm with marginal notation made by an early modern reader in a very early fifteenth-century practical manuscript. Medical recipe collection, England, fifteenth century. London, Wellcome Library, MS 5262, f. 38v. Public Domain Mark.

yourself censors cannot now be satisfactorily reconstructed. Their interactions with fifteenth-century manuscripts were mostly confined to X marks, cancellations, and erasures. We have only a few names to place alongside these reader marks, and even those are difficult to assign with absolute certainty. What we can say, however, is that early modern readers' censorship efforts were almost entirely directed toward categories of knowledge associated with women. And, moreover, that their discomfort with recipes relating to childbirth, charms, and healing rituals represents a significant shift in attitudes since the fifteenth century, when such material circulated openly in manuscript. Further, we should note that they censored charms and fertility rituals in manuscript collections even as similar material circulated in print. This shift in early modern readers' attitudes toward manuscripts and magic cannot be reduced to any single factor. The concept of secrecy as outlined in printed books, the inquiries of Elizabethan church officials, and the crescendo of witchcraft allegations in the later sixteenth century all combined to make these early modern readers especially wary when encountering what looked like women's superstition in the pages of "old English written books" whose contents might be secret, given that they had never before appeared in print.

This is not to suggest that early modern women lacked real knowledge or expertise that they might have described as "secret," nor is it to imply that women played no part in shaping the preservation and circulation of "women's knowledge." By the later sixteenth century, a woman like Catherine Tollemache, matron of Helmingham Hall and owner of a fifteenth-century practical manuscript, had the means to compose her own manuscript recipe collection filled with knowledge she had accumulated and mastered herself.[99] Indeed, Elaine Leong has shown that seventeenth-century England nurtured many women like Tollemache, who cultivated medical expertise that they recorded in household manuscript collections.[100] Yet, despite a steady increase in female authorship of manuscript recipe books, few ever published their recipe knowledge in print—Hannah Woolley being the notable (and well-studied) exception.[101] Even as women took pains to set down and preserve their expertise in writing, print remained the realm of male expertise while manuscript retained its associations with femininity.[102] We can trace this dynamic back to the narrator of *The gospelles of dystaves*, back to Master Alexis, and even back to Thomas Lupton and his "thousand notable things." In all of these books, publication was the means by which men solidified their position as masters of discovery, taking credit for the knowledge of "worthy doctoresses" or "poore women," whose expertise absolutely did shape the economy of natural knowledge in early modern England.

Conclusion

This chapter's examination of wildly successful printed books alongside censored manuscript recipe collections has shown, perhaps more than anything else, that readers' attitudes toward "women's knowledge" were as slippery as the category itself. Knowledge that might be characterized as women's secrets in one book could be couched as learned philosophy in another. What was licit in print was suspect in manuscript. As a result, early modern readers learned to make value judgments regarding the surety or safety of the same kinds of knowledge depending on how it circulated and in what medium. The category of "women's secrets" could be constructed and deconstructed with ease, according to authors', publishers', or readers' needs. In effect, this meant that when readers learned to parse from among nearly identical versions of the same magical charms according to the media in which they circulated, they were learning to read femininity as one of the most important factors in determining the value and security of natural knowledge.

Which is all to say that this chapter has shown the futility in trying to identify "women's secrets" as they circulated in manuscript and early print. In attempting to do so, we run the risk of instantiating that which was only ever a discursive category, retroactively imposing a logic that was only intermittently, haphazardly applied by historical actors. In attempting to make sense of misogyny, we risk building in a coherence that was never really there. There was no such thing as "women's knowledge" in fifteenth- and sixteenth-century England. There was only ever the threat and thrill of it. Readers (most of them male) navigated the tension between that threat and thrill, seeking out what they understood to be precious knowledge because it supposedly came *from* women, even as they denigrated it for the same reason. In the end, commercial publishing added another set of terms to the existing binaries that structured early modern thought. To neat oppositions between man and woman, good and evil, literate and illiterate, the sixteenth-century reader could add print and manuscript.[103]

7

Englishing Medicine and Science

By the time Thomas Buttus began the laborious process of copying recipes into his "book of medicines for diverse and sundry diseases" in 1564, he and the rest of the English reading public were inundated with printed books, broadsides, and pamphlets filled with medical recipes, prognostications, herbal lore, and agricultural instructions.[1] Dozens of editions of practical books appeared at the bookshop every year, their titles proclaiming the discovery of new, secret, or "never before published" knowledge. Each of these new editions represented a choice for a discerning reader like Buttus: would he select this or that medical recipe book? Purchase this or that almanac? Subscribe to the directions in this or that manual on horsemanship or husbandry? These daily acts of discernment honed readers' sense of their own expertise, their ability to make value judgments, and their confidence assessing the utility of information. In the decades since 1485, when William Machlinia published the first practical book in English, consumers had become capable adjudicators of natural knowledge—though Thomas Buttus probably thought himself more capable than most. He was, after all, the middle son of Sir William Buttus (or Butts), personal physician to Princess Mary before his death in 1545.[2] After years of observing his father at work, the younger Buttus presumably felt he could readily identify the best, most authoritative medical knowledge. And so, when he sat down with fresh quires of paper to compile recipes for his "book of medicines" in 1564, he was careful to copy only recipes that were "by late experience proved" and—perhaps surprisingly—those which were "taken out of diverse olde English books."[3]

We might expect that the son of a well-known English physician would recognize the value of trying and "proving" medical knowledge to judge its efficacy, but Buttus's personal history might have made him desirous of recipes utilizing new and exotic ingredients rather than those from "old English books." In 1564, Buttus could count himself just one among a handful of living Englishmen who had seen North America, having sailed to Newfoundland with Captain John Rut in 1527.[4] Yet Buttus's recipe book suggests no fascination with marvelous cures sourced from New World plants and minerals. Instead, the collection makes clear that Buttus was totally enamored of recipes that came from manuscript sources with what he perceived to be a distinguished pedigree. He admitted as much in an addendum to his remedy book. After copying 146 medical recipes in 1564, arranging their titles neatly in his table of contents, another batch of "old English books" presented themselves. The next year he added another nineteen recipes to his collection, as well as a note: "Hereafter followeth diverse other very good medicines . . . gathered out of diverse old English books, which came unto hand since the time that the other [recipes] were written, by means whereof they could neither be orderly placed, nor numbered, but yet they be all such as the worst is too good to be left unwritten."[5] Thomas Buttus, son of a physician at Henry VIII's court, fastidious organizer of knowledge, found the recipes in "diverse old English books" so compelling that he simply couldn't leave them out of his manuscript collection, no matter what it cost him in terms of aesthetics or order. They were simply "too good to be left unwritten."

Buttus's drive to collect and copy knowledge "gathered out of diverse old English books" is striking, but it is not singular. Recipe collecting and recipe book authorship were all the rage among sixteenth- and seventeenth-century Englishmen and -women. Though studies of these early modern collections have tended to focus on their value as evidence of household experimentation and manufacture, we should also recognize early modern recipe books as vehicles for passing on a great deal of natural knowledge from medieval sources.[6] For example, George Walker, an early sixteenth-century landowner from East Farnedowne, Northamptonshire, took pains to partially reproduce the pictorial prognostications by dominical letter (discussed in chapter 2) in his sixteenth-century practical manuscript. Like Buttus, Walker must have had a fifteenth-century manuscript as his source, though we have no way of knowing exactly which manuscript he copied when he sketched those icons depicting crops or livestock.[7] We can only infer that Walker had an interest in preserving older traditions of natural knowledge that he found in "diverse old English books."

Though Buttus's and Walker's manuscript sources are now lost to us, it is possible to reconstruct sixteenth-century readers' interest in very old natural knowledge through the hundreds of reader marks, commentaries, notes, and signatures scattered throughout the margins of fifteenth-century manuscripts.[8] In total, 112 of the practical manuscripts examined for this book contain early modern reader marks, ranging from added recipes, indices, or tables of contents to the addition of inventories or correspondence.[9] These marks make clear that fifteenth-century manuscripts were read and reread well into the era of print, long after much of the knowledge contained within them was widely available in published editions.[10] In fact, as this chapter will argue, print seems to have made early modern readers *more* interested in manuscript sources. Because printers had convinced these readers to value natural knowledge that was "practiced" or "proved" through experience (as noted in chapter 4), those same readers may very well have assumed that handwritten recipes were vetted more thoroughly than those compiled by unscrupulous printers. Likewise, readers inspired by books of secrets and interested in uncovering "hidden" knowledge themselves may have viewed manuscripts as opportunities for discovery. Buttus was likely influenced by both trends. Not only did he select recipes for his remedy book that had been "by late experience proved," he also owned and treasured a copy of "the secretes of the Reverend father Mr Alexis of piedmont," so much so that he specifically mentioned it in his will.[11]

Yet even if Buttus was driven to collect knowledge that had been "proved" and to search for "secrets" in unprinted sources, neither impulse wholly explains why recipes from "old English books" were "too good to be left unwritten." What made older English knowledge so valuable? And how did Buttus come to recognize his sources *as old*? This chapter will seek answers to these questions by piecing together the scattered evidence of early modern readers' notes, initials, and marks in fifteenth-century manuscripts, reconstructing how those readers approached century-old collections of natural knowledge. As we will see, identifying a manuscript as both old *and* English gave it a special cachet within a culture obsessed with defining Englishness through attention to the past. Indeed, I show that practical manuscripts played an important and largely unacknowledged role in the Elizabethan project of constructing a national English identity. Not only did fifteenth-century manuscripts turn middling bureaucrats into amateur antiquarians and vernacular humanists, eager to collect "diverse old English books," these manuscripts also played a part in convincing early modern readers that there had once existed a distinctly English tradition of engaging with the natural world—a tradition that some in Elizabethan England were eager to defend.

Manuscripts in a World of Print

Though the number of printed practical books for sale in English bookshops grew exponentially in the century following the arrival of the press, during all that time, early modern readers never stopped reading manuscripts. If anything, the number of manuscripts in circulation in Tudor England may have grown even faster than the number of printed books.[12] More people could write, and paper was more widely available, making it easier for sixteenth-century men like Thomas Buttus, or women like Catherine Tollemache, to compile recipes in their own manuscript collections.[13] Indeed, Elaine Leong has documented a precipitous rise in the composition of household recipe collections in England over the later sixteenth and seventeenth centuries.[14] Nothing about the dominance of print turned readers away from the affordances of manuscript as a medium for collecting and assessing natural knowledge.

This is not to say, however, that early modern readers hadn't grown accustomed to the format and function of printed books. When they created their own manuscript collections, or when they amended older manuscripts from the era before print, they brought these books up to date, creating a table of contents, correcting odd Middle English spellings, adding an index, or dividing a text with clearly marked headers.[15] In some cases, sixteenth-century annotators of fifteenth-century manuscripts even dated their marginalia, leaving signatures reminiscent of printers' colophons.[16] And very often, early modern readers used printed books to update or analyze the natural knowledge they found within older manuscript collections.[17] For example, the later sixteenth-century owner of Huntington MS HM 58 must have had several printed medical books to hand while he read and annotated this fifteenth-century copy of the Middle English *Agnus castus* herbal. On the first leaf of the book, this reader copied a pleasing verse from the Bible—"The Lord hath created Medicine of the earth, and he that is wise will not abhor it"—which he had copied from the title page of *An hospitall for the diseased*, published for the first time in 1578.[18] Likely this same reader had a printed edition of the *Agnus castus* available to him—perhaps one of the many editions of Banckes's herbal—whose text he systematically compared with the version in his manuscript to note where his copy was deficient.[19]

For many early modern readers, like Thomas Dib, older manuscripts were simply one part of a larger information economy that included printed books and more contemporary manuscript sources. Only by working across these varying media could one hope to ascertain which bits of knowledge were the most authoritative and effective. And that meant collecting quite a lot of books. Dib was owner of at least three fifteenth-century practical manuscripts, though the inventories and notes he added to those manuscripts' flyleaves suggest he may have owned many more that have either been lost or have yet to be identified.[20] The first of Dib's fifteenth-century manuscripts is British Library MS Sloane 1764, the same manuscript once owned by Peter Cantele of Toft Monks, who served as our guide in chapter 1. After Cantele, the manuscript passed into the hands of Thomas Say, who also owned the second of Dib's manuscripts, British Library MS Sloane 121.[21] When Dib got his hands on these manuscripts in the mid-sixteenth century, he rebound both of them in old pieces of parchment cut from a medieval antiphonary, a genre of medieval religious manuscript intended to guide monks through the liturgy of the Mass—exactly the sort of manuscripts, in other words, that would have had no value save as scrap material after the English Reformation.

Onto the outer parchment wrapper of Sloane 1764, Dib made a list of books he wished to give "to my cousin": a copy of "vigoes little practyk," "an old book with pythagoras sphere," and "an other old book of colors & medicines."[22] "Vigoes little practyk" was likely an edition of a book of surgical recipes published by Robert Wyer in 1550, one of several of his editions that claimed to have been "proved" by contemporary authorities.[23] But the two "old books" that Dib intended to send to his cousin were probably fifteenth-century manuscripts. One of these must have contained a popular Middle English prognostication known as "Pythagoras's sphere," which supposedly determined whether a sick patient would live or die.[24] The other "book of colors & medicines" probably contained instructions for making colors and inks alongside medical recipes, perhaps very much like British Library MS Sloane 2584, discussed at length in chapter 3.[25]

Thomas Dib, like Thomas Buttus, clearly distinguished between "old books" and new, though both had value to him. In the second of his manuscripts, British Library MS Sloane 121, Dib made several more notes about manuscripts and printed editions, though it is hard to say whether these notes comprise a wish list or an inventory: "A book *de natura et essentia Angolorum* in black friars," "A book of natures working in herbs," "Escalapius physics," "Anglographia, a book in 8to [octavo] printed by Plantyne" (probably Christophe Plantin of Antwerp), and "A book in French called the Good steward of distillations & showing how to make a furnace that shall give diverse degrees of heat."[26] His bibliographic notes continue on the back cover of the manuscript as well, where he wrote summaries of the second book of Heinrich Cornelius Agrippa's *De occulta philosophia*, expounding on "the many & marvelous works which only are wrought by the mathematical Artes."[27] It seems the "occult sciences" were a special interest of Dib's, so much so that he composed his own manuscript collection of various prognostications and horoscopes—a collection that calls to mind the notebooks of the well-known Elizabethan astrologer-physician Simon Forman—and bound it together with his fifteenth-century manuscript to form what is now Sloane 121.[28]

Dib's collection of manuscripts and printed books wasn't just for show. He was clearly a voracious reader, but as he read, he constantly compared the knowledge he found on the page with knowledge he had garnered from experience. Those careful comparisons are documented in the comments he added alongside recipes in both of these practical manuscripts. In the left margin of one page in Sloane 121, for example, next to a recipe that lists ingredients for preparing a "clyster" (or glister), Dib added notes to remind himself of best practices: "not[e] soap is necessary in glister," and in the right margin, he added, "note that oil might never be put in clister . . . if they be in dysentery which is a flux of the body."[29] Alongside a recipe "For the stone" in Sloane 1764, Dib amended the recipe with his personal experience: "this hath been oft proved by diverse practitioners that have used almost no other purge. But it works strongly & it will work upward & downward. But to make it work downward walk a good pace a pretty while after you have biben [drunk] it."[30] In these two annotations, we see a man invested in refining the knowledge that had come to him in textual form, using his experience to do so. His commentaries bring to mind Nicholas Neesbett, the mid-fifteenth-century practitioner from York whose "Sururgia," discussed in chapter 3, was also a vehicle for refining a textual record of healing practice with knowledge born from experience.

But unlike Neesbett, Dib had more than experience to guide him. He also had a considerable library of printed books *and* manuscripts. If the lists he kept in the front pages of his manuscripts are any indication, it would seem he could access a huge number of scientific, medical, and mathematical texts, which meant he could compare one copy of a recipe or horoscope with multiple others, before finally comparing those multiple versions to his own experience. For example, in Wellcome MS 404, the third of Dib's fifteenth-century practical manuscripts (also wrapped in parchment from the same medieval antiphonary), Dib left notes that reveal exactly this process.[31] Underneath a recipe for making "pitched cloths" (or bandages soaked in a sticky substance), Dib wrote, "the quantities are these as I find in another old copy [of the same recipe]: iiij. li of [illegible] iij. quartons of resin, iij. quartons of wax, a quarton of turpentine."[32] On the following page, Dib added another marginal note next to a recipe "For to do away the ache & the bruised blood of wound: I find it in many copies."[33] Dib certainly seems to have amassed a collection of "old books" (probably manuscripts) large enough that he could have compared one "old copy" of this recipe for sticky bandages to another, equally old copy. He was someone who could closely read and compare a variety of printed and handwritten texts and then examine them in the light of his experience, a skill that Jennifer Rampling has called "practical exegesis."[34] For Dib, trying-and-testing knowledge meant reading widely just as much as it did tinkering with ingredients in the stillroom.

We can glean something about Thomas Dib's reading habits, his interest in techniques of healing, and his careful research methods from the notes he left in his three fifteenth-century manuscripts. But nearly everything else about him—and, indeed, everything else about most early modern annotators of practical manuscripts—remains a mystery. It just so happens that William Thorowgood recorded the birth of his daughter, Ann, in Cambridge University Library MS Dd.10.44, and for that reason we know he had possession of the manuscript in August 1561. Because Thorowgood was a prominent Draper, he is easy enough to locate in records of London city governance, but the same cannot be said of the two men to whom Thorowgood lent the manuscript, John Young and John Tight. Though they also added their signatures to the book's pages, above an *ex libris* note reminding future readers that the book belonged to "William Thorowgood only," these marks tell us nothing of their profession or place of residence.[35] Though scores of early modern readers left marks in 112 of the fifteenth-century manuscripts studied for this book, most of these figures remain anonymous, or, like Young and Tight, are no more than the names they scribbled in their manuscripts' pages.

Of the early modern annotators whom we can securely identify, many lived in London and many were members of the medical profession. Thomas Dib may have been a medical practitioner, given his interest in comparing instructions for producing medical bandages and preparing purgatives. He likely shared these interests with John Smerthwaite, another owner and annotator of an early fifteenth-century remedy book, who was also a member of the London Barber-Surgeons.[36] Smerthwaite very thoughtfully amended Wellcome Library MS 406 with a dated colophon, so we know that he spent some time in January of 1511/1512 copying recipes for "text ink" and to "stanch blood."[37] Sir John Aylyff, also a member of the London Barber-Surgeons and one of Henry VIII's personal surgeons, owned British Library MS Sloane 1, a nearly five-inch-thick fifteenth-century copy of Guy de Chauliac's "Treatise on Fractures" and "Treatise on Wounds," which he made available to other members of his profession. Aylyff's name, along with the names of two of his fellow barber-surgeons, John Jonson and William Rewe, appears on the last leaf of the manuscript, as does a note under Aylyff's name stating that he had once been "surgeon to the king Henry the eighth."[38] Finally, the last leaf of

Bodleian MS Bodley 483, another fifteenth-century practical manuscript, contains a note about the "act made the third year of the reign of king henry the viii for the establishing of the physicians and surgeons"—a mark that suggests a professional's interest in the official incorporation of these two classes of medical practitioners.[39]

It isn't particularly surprising that John Smerthwaite and John Aylyff, both of whom practiced medicine under Henry VIII, would be interested in medical manuscripts created in the fifteenth century, perhaps even by their forebears in the surgical profession. What is noteworthy, however, is that early modern readers' interest in these old sources actually intensified over the course of the sixteenth century. As time passed and the years marched on, men like Thomas Buttus came to believe that hundred-year-old manuscripts contained knowledge that was *superior* to that available in contemporary books, that their recipes were, as Buttus put it, "too good to be left unwritten."[40] Sometime in the mid-sixteenth century, John Rice paid twenty-five pence for Royal MS 17 A.xxxii, a mid-fifteenth-century manuscript with a prognostication, an herbal, a collection of medical recipes, and a treatise on "Medicines for horses." Though every one of these genres of practical knowledge had been issued in print by the 1520s, most in small octavo editions that cost just a few pence, Rice paid several times that for the same knowledge in a decades-old manuscript.[41] By the end of the sixteenth century, the value of practical knowledge in Middle English manuscripts had risen even higher. The physician-astrologer Simon Forman wrote a note at the front of Bodleian MS Ashmole 1396, a beautiful early fifteenth-century copy of Lanfranc of Milan's *Chirurgia magna* in Middle English, that "this book is worth x li [ten pounds] for th'excellence of the remedies therein contained."[42] Ten pounds was a princely sum in sixteenth-century England, and yet, for Forman, centuries-old medical knowledge was worth its weight in gold.

What was it that convinced men like Buttus, Rice, and Forman of the superior value of century-old books as compared to printed editions? Were they immune to the marketing tactics of publishers who had spent decades encouraging readers to buy editions of herbal lore or medical recipes? Did they not believe publishers who claimed their printed editions were more complete and better organized than what was available in manuscript?[43] If they didn't, they should have. As Elizabeth Eisenstein famously argued, print's capacity to reproduce and disseminate a single, stable text should have made that medium the obvious choice for those interested in acquiring, comparing, and assessing knowledge.[44] Certainly the corrector of the *Agnus castus* herbal in Huntington MS HM 58 whom we met earlier in this chapter must have quickly realized that his manuscript version was deficient when compared to the multiple printed versions available for sale in English bookshops. Yet this early modern reader still took the time to annotate his manuscript copy, noting the entries on *Genestula* and *Lactuca leporita* that were missing.[45] The text in his manuscript was not as reliable as that available in print, but something about the manuscript as an epistemic object made it worth keeping—even correcting.

On the one hand, printed editions of natural knowledge were often more accurate. But on the other hand, they were also tainted by the commercial nature of the publishing industry. Early modern compilers of recipes and herbal lore were sometimes mistrustful of the natural knowledge they found in printed books because, as Leonard Digges had put it in *A prognostication of right good effect*, publishers worked for "profit only."[46] Yet, that alone cannot explain early modern readers' interest in "old English books." If the profit motives of English printers were the problem, these readers could have turned to contemporary manuscript collections or friends and neighbors to collect knowledge for their households—as, no doubt, many of them did.[47] But just as many seem to have sought out fifteenth-century practical manuscripts not in spite of their age but *because* of it. When these early modern readers scoured century-old books for natural knowledge, they were looking for more than tried-and-true recipes. They were amateur antiquarians, looking to recover traces of a past that had been lost, acting on the same impulses that drove much more famous collectors of old English manuscripts, like Matthew Parker, John Stow, John Dee, and Sir Robert Cotton. Which is to say, the only way to understand early modern readers' interest in fifteenth-century practical books is to understand how the past and present were bound up together in post-Reformation England.

Amateur Antiquarians

Right from the outset, the English Reformation was built on claims to historical precedent. When Henry VIII wanted to divorce Catherine of Aragon in the late 1520s, he sent his ministers off to check the chronicles of medieval churchmen like the Venerable Bede, Matthew Paris, or William of Malmesbury to find some historical authorization for his refusal to acquiesce to Rome.[48] Very old manuscript sources were thus critical to Henry's Reformation, and yet, in his zeal to break Rome's hold over the church in England, Henry oversaw the destruction and dispersion of the most ancient manuscript collections in England. There is no telling how many manuscripts were destroyed, scattered, or sold following the dissolution of the monasteries, nor how many more were burned up in the fires of Edward VI's reign.[49] And it wasn't just manuscripts that were lost when Cromwell's men dissolved the nearly 900 monastic institutions that were scattered across the English polity. In destroying the edifices of the old church institutions, reformers also severed many of the ties that bound the English past to its present. Because the ruins of abandoned abbeys and cloisters remained so visible across the English countryside, the faithful were constantly reminded that their society had changed markedly in just the span of a few years.[50] As Margaret Aston argued years ago, the ruins of the monasteries became physical markers of even greater losses that didn't mar the landscape: the loss of ritual, the loss of tradition, and, most significant, the loss of knowledge.[51]

When Henry's daughter Elizabeth ascended to the throne in 1558, following five years of a return to Catholicism under Mary, the Church of England resumed its commitment to severing ties with the old traditions and rituals of the medieval church. But Elizabeth's ministers also recognized that the new church needed a history, and that the construction of this history could prove beneficial to the Protestant cause. Like her father, Elizabeth sent ministers looking for manuscripts, but in her case, the search was inspired not by marital strife but by a group of Lutheran church historians from Germany known as the Magdeburg Centuriators.[52] These men were engaged in a massive project to write a universal church history that they believed would affirm the Protestant church as true inheritor of St. Peter's legacy. They hoped that Elizabeth would give them access to documents from the very early English church, and she did. Elizabeth sent for her new Archbishop of Canterbury, Matthew Parker, who scoured the country for important religious manuscripts that had been scattered by the dissolution. He wrote letters to priests across England and turned up at cathedrals and even private estates asking to see very old books.[53] When all was said and done, Parker had compiled an extraordinary collection of English religious manuscripts and given English theologians and historians new fodder from which to compose a history of their church. Quickly, interest in old books spread, and thanks to the dissolution of the monasteries, many beyond Parker's circle found that they, too, could collect ancient manuscripts: men like the historian John Stow or the queen's astrologer and alchemist, John Dee.[54] By 1586, Dee and Stow, together with the historian William Camden, were meeting regularly as founding members of the Society of Antiquaries.

The rise of antiquarianism in Elizabethan England is well known, but it is not a story that has yet been associated with amateur collectors of run-of-the-mill vernacular manuscripts. Yet the same spirit that drove Parker,

Stow, Dee, and Camden to collect treasures of Old English literature or theology also inspired anonymous collectors like the owner of Huntington MS HM 19079. On the first leaf of that manuscript, a copy of the Middle English translation of Gilbert the Englishman's *Compendium of medicine*, this reader left a note for later scholars: "this book was written the year of Christ 1100 as may be seen in the second side of the fifteenth leaf."[55] The passage that convinced this reader of the great antiquity of his manuscript was the following explicit, written by the manuscript's fifteenth-century scribe: "In his hand for crooking of his fingers *deo gracias*. A.M.C." Convinced that A was an abbreviation for the Latin *anno*, or year, and that the initials MC were the Roman numerals for 1100, the later sixteenth-century owner of this manuscript added a note next to the explicit: "Anno 1100 this book was written as appeareth here under."[56]

But the manuscript wasn't written in 1100. The original Latin text of the *Compendium of medicine* wasn't completed until at least the third decade of the thirteenth century, and Huntington MS HM 19079 probably wasn't composed until around 1425.[57] Yet the inaccuracy of this reader's estimate hardly diminishes the significance of his effort. He misread a confusing passage in his manuscript and misdated it as a result, but even learned antiquarians were prone to similar kinds of mistakes.[58] What matters is that this anonymous early modern reader was on the lookout for evidence that would help him place the manuscript in a time and place that was alien to his own. Whereas fifteenth-century compilers had happily appended the names of Hippocrates and Galen to Middle English remedies, stitching together past and present and eliding the events of history that made fifteenth-century England so very unlike ancient Greece or Rome, this early modern reader perceived the value of situating knowledge within its historical context. He could only do so because the English Reformation had made the events of his lifetime into history.

Though historians have tended to associate an awareness of the past *as past* with learned humanists, in fact, this awareness was shared by everyone in England who watched as vines crept up the walls of dilapidated monasteries or pigeons set to roost in the unused bell towers of abbey cathedrals. Because this awareness was so widely shared, even members of the minor gentry or middling sort became interested in collecting manuscripts that preserved something of the past they saw around them. We might say this is what happened to the men who would become the most famous and devoted antiquarians in England, each of whom began their lives as sons of tradesmen. John Stow's father was a Tallow Chandler, John Dee's father was a Mercer, and William Camden's father was a member of the Painter-Stationers.[59] None of them were born with scholarly reputations. Instead, they made them. Their collections of very old English books were eventual proof of their status as scholars and gentlemen.

If collecting knowledge from old manuscripts was a means of achieving social *and* intellectual status in sixteenth-century England, then we can understand why Henry Dyngley (or, alternatively, Dyneley) devoted himself to acquiring his own manuscript collection, even if it was decidedly less learned than those of Dee, Camden, or Stow. Like those men, Dyngley had no impressive intellectual pedigree. His was one of those families that had only sprung themselves into the minor gentry in the aftermath of the Black Death, when they took possession of the manor of Charlton in the parish of Cropthorne, Worcestershire. For the next century and a half, the manor passed in an orderly fashion from son to son until, in 1541, Henry Dyngley inherited it from his father, John.[60] Once lord of the manor, it seems that Henry Dyngley developed a taste for old manuscripts.[61]

Though Dyngley left no dated reader mark in Wellcome Library MS 5262, it may well have been Dyngley's first acquisition. The manuscript is a beautiful late fourteenth-century collection of medical recipes, into which Henry inscribed his name and initials. It was likely created at Winchcombe Abbey in Gloucestershire, a Benedictine foundation just twenty miles or so from Dyngley's manor at Charlton—and one that was dissolved in 1539, just two years before he inherited the manor.[62] Though we cannot know how or when Dyngley got his hands on the book, someone who possessed the manuscript after the dissolution took care to partially obscure the illustrations of saints on the manuscript's opening pages using the same technique found in Bodleian MS Rawlinson C.506, another of Dyngley's manuscripts, and one we can securely place in his hands before 1547.[63] That manuscript of over 300 small paper pages, crammed with recipes, charms, verses on bloodletting, equine medicine, and treatises on planting, grafting, fishing, and hawking, may also have come to Dyngley as a spoil of the dissolution. Before Dyngley owned the manuscript, it passed through the hands of "humfridus harrison Capellanus" (Humfrey Harrison, chaplain), who left a reader mark on the manuscript's final page. This may well be the same Humfrey Harrison who was vicar of Alstonefield in Staffordshire in the later fifteenth century.[64] And, because Alstonefield was owned by the Cistercian abbey of Combermere, it is possible that the chaplain's possessions reverted to the Crown upon the abbey's surrender on 27 July 1538.[65] By 1547 it was in the hands of Henry Dyngley, who added a new recipe "for migraines" to its pages, which he signed "by me henry

Dyngley of Charleton in the parish of Cropthorne written by me the .14. day of august anno domini 1547 I being of the age 32."[66]

Over the next two decades Dyngley amassed another three fifteenth-century medical manuscripts as his fortunes and family increased. By 1550, Dyngley had married Mary Neville, daughter of Sir Edward Neville, a courtier close to Henry VIII. By 1553, when he was appointed Sheriff of Worcestershire for the first time, he was also the proud father of three sons: Francis, the eldest, followed by George and Henry. And by 1554, Henry had managed to procure yet another medical manuscript, Trinity College Cambridge MS O.8.35, a professionally produced vernacular medical textbook, with recipes and treatises on uroscopy, on the four humors, on anatomy, and on "simples." Just as in his other manuscripts, Dyngley used blank space in the Trinity College manuscript to add extra recipes. In 1557, he copied a list of "waters" onto the last page of the book, which he again signed and dated, "Henry Dyngley anno Christo 1557 xxx may at Adyngton in Buckinghamshire."[67] Around 1560, Dyngley began consolidating his landholdings in Cropthorne, bringing suit against the Dean and Chapter of the Cathedral at Worcester for rights to common lands pertaining to his manor at Charlton.[68] That same year, he acquired both a new daughter, Mary, and another medical manuscript, British Library MS Royal 17 A.xxxii, which he again signed and dated over several folios.[69] The following two years brought another two daughters in quick succession, Barbara and Alice, the last of Henry's nine children.[70]

Perhaps it was the births of Mary, Barbara, and Alice in Dyngley's late forties that inspired him, sometime after 1563, to fill blank leaves toward the end of the Royal manuscript with a series of verses which he titled "A godly exhortation for a father to his children." The poem, published in John Foxe's *Actes and monuments* and attributed to the Marian martyr Robert Smith, was originally written as a macabre invocation for Smith's children to remember their father "in prison and in payne" and to reflect on their own mortality.[71] In Dyngley's hands, however, the poem took on a different tone. It was still a reflection on a father's mortality and the ephemeral nature of familial bonds, but where the printed version in the *Actes and monuments* ended with an extended reflection on the horrors of Catholicism and an admonition to reject that errant "whore of Rome," Dyngley chose not to include those verses in his copy. Instead, he closed his poem thus: "I leave you here a little book to look upon / to see your fathers face when he is dead and gone."[72]

In 1564, just one year after "A father's exhortation for his children" appeared in the first English edition of Foxe's *Actes and monuments*, Dyngley began work on another "little book" to leave for his children when he was dead and gone: Wellcome Library MS 244, a manuscript intended to collect and organize the knowledge he had garnered from Middle English sources. Begun in Dyngley's hand, the large paper manuscript opens with a *kalendarium*. Though it is devoid of most of the saints' days that filled medieval examples of the genre, it features Dyngley's own additions of astrological information for every month of the year.[73] On what is now the final leaf of Wellcome MS 244, Dyngley wrote notes about the weather ("The year of our lord god 1564 was the coldest spring & the windiest that ever I did see & it was the goodliest year of blooming of all manner of fruit trees"), followed immediately by a copy of a prognostication that would help him forecast future weather events.[74] The prognostication was a version of the "Book of Thunders," one of those discussed at length in chapter 2, predicting weather and crop yields according to whether one hears thunder in a given month. And though Dyngley did not leave a note for future readers about the value of the "diverse old English books" he had surely consulted, he did provide a disclaimer to the reader about the value of the recipes he'd collected: "Where so ever you see this character HD stand in the margin of this my book, against any medicine, oil, ointment, salve, plaster, trete, powder, syrup, electuary, unguent, water, or any other thing contained within this book, that have I, Henry Dineley proved without doubt & none other have I myself proved."[75]

In 1573, at nearly sixty years old and almost a decade after he composed this disclaimer about recipes marked with an HD, Dyngley was still annotating old manuscripts with his initials. Alongside a vernacular treatise on reproduction in Trinity College Cambridge MS R.14.52, Dyngley wrote, "Note H D 1573." And, at the back of the same massive (and beautiful) manuscript, he updated a running list of the dates of Easter with a comment that in 1573 "easter was the xxii day of march."[76] It is the latest date that appears in any of his manuscripts. Sixteen years later, in 1598, Henry Dyngley died at the age of seventy-four.[77] Over a lifetime in pursuit of natural knowledge, Dyngley left initials, notes, and dates in old manuscripts as a system of annotation by which he could keep track of those recipes he had "proved without doubt" and those he hadn't. He then began to compose a new manuscript collection, Wellcome MS 244, into which he could copy the best knowledge collected from "diverse old English books," as his contemporary, Thomas Buttus, had put it. But Dyngley went one step further than Buttus. He didn't just read old recipes, evaluate their contents, and copy them into a new manuscript. He preserved the original sources of his new knowledge, too. He built an archive as well as a reference book, and in so doing, Dyngley joined the ranks of Elizabethan

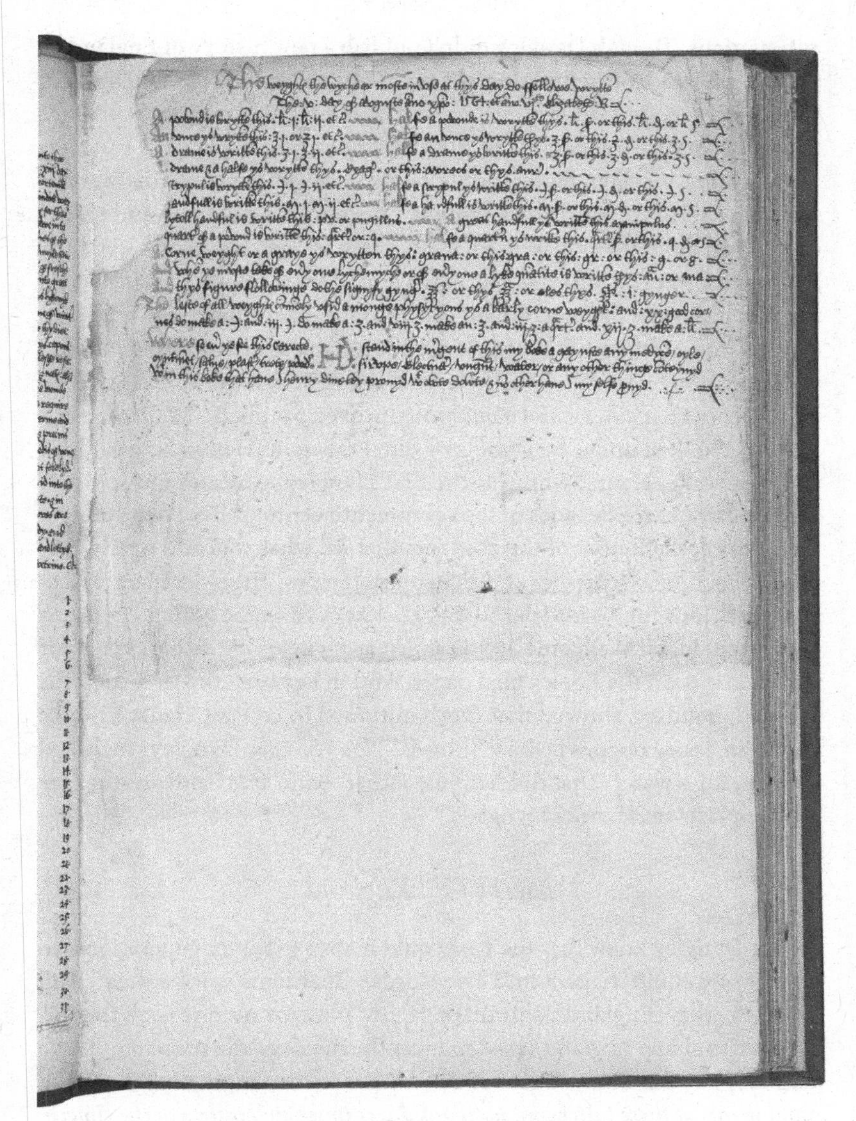

FIGURE 7.1. A page of Henry Dyngley's notes about apothecaries' weights and measures and about his mark of approbation, HD. Dinely/Dyneley (or Dingley/Dyngley), Henry (& others). London, Wellcome Library, MS 244, f. 4. Wellcome Collection. Public Domain Mark.

antiquarians. Though Dyngley didn't publish a new history of England or acquire a formidable reputation as a *magus*, he shared with Stow, Dee, and Camden a recognition that the pursuit of knowledge was an important means of social advancement in Elizabethan England.[78]

In his lifetime, Dyngley made all the right moves to elevate his family's status: he went to court to consolidate his family's holdings and he accepted two different monarchs' offers to serve as Sheriff of Worcestershire. His manuscript collecting was an extension of these efforts to advance his family's interests. As he wrote on one of the pages of his many fifteenth-century manuscripts, his books would be available one day for his children to "look upon to see [their] fathers face" when he was gone.[79] That verse, drawn from Foxe's *Actes and monuments*, proved prophetic. Wellcome MS 244 was "looked upon" by Dyngley's son, Francis, as well as his grandson and namesake, Henry, who was the first Dyngley to attend university.[80] Over the first three decades of the seventeenth century these two younger Dyngleys copied most of the contents that fill what is now a nearly 350-page collection of remedies and alchemical recipes. The elder Henry's disclaimer to look for his initials, HD, as a marker of a recipe's quality remains at the opening of Wellcome MS 244, just as Dyngley's weather report for 1564 remains on the book's final pages. And in between those entries, his son and grandson showed that they continued to consult Henry's books and value those recipes he had "proved." The younger Dyngleys included a recipe for a water "that did help my father Anno 1541" and another for an "Emplastrum Henrici Dyngley."[81]

Merry Old England

Henry Dyngley knew that the past could matter to future generations, so long as one could shape it into a new order. That same spirit animated all of Elizabethan England's antiquarians, but perhaps no one took the imperative to shape England's past to meet the needs of the present quite so seriously as John Foxe, author of the famous Protestant history *Actes and monuments of these latter and perillous dayes touching matters of the church*, first published in English in 1563. The historian Matthew Phillpott has shown that Foxe drew from a number of manuscripts collected by Matthew Parker and his circle for his *magnum opus*, but instead of publishing these sources as stand-alone translations or transcriptions (as Parker did), he presented these sources in a narrative history that made the distant past directly relevant to Foxe's present.[82] Like other Protestant church historians, Foxe argued that the early church was the "true" church, and that everything that came after the expansion of the medieval papacy was degradation and corruption. For that reason, the English Reformation, which had purged the English church of the false doctrine of popery, was not so much a break from the past as a return to it.[83] To prove it to his readers, Foxe published transcriptions of medieval letters and other documents within the *Actes and monuments*, many of them reproduced in their original Latin and in English translation.[84]

Foxe's book—hugely expensive though it certainly was—seems to have made quite an impression on the Elizabethan collectors we've met in this chapter. Henry Dyngley was clearly inspired by Foxe's account of the martyrdom of Robert Smith, so much so that he copied an excerpt of Smith's "A father's exhortation for his children" in one of his fifteenth-century manuscripts.[85] Thomas Buttus, the sixteenth-century compiler we met in the introduction to this chapter, not only owned a copy of Foxe's great martyrology but also kept a picture of the author hanging in his parlor.[86] And, though it is surely a coincidence, both Buttus and Dyngley began compiling their own early modern recipe books sourced from "diverse old English books" just one year after Foxe published the first English edition of his great history of the church, filled with knowledge sourced from manuscripts and ancient documents.[87]

To a greater extent than almost any other publication that emerged from the wave of antiquarian collecting in the early Elizabethan era, it seems that Foxe's martyrology played an important role in demonstrating to sixteenth-century readers like Dyngley and Buttus that attention to historical documents and original sources was the only way to ascertain the veracity and legitimacy of knowledge. But even more than that, Foxe's attention to the material evidence of the past—the letters of bygone Popes or the testimony of Lollard heretics—facilitated what Alexandra Walsham has described as the emergence of a "Protestant consciousness" in England, anchored by a belief in divine providence, or the idea that God's hand could be seen at work in the world, shaping events to reflect his divine will.[88] In Foxe's narrative history, readers could discern God's hand shaping events in the past such that they ordained the rise of the Protestant church in the present. Foxe was adamant that readers understand that the new religion had not "sprung up" in the last twenty or thirty years, but had rather "been spread abroad in England, by the space of almost CC [200] years, yea and before that time."[89] He devoted considerable space to John Wyclif and the Lollard movement of the later fourteenth and fifteenth centuries, recounting stories of the English men and women who had tried to reject the corrupting influence of Rome and restore the true church in England.

In Foxe's telling, the English church had a history that set it apart, not just from Rome but from the Protestant church elsewhere in Europe, too. Though his history took into account events beyond the English channel, and though it was intended to contribute to the broader project of a "universal" church history begun by those historians from Magdeburg, his book played at least some part in developing among the English faithful a belief that God had bestowed his special providence on England, a nation of the elect.[90] That conviction was only strengthened by the geopolitical realities of early modern Europe, where confessional conflicts were also political conflicts. Protestants in Elizabethan England knew that the Catholic powers of both Spain and France were pitted against them. Given the might of these two foes and the constant threat of Catholic interference in English politics, it is no wonder that worshippers in London or York began to define their religious identities in terms of their Englishness. What made Foxe's work so critical for cementing this nascent concept of English Protestantism was that it showed readers how to trace that English religious identity back into the past, locating the threads of divine providence that preordained the triumph of the Church of England in documents rescued from archives and libraries.

That English antiquarianism was driven by nostalgia and a nascent English nationalism is of no surprise to historians.[91] These dual forces inspired church historians like Foxe and historians of Britain like Stow and Camden, just as they drove Sir Robert Cotton's desire to possess manuscripts filled with Old English poetry and legal documents. But could the same be said of the amateur antiquarians who collected practical manuscripts? It is one thing to cling to Englishness in a church founded on the premise that parishioners should be able to pray with an "Our Father" rather than a *Pater noster*, and to seek out very old English documents to tell a story of the *English* church or the *English* nation. It is another thing entirely to seek Englishness within the domain of natural knowledge, which, as this book has demonstrated, very often entered England from elsewhere. That was true in the fifteenth century when Peter Cantele commissioned a practical manuscript with texts first composed in Salerno and Burgundy, and it was true for much of the sixteenth century, when English publishers scrambled to translate and print new works on European plant life, on the interior of the human body, or on the movements of the heavens produced by continental scholars for continental presses. As we have seen, even those printed practical books derived from English manuscripts nearly always originated in earlier, Latin sources.

But for Elizabethan readers steeped in a culture of antiquarianism, with budding notions of English exceptionalism, none of that mattered. Despite the fact that practical manuscripts' contents were as likely to originate in

France, Spain, or Italy as in England, collectors like William Gale, a London barber-surgeon, saw in them evidence of English excellence. Sometime around 1578, Gale came into possession of the large manuscript that is now Bodleian MS Ashmole 1505 and inscribed his name and that date on the manuscript's first leaf. He had inherited the manuscript from Thomas Gale (no relation), another member of the London Barber-Surgeons and author of a printed practical book: *Certain works of chirurgerie*, published in 1563.[92] Before that, the manuscript had belonged to Richard Cler, rector of St. Pancras in Winchester, and before that, it had probably been created for Robert Brooke, the man who had served King Henry VI as his master distiller.[93] The manuscript created for Brooke in the second quarter of the fifteenth century contains the only extant Middle English translation of the French physician Bernard of Gordon's early fourteenth-century Latin recipe collection, known as the *Lily of Medicine* (*Lilium medicinae*).

Not surprisingly, the manuscript was a lavish production befitting a member of the king's court, with illustrations and penwork flourishes adorning its hundreds of medical recipes. One of its illustrations, in the bottom margin of the first leaf of the manuscript, features a portrait of Brooke alongside an illustration of the original author of the medical treatise, Bernard of Gordon. Both figures gesture toward a lily at the far left. A banderole (sort of like a medieval speech bubble) issuing from the finger of the figure closest to the lily reads "*Ecce nomen huius libri*" (Behold the name of this book), that name being the *Lily of Medicine*. Another banderole gives the name of the treatise's author: "*Bernardus auctor.*" Finally, all the way to the right of the illustration, in handwriting that looks different from that of the original scribe, someone—perhaps even Robert Brooke himself—has written the name "Brock."[94]

When William Gale began to assess the value of this medical manuscript in the late sixteenth century, it was this addition of the name "Brock" that tripped him up. Thinking the name "Brock" to be the surname of the "Bernardus" on the left, Gale inscribed the following in bold letters on the manuscript's second leaf: "*Barnardus Brooke est auctor huius libri. 1303*" (Bernard Brooke is the author of this book. 1303).[95] Like the owner of Huntington MS HM 19079 whom we met earlier in this chapter, who dated that manuscript to the year 1100, William Gale also found it worthwhile to date his recipe collection. In this case, however, Gale got the date right. The *Lily of Medicine* was composed in 1303, as anyone who had read one of the many published editions of the work in its original Latin would have known.[96] Gale must have recognized the text for what it was, perhaps because he had one of those printed Latin editions of the work to hand. But though he ascertained the date of its composition, he was totally misled by the illustration of the manuscript's patron. With that illustration as his evidence, Gale made the French physician Bernard of Gordon into an Englishman: one Bernard Brooke. Perhaps he did so because he read another note left by a different early modern reader in his manuscript, explaining that Brooke had been "Master of the king's stillatories & maker of his excellent waters."[97] It wasn't hard for Gale to imagine that this eminent English physician was the same Bernard who authored the *Lily of Medicine*.

Gale's misattribution seems a harmless enough mistake. So what if he didn't know the last name of the "*Bernardus auctor*" illustrated on his manuscript's opening page? The point is not that Gale was a deliberate manipulator of English medical history, but rather that men like Gale felt the need to identify an *English* medical history at all. Just as church histories like Foxe's *Actes and monuments* or national histories like those authored by John Stow were clearly the product of confessional and political divides in post-Reformation Europe, so too was the history of medicine new ground on which English authors could build the case that God had specially provided for the English. And never was this effort more necessary than in 1578, when William Gale took up his pen to add his *ex libris* and note of attribution to the flyleaf of MS Ashmole 1505. Just one year prior, the English merchant John Frampton, who had spent years in Spain, published his translation of the first volume of Nicolás Monardes's manual on the *materia medica* of the Americas, titled *Joyfull newes out of the newe founde worlde*.[98] Three years later, Frampton would publish a complete translation of all three of Monardes's original volumes (published in Seville in 1565, 1569, and 1574, respectively), in which he touted the "Trees, Plants, Herbes, Roots, Juices, Gums, Fruits, Licours & Stones . . . of great medicinal virtues" that the Spanish had harvested from their territories in New Spain.[99] In both editions of *Joyfull newes out of the newe founde worlde*, English readers were confronted with the unsettling reality that the Spanish were benefitting from medicines about which the English knew nothing. The stated aim of Frampton's translation was to remedy this lacuna. He wrote in the book's epistolary dedication that the work would "bring in time rare profit, to my Country folks of England, by wonderful cures of sundry great diseases, that otherwise than by these remedies, were incurable."[100]

There were some in England who embraced Frampton's promise that Spanish medicines would "bring in time rare profit" to the English, but Timothy Bright was not one of them.[101] In 1580, the same year that Frampton published the full translation of Monardes's three volumes and just after matriculating with his MD from Cambridge, Bright composed *A treatise: wherein is declared the sufficiencie of English medicines, for cure of all diseases, cured with medicine*, the first of several well-known medical works he would author over his lifetime.[102] In this treatise, Bright wrote that he hoped "this my enterprise shall be a means to provoke others to inquire after the medicines of our own country's yield, and more care to put them in practice."[103] His aim was not only to discount the efficacy of the medicines flowing into English apothecary shops by way of foreign merchants, but also to demonstrate to his readers the utility of native English plants. He divided his treatise into two parts: the first presented the case against foreign medicines through logical argument, backed by well-respected and sometimes ancient authorities. The second and much shorter part was akin to an herbal or recipe collection. It offered instructions for using English plants to cure "all diseases, cured by medicines." As Bright emphasized, "England aboundeth plentifully with all things necessary for thy maintenance of life, and preservation of health."[104]

Bright had no problem finding support for his argument against the importation of foreign medicines. He cited Pliny, who warned readers of the *Natural History* not to trifle with medicines from Arabia or India but to use only those that were universally available.[105] As a more recent authority, he noted that the great German botanist and physician, Leonhart Fuchs, had praised the virtues of native German plants over those imported from "strange and far countries."[106] He might have cited many more. As Alix Cooper has shown, over the course of the sixteenth and seventeenth centuries, German, French, and Dutch physicians made arguments similar to Bright's in herbals and other medical books that championed indigenous plants over exotic ones.[107] Indeed, the German physician Theophrastus von Hohenheim, better known as Paracelsus, previewed much of Bright's rhetoric in his own diatribe against foreign medicines, the *Herbarius*, written piecemeal over Paracelsus's lifetime but not published until 1570. In that text, the firebrand German doctor had railed against those who "want to prepare medicines from across the seas while there are better remedies to be found in front of their noses."[108]

Yet, though the authorities cited by Bright (and a few others he didn't cite) emphasized the benefits of local medicines and the foolishness of traveling halfway around the world to retrieve *materia medica*, they did not invoke what Bright knew to be the definitive argument against New World pharmacopeia: the doctrine of divine providence. Bright, who would later publish an abridged edition of Foxe's *Actes and monuments* in 1589, and who eventually gave up a career in medicine for a career in the Church of England, knew that God had provided for the English.[109] And if God's providence for the English could be trusted—and of course it could—then his countrymen and -women had no business importing medicines. First, as Bright reminded his readers, medicine was invented before navigation. The all-knowing divine would not have intended the English people to remain deficient in the medicines they needed while awaiting the invention of the compass or sextant.[110] Second, God would never have allowed the unconverted peoples of the Americas access to life-saving medicines while denying God-fearing Englishmen the same. "Nay," Bright wrote, "which is more absurd, that the health of so many Christian nations should hang upon the courtesy of those Heathen and barbarous nations, to whom nothing is more odious than the very name of Christianity?"[111]

Bright's belief that God would provide everything necessary to preserve and maintain health wasn't new in the sixteenth century. That same belief animated the fifteenth-century herbals and recipe collections examined in chapter 3, which purported to reveal and exploit the divinely appointed "virtues" of plants, herbs, and stones. Yet these earlier texts did not suggest that the divine order of nature might be regional or even local in character, nor did they focus only on English plants. Rather, because their ultimate sources were Latin and Arabic texts, many Middle English herbals included ingredients that were native to warmer climes. The Middle English translation of the *Circa instans* herbal, for example, begins with "aloe," just like the Latin original, and the Middle English *Agnus castus* includes a lengthy entry on "mastic," described (in the 1525 printed edition) as "a gum of a tree growing in a part of the country of Greece."[112] Yet by the time of Bright's writing in 1580, providential thinking had taken on a different character. The confessional divides of post-Reformation Europe and European contact with the New World had given rise to a new sense that God's providence was not meted out equally.

The Spanish were somehow benefiting from New World drugs, both medicinally and financially, despite the fact that they had rejected the doctrine of the one, true (Protestant) church. There could only be one explanation: English medicines must be as beneficial for English bodies as foreign ones were for the Spanish. Bright argued that "home medicines of our country yield, of equity must necessarily perform the same to us, which their medicines do to them. Else I would know why we should be inferior unto them, or one nation more privileged that way than another, the need being common, and the providence of God all one."[113] Bright conceived of pharmacopeia as regional in character: "as the Indian, Arabian, Spaniard have their Indish, Arabian, and Spanish medicine, so also the German hath his, the French man his, and the English man his own proper, belonging to each of them."[114] In Bright's treatise, medicine differed not just according to environmental or astral influences, as ancient and medieval theorists had believed, but rather according to what we might call national character—which in 1580 also meant religious identity.[115] The confessional divisions that had fractured post-Reformation Europe made it possible to assert that God's providence recognized national borders.

And just as Bright would later abridge Foxe's history to make the story of God's intentions for the English church more legible to English readers, so too did he recognize that readers would need a means of comprehending God's intentions for the English in the natural world, too. He couldn't just argue for the existence of a specifically *English* medicine, he would need to give readers the means to test his claims. And so, he did as so many others had done since the arrival of the printing press in England: he published medical recipes and herbal instructions. He suggested that his readers use these recipes, collected on the final ten pages of his treatise, to experiment with local plants and herbs since, as he wrote, "the causes and effects of our bodies [are] of all arguments the most forcible to establish and overthrow anything to be decided by reason."[116] As recipe collections go, however, Bright's left much to be desired. None of his recipes were organized according to their main ingredient, as herbals had been organized since the time of Dioscorides, nor did his publisher employ any of the usual paratexts to aid a reader in discovering the right cure for a particular disease. There are no headings, no judicious use of blank space to separate recipes, and no index or table of contents to serve as finding aid. Anyone who had purchased one of the scores of English herbals available for sale alongside Bright's treatise would have recognized the wormwood, angelica, herbgrace, horehound, pimpernel, juniper berries, and marigold in Bright's treatise from other, far more useable texts.[117]

But, like other authors of practical books, Bright recognized that English readers would have their choice of which recipe collections to buy and which cures to try. When a new book with new cures from the New World arrived on the English market, he thought he'd give those readers a reason to choose *English* medicines over those from foreign merchants. He recognized that a practical book was an effective means of persuading these readers, even if he could not be sure that they would value wormwood over tobacco or sassafras. All he could do was make his case for the sufficiency of English medicine in the popular press, just as he worked to bring Foxe's history of the English church to a wider readership. And, critically, he did both because he believed that God had provided especially for the English, and that history and nature would prove it. Historian Alexandra Walsham has defined the doctrine of providence in early modern England as a set of beliefs anchored in these two complementary planes: "History was the canvas on which the Lord etched His purposes and intentions; nature a textbook and a laboratory in which He taught, demonstrated, and tested His providence."[118]

Timothy Bright knew that history and nature were dually reflective of divine will, but so, too, did many collectors of fifteenth-century practical manuscripts. Those manuscripts filled with faintly archaic medical recipes, herbal instructions, and prognostications were nature and history wrapped up neatly in a package that was wholly English in character. That at least partially explains why sixteenth-century collectors valued them so highly, and it may certainly help us understand why John Gerard, surgeon to the queen and keeper of William Cecil's gardens, acquired British Library MS Sloane 7, an early fifteenth-century manuscript featuring a copy of the *Agnus castus* herbal, to which he added lists of herbal ingredients, as well as his signature and the date 1577.[119] Twenty years later, John Gerard would go on to author *The herball, or Generall historie of plantes* (1597), a work that really did give English readers access to specialized knowledge about the benefits of native English botanicals and would remain the definitive source for information on English flora for a century more, especially after it was revised by Thomas Johnson in 1633. By providing a comprehensive source for identifying and using English botanicals, Gerard in some ways fulfilled the mission of Timothy Bright.

Yet even though Gerard was a champion of English flora, he was also—like most authors of practical books—more than happy to borrow from continental authors: namely, Matthias L'Obel and Rembert Dodoens.[120] In Deborah Harkness's telling, the making of Gerard's herbal was an act of plagiarism and trenchant anti-foreign sentiment. Gerard was supposedly suspicious both of foreign naturalists living in London and of foreign species of plants growing in his garden—perhaps a bit like Timothy Bright.[121] Sarah Neville has attempted to rehabilitate Gerard's reputation, arguing that he was no plagiarist but was instead merely an anthologizer of other printed herbals, in addition to being a noted botanical expert in his own right.[122] Yet both can be true: if Gerard was both suspicious of foreign plants and unconcerned about borrowing from foreign authors, that would hardly make him an anomaly in Elizabethan England or in the market for practical books. Indeed, Gerard's making of *The herball* may represent the culmination of techniques we have seen throughout this book. Gerard drew from his experience with English plants, from the texts of his competitors, and—perhaps—from a Middle English herbal in a centuries-old manuscript. When he left an index of herbal names and his signature in that manuscript, perhaps he hoped, like Henry Dyngley,

William Gale, and Thomas Buttus, to recover old English knowledge for a new English future.

Conclusion

Timothy Bright's appeals to seek only after English medicines fell on deaf ears. There was no way to put the cork back in the bottle, no way to convince the English of the obsolescence of tobacco or sassafras. In that respect, Bright's treatise reads almost like the coda to a story that was already wrapping up by the later sixteenth century. The insular, derivative culture of natural knowledge exchange that predominated in fifteenth- and sixteenth-century England—the culture closely examined in this book—was cracking open under Elizabeth. In the seventeenth century, England laid the foundations for a mercantile network that would eventually stretch across the globe. By 1680, no one in England would have seriously considered eschewing the natural bounty of the New World out of some misguided loyalty to native English species, provided by God expressly for the English. To the contrary, confidence in God's special providence was all the assurance Englishmen and -women needed as they moved into the world and claimed it as their own.

When looked at another way, however, Bright's insistence on an *English* mode of interaction with the natural world—one that was markedly different from that of France, Germany, Italy, or Spain—did live on, though not as Bright could have imagined. Following the constructivist turn in the history of science, narratives of the so-called scientific revolution have occasionally—and perhaps unintentionally—argued that *Englishness* played a critical factor in the development of modern methods for understanding the natural world. Steven Shapin has shown us how thoroughly the experimental method was imbued with the conventions of English polite society, while Peter Dear has argued that experimentation as a governing philosophy for interpreting natural phenomena depended on English Protestantism.[123] I do not mean to suggest that either Shapin or Dear believe that science is somehow an *English* way of knowing the world. Rather, I mean to suggest that there is an element of Timothy Bright or Thomas Buttus in all our historicist attempts to mine the past for evidence that can help us make sense of our present. We may sometimes be guilty of "Englishing" the history of medicine and science, just as our early modern historical subjects did. The trick is to resist the temptation, as Bright and Buttus could not, to ignore the extent to which English ways of knowing had always depended on knowledge from faraway places.

Conclusion

In 1605, Francis Bacon, father of English science, addressed a critique of English learning to King James I/VI. In it, he insisted on the value of observation and experience over the wisdom of ancient authorities, diminished and occluded by centuries of scholasticism. The men of his age had "withdrawn themselves too much from the contemplation of Nature, and the observations of experience." The poor state of knowledge might be overcome, according to Bacon, if only "contemplation and action may be more nearly and straightly conjoined and united together."[1] This is the founding narrative of English science: Bacon issued a call to temper the authority of natural philosophy with knowledge born from experience, and the experimental method was born. How this happened, precisely, and by whom, has been the preoccupation of historians of early modern science ever since. A number of historians have echoed Bacon's praise for ordinary practitioners of the "arts mechanical" and argued that it was regular folks—artisans, printers, instrument makers, or navigators—that injected experiential knowledge rather than book-learning into conversations about the natural world.[2] But regardless of whether one pins the origins of experimental science on anonymous craftspeople or on the likes of Hugh Plat, Francis Bacon, or Robert Boyle, there is widespread consensus that the spark that generated new ways of understanding the natural world was exactly as Bacon described: a rejection of a stagnant textual tradition had to precede the rise of empiricism.

This book has done its best to show that even among the "ordinary people" that Bacon credits with an ingenuity uninhibited by the weight of authority, textual tradition still meant quite a lot. The readers we've encountered in the pages of this book didn't pore over editions of Aristotle or Hippocrates, but they did read practical manuscripts and books that distilled the theories of these thinkers into bite-size recipes or instructions. Their encounters with the natural world were just as shaped by these textual encounters as were those of the natural philosophers whom Bacon accused of having "withdrawn themselves too much from the contemplation of Nature." Ordinary readers' curiosity and inquisitiveness about the natural world didn't develop *in spite of* their dependence on almanacs, prognostications, medical recipe collections, or herbals but rather *because of* them. These practical books were very often the vehicles through which new ideas about the heavens or the human body made their way into the homes of merchants, well-to-do farmers, village priests, or curious matrons. When that happened, these new ideas weren't presented with the solemn gravitas of a learned philosophical treatise. They were injected into an already crowded and contentious information economy, and readers had to make sense of them alongside all sorts of other competing texts. They had to hone their critical faculties to be able to assess the utility and validity of all sorts of natural knowledge circulating in cheap pamphlets.

That isn't how Bacon would have seen it. No doubt he would have been skeptical of the critical faculties of anyone reading prognostications or recipe collections. He didn't trust the predictions of astrology, which he lumped together with "natural magic and alchemy" as the three "sciences" that have more to do with the "imagination of man than with his reason." He had no patience for the medicine of "empiric physicians," whom he accused of having "a few pleasing receipts, whereupon they are confident and adventurous, but know neither the causes of diseases, nor the complexions of the patients, nor peril of accidents, nor true method of cures."[3] He clearly worried that English readers' minds were being dulled, not honed, by the claims of charlatans offering knowledge of the future or of the human body. He might have called ordinary English readers credulous rather than critical—and certainly as regards many English readers, he wouldn't have been wrong.

Bacon's rejection of these pseudo "sciences" became its own sort of prognostication. Over time, the rituals and healing practices we have encountered in fifteenth- and sixteenth-century practical books would cease to have meaning for English readers. Baconian science would triumph over superstition. The question of why and how empiricism triumphed over belief has figured prominently in the work of historians and sociologists, beginning with Max Weber in the early twentieth century and culminating with the work of Keith Thomas. In his magisterial *Religion and the Decline of Magic*, Thomas concluded that the causes of English "disenchantment" were many. Certainly the new, empiricist science of Bacon and his later contemporaries played a role, in that the new philosophy demanded that natural knowledge be demonstrated to be believed. But for Thomas, far more important than the philosophy of the New Sciences was a general change in how English people understood their position in the world. If astrology, healing charms, herbal lore, and prognostications had essentially been "ineffective rituals employed as an alternative to sheer helplessness in the face of events," there was no need for these rituals once English people decided they were not, in fact, helpless. By the seventeenth century, Thomas asserts that English people had found a "breath-taking faith in the potentialities of human ingenuity."[4]

There can be no question that practical books played a role in fostering that "breath-taking faith" in humanity's ability to find solutions to problems that had plagued them since time immemorial. Over the two centuries between 1400 and 1600, English people found they had access to a wealth of information, right at their fingertips. They could collect recipes in manuscripts or, later, purchase dozens of editions of practical books at the bookshop. Here were pages and pages of texts that showed all sorts of people trying all sorts of things to harness the power of nature in medicines or to better understand the workings of the heavens to prevent the sorts of calamities that would wreak havoc in a person's life. What could be better proof of human ingenuity than a printed recipe collection chock full of remedy after remedy, each one an attempt to find a solution to the intractable problem of disease, injury, and suffering? These books encouraged readers' ingenuity, too. Printed prognostications, herbals, and medical recipe collections put experiences that readers were intimately familiar with on the page: bad harvests, foul weather, sickness, aging, suffering. The sheer variety of solutions to these problems offered in practical books forced English people to think about discrepancies between what they experienced in the day-to-day, and how they read about it in books. And there were quite a lot of discrepancies, because practical books (whether in manuscript or in print) didn't exactly present natural knowledge systematically or as part of a congruous whole. There was rarely much in the way of framing to guide a reader through a clearer understanding of the theories or principles on which these texts depended. There was no single, unifying framework through which to understand these collections of natural knowledge. And so English people were largely on their own to work out what *they* believed, what *they* valued.

We have charted that process throughout this book. Though the natural knowledge contained in practical books was often quite old, and though it usually capitalized on the perceived authority of textual tradition, that tradition was always open to revision. In the manuscript culture of fifteenth-century England, compilers invented a pedigree for texts that had none, tacking on the names Hippocrates or Galen to Middle English recipe collections that had more to do with "empiric physicians" than ancient authorities. On the other hand, empirics like Nicholas Neesbett had a healthy respect for the authority of textual tradition, which made them all the more eager to put their experiential knowledge in writing. In the crowded and competitive print market of the sixteenth century, English readers readily adjusted their expectations about the novelty, originality, legitimacy, and secrecy of natural knowledge based on the marketing slogans of printers, the influence of religious doctrine, or even the material contexts in which they encountered recipes, prognostications, and the like. Though their reassessments of the same old medical recipes, prognostications, and herbal texts don't seem especially *critical* when compared to those of learned humanists, ordinary readers' engagement with this natural knowledge encouraged them to think of books as sites where authorities could be challenged and discoveries could be made. That recognition alone is significant for understanding why English people so readily embraced an empiricist philosophy that depended on individual persuasion through firsthand witness, nearly always mediated through texts.[5]

But while readers did learn to assess and evaluate what they read in almanacs, medical recipe collections, books of secrets, and herbals, that doesn't mean that English readers necessarily adopted the same perspective as Francis Bacon. Natural magic, astrology, and the remedies of empirics won out among ordinary English readers just as often—and probably more so—as the cutting-edge medicine or science on offer from men like Andreas Vesalius or Nicolaus Copernicus. There is a reason why Thomas Gemini and Thomas Digges both used old texts to sell these revolutionaries' new ideas.[6] Gemini and Digges the younger recognized that English readers might more readily adopt Copernican astronomy or Vesalian anatomy if they were accompanied by a centuries-old prognostication or a thirteenth-century surgical treatise. The sixteenth-century readers I have examined in this book were not so very different from today's media consumers. While we may turn to Google or social media instead of practical books for information, we, too, are more likely to accept new information if it builds from established narratives. And, as we know all too well, when people are offered a variety of solutions to difficult problems, when they are given access to a world of information about the human body and the natural world, not all of them will choose that which is rational or empirically sound. They may, instead, choose what Bacon described as knowledge having more to do with the "imagination of man than with his reason."

I have argued throughout this book that readers developed their critical faculties by selecting from competing knowledge claims and assessing the quality of natural knowledge available for copy or for sale. Even so, a facility for criticism is not the same as a commitment to the principles of Baconian science. Not every fifteenth- or sixteenth-century reader of a practical book or manuscript was a budding empiricist. Indeed, though practical books are full of hands-on knowledge and though they were certainly read by practitioners and artisans, they don't fit a narrative that shows *practica* triumphing over *theorica*.[7] Though these books are defined by their association with the *practices* of healing, reading the heavens, or manipulating natural matter, for the most part their contents are no more empirically grounded than a scholastic treatise on humoral medicine. The contents of Richard Banckes's medical recipe collection or Timothy Bright's treatise in defense of English medicine have nothing whatsoever to do with experimentation. If readers did try and test the recipes they found in those books, they would have been no closer to a holistic understanding of the human body than the early fifteenth-century readers who carefully copied the *Pater noster* onto a sage leaf as a cure for fever.

But practical books still matter to the history of medicine and science. Their value lies in their ubiquity. For more than two centuries, the same texts were read and reread in relatively costly manuscripts, and then in less expensive paper remedy collections, and then in cheaply printed pamphlets. What mattered to fifteenth- and sixteenth-century readers was not practical books' accuracy but rather their accessibility. Because practical manuscripts and printed books made natural knowledge of all kinds available, English readers came to see themselves as full participants in a culture of knowledge exchange and inquiry. Though these readers turned their critical acumen on texts that were often centuries old, which were neither innovative nor particularly effective by modern standards, the very fact of being able to *choose* what sort of knowledge to implement in moments of suffering, anxiety, or insecurity had a tremendous effect on readers' belief in human ingenuity. The power of practical books lay in their promise. All who turned their pages did so in the belief that nature could be tamed, health could be managed, and life could be made more predictable, but only if they continued to pursue natural knowledge wherever it could be found.

Acknowledgments

Like practical manuscripts and printed editions, this book collects the wisdom of many scholars who came before me. I am grateful to acknowledge these scholars and sources of support here. Over a decade ago at the University of Alabama, James Mixson set me on a course toward the study of vernacular manuscripts, and Dan Riches encouraged me to follow that course further. At Rutgers, James Masschaele taught me Latin paleography, and after a semester of thirteenth-century English court hand, Middle English medical manuscripts came as a relief. Leah DeVun inspired me to reflect on how the past bears on the present while at the same time holding me to the highest standards of historical scholarship. Finally, Alastair Bellany was a generous mentor who read innumerable drafts and offered comments—sometimes within the span of a single day!—that gently nudged me in the right direction. I remain deeply grateful for Alastair's guidance and support. The scholars with whom I shared ideas, office space, and delicious lunches at the Princeton Society of Fellows have each left their imprint on this book. I wish especially to thank Yelena Baraz, Michael Gordon, Joshua Freeman, Amanda Lanzillo, Matthew Larsen, Natalie Prizel, Aniruddhan Vasudevan, and Beate Witzler for insightful suggestions, probing questions, and, most of all, friendship. I was fortunate to find a second home at Princeton in the History Department and Program in the History of Science, where both Tony Grafton and Jenny Rampling offered sage advice about manuscripts, readers, and natural knowledge that has improved this book in every way. Thanks also to Pamela Smith for igniting my interest in recipes during my time on the Making and Knowing Project, and to Ann Blair, who encouraged my book-historical pursuits from the outset.

I am grateful to those who generously gave of their time to read the book manuscript at various stages and in various forms. Mary Fissell, Hannah Marcus, and Leah DeVun read the manuscript in its entirety and whipped this book into shape. Tony Grafton, Elaine Leong, Jenny Rampling, Michael Gordon, Alastair Bellany, the Fellows of the Princeton Society of Fellows, members of the Princeton Early Modern History Workshop, participants in the Princeton History of Science Program Seminar, and members of the Early Modern Science Working Group at the Consortium for the History of Science, Technology, and Medicine read individual chapters, asked questions, and gave thoughtful feedback that improved my thinking tremendously. Finally, my thanks go to the anonymous readers for the University of Chicago Press.

For financial support while researching and writing this book, I am grateful to acknowledge the American Council of Learned Societies, the Andrew W. Mellon Foundation, the Medieval Academy of America, the Richard III Society, the Rare Books School at the University of Virginia, the Folger Shakespeare Library, Rutgers Graduate School and School of Arts and Sciences, and the Princeton Committee on Research in the Humanities and Social Sciences. Subvention funding was generously provided by the Paul Oskar Kristeller Fellowship from the Renaissance Society of America and the Clark-McClintock-Gershenson Memorial Fund from the Rutgers University History Department. In addition, I am enormously grateful to the Special Collections librarians and curators who fielded a flurry of manuscript requests and shared their expertise with me at the following institutions: the Bodleian Library, University of Oxford; the British Library; the Cambridge University Library; the Folger Shakespeare Library; the Henry E. Huntington Library; the Morgan Museum and Library; the National Library of Medicine; Trinity College Library, Cambridge University; and the Wellcome Library. Portions of chapter 3 have appeared in the journal *Social History of Medicine* and portions of chapter 4 in the *Journal of British Studies*, and I thank the editors of those journals for permission to reproduce sections of those articles here. Karen Merikangas Darling, Fabiola Enriquez Flores, Marianne Tatom, and the design, marketing, and production teams at the University of Chicago Press deserve special kudos for seeing this book through to the finish.

I have been fortunate in a career that has taken me far from home to find friends who have been like family: Corey and Jennifer, Pierre-Luc and Emilie, Zach and Shannon, Carolyn and Jonathan, Kat, Merissa and Nikola, Matthew and Lauren, Paris and Iggy, and Amanda and Matthew. In the time it took to research and write this book, my two children, Hansen and Lucy, have grown from toddlers into teenagers. I thank them for their patience, love, and, especially, for their enthusiasm for "Mom's book." My sister, Christina Buckner, indulged my obsessions with *Lord of the Rings* and medieval history as a teenager and probably did more to steer me toward a career reading English manuscripts than she'll ever know. My in-laws, Richard and Sharon Reynolds, and my late father, Frank Buckner, lovingly cared for Hansen and Lucy every time I jaunted off to read manuscripts in English libraries. My mother, Diane Buckner, deserves special mention. She accompanied me on my first research trip to London and ever since has been babysitter, research assistant, cheerleader, and home chef whenever I needed her. I could not have written this book without their collective support. But my deepest thanks are reserved for my husband, Justin Reynolds. Over nearly twenty years and a total of four post-grad degrees between us, homes in five states, at least four career changes, and two kids, he has never wavered in his support of my academic dreams—even when the pursuit of those dreams has tested our collective resilience.

Notes

Introduction

1. I recognize that in naming an internet search engine and popular website I am dating this book. Readers should replace "Google" and "YouTube" with whatever information resource seems most relevant.

2. Tamara Atkin and A. S. G. Edwards, "Printers, Publishers and Promoters to 1558," in *A Companion to the Early Printed Book in Britain, 1476–1558*, ed. Vincent Gillespie and Susan Powell (Cambridge: D. S. Brewer, 2014), 28. For an excellent example of a transnational comparison of vernacular medical publishing, see Sandra Cavallo and Tessa Storey, "Regimens, Authors and Readers: Italy and England Compared," in *Conserving Health in Early Modern Culture*, ed. Sandra Cavallo and Tessa Storey, *Bodies and Environments in Italy and England* (Manchester: Manchester University Press, 2017), 23–52.

3. Andrew Wear notes that both John Caius and Thomas Linacre chose not to publish their Latin editions with English printers, but rather used continental print houses; see *Knowledge and Practice in Early Modern English Medicine, 1550–1680* (Cambridge: Cambridge University Press, 2000), 41.

4. John Foxe, *Actes and monuments of these latter and perillous days*, STC 11222 (London: John Day, 1563), sig. B.v *b*.

5. Though she contends that an experimental culture emerged in London before Bacon conjured it as Salomon's House, Deborah Harkness essentially follows this model as an explanation for the rise of the New Sciences in England in *The Jewel House: Elizabethan London and the Scientific Revolution* (New Haven, CT: Yale University Press, 2008).

6. Francis Bacon, *The tvvoo bookes of Francis Bacon, of the proficience and aduancement of learning, diuine and humane: to the king*, Princeton University Scheide Library 2015-0292N (London: Henrie Tomes, 1605), sig. F.iv.

7. Oronce Fine, *The rules and righte ample documentes, touchinge the vse and practise of the common almanackes*, trans. Humfrey Baker, STC 10878.7 (London: Thomas Marshe, 1558). Fine's original French "how-to" manual, *Les canons & documens tresamples, touchant lusaige & practique des communs almanachz, que l'on nomme ephemerides*, was published by Simon de Colines in Paris in 1543, 1551, 1556, and 1557.

8. Fine, *The rules and righte ample documentes*, trans. Baker, sig. A.ii *b*–A.iii *a*.

9. Elizabeth L. Eisenstein essentially made exactly these arguments for the supe-

riority of printed scientific books in *The Printing Revolution in Early Modern Europe*, 2nd ed. (Cambridge: Cambridge University Press, 2012), 46–101.

10. Fine, *The rules and righte ample documentes*, trans. Baker, sig. A.ii *a–b*.

11. Fine, *The rules and righte ample documentes*, trans. Baker, sig. A.iii *b*.

12. This argument, a corrective to the "print revolution" thesis of Elizabeth Eisenstein, is best expressed in Adrian Johns's *The Nature of the Book: Print and Knowledge in the Making* (Chicago: University of Chicago Press, 2000), 6–11 passim.

13. On Shapin's original insight into print's role in the construction of scientific authority, see his canonical article "Pump and Circumstance: Robert Boyle's Literary Technology," reprinted in Steven Shapin, *Never Pure: Historical Studies of Science as if It Was Produced by People with Bodies, Situated in Time, Space, Culture, and Society, and Struggling for Credibility and Authority* (Baltimore: Johns Hopkins University Press, 2010), 89–116. Adrian Johns writes about Brahe's astronomical table in *The Nature of the Book*, 14–19, as a counterpoint to Eisenstein's treatment of the same material; see, for comparison, Eisenstein, *The Printing Revolution in Early Modern Europe*, 235–45. See also Sachiko Kusukawa, *Picturing the Book of Nature: Image, Text, and Argument in Sixteenth-Century Human Anatomy and Medical Botany* (Chicago: University of Chicago Press, 2012).

14. My database of printed practical books is one of several Digital Appendices available at https://readingpractice.github.io.

15. Italics are mine. Roger Chartier, *The Order of Books: Readers, Authors, and Libraries in Europe between the Fourteenth and Eighteenth Centuries*, trans. Lydia G. Cochrane (Stanford, CA: Stanford University Press, 1994), 8.

16. See, for example, Ethan H. Shagan, *Popular Politics and the English Reformation* (Cambridge: Cambridge University Press, 2003). For an alternative view that very little *did* change at the level of parish life after the English Reformation, see Christopher Haigh, *English Reformations: Religion, Politics, and Society under the Tudors* (Oxford: Clarendon Press, 1993).

17. Eamon Duffy, *The Stripping of the Altars: Traditional Religion in England, 1400–1580* (New Haven, CT: Yale University Press, 1992); Tessa Watt, *Cheap Print and Popular Piety, 1550–1640* (Cambridge: Cambridge University Press, 1991); see also Ian Green, *Print and Protestantism in Early Modern England* (Oxford: Oxford University Press, 2000).

18. Elaine Leong, *Recipes and Everyday Knowledge* (Chicago: University of Chicago Press, 2018). A notable exception is Jennifer M. Rampling, *The Experimental Fire: Inventing English Alchemy, 1400–1700*, Synthesis (Chicago: University of Chicago Press, 2020).

19. Pamela O. Long, *Artisan/Practitioners and the Rise of the New Sciences, 1400–1600* (Corvallis: Oregon State University Press, 2011), 9 and passim. There was, of course, some scientific patronage at the English court, particularly of alchemists and astrologers; see Rampling, *The Experimental Fire*, 64–73; Hilary M. Carey, *Courting Disaster: Astrology at the English Court and University in the Later Middle Ages* (Hampshire, UK: Macmillan, 1992).

20. Danielle Jacquart, "Theory, Everyday Practice, and Three Fifteenth-Century Physicians," *Osiris* 6 (1990): 140–60.

21. Vivian Nutton, "'A Diet for Barbarians': Introducing Renaissance Medicine to Tudor England," in *Natural Particulars: Nature and the Disciplines in Renaissance*

Europe, ed. Anthony Grafton and Nancy G. Siraisi (Cambridge, MA: MIT Press, 1999), 275–76.

22. Pamela H. Smith, *The Body of the Artisan: Art and Experience in the Scientific Revolution* (Chicago: University of Chicago Press, 2006), 6–8, 20 and passim.

23. Pamela H. Smith, *From Lived Experience to the Written Word: Reconstructing Practical Knowledge in the Early Modern World* (Chicago: University of Chicago Press, 2022), 117–37.

24. See, for example, Hannah Bower's *Middle English Recipes and Literary Play, 1375–1500* (Oxford: Oxford University Press, 2022), which treats many of the same manuscripts studied in this book as literary texts.

25. Rossell Hope Robbins and H. S. Bennett established scholarly interest in Middle English medicine and science in early exploratory articles: Rossell Hope Robbins, "English Almanacks of the Fifteenth Century," *Philological Quarterly* 18, no. 4 (October 1939): 321–31; H. S. Bennett, "Science and Information in English Writings of the Fifteenth Century," *Modern Language Review* 39, no. 1 (1944): 1–8; Rossell Hope Robbins, "Medical Manuscripts in Middle English," *Speculum* 45, no. 3 (July 1, 1970): 393–415.

26. Linda Ehrsam Voigts and Patricia Deery Kurtz, *Scientific and Medical Writings in Old and Middle English: An Electronic Reference* (Ann Arbor: University of Michigan Press, 2000), http://cctr1.umkc.edu/search. Nearly all the manuscripts featured in this book have been catalogued at least in part by Voigts and Kurtz. Other critical reference works consulted for this project are Keiser, *MWME*, and Peter Murray Jones, *Medieval Medicine in Illuminated Manuscripts*, rev. ed. (London: The British Library, 1998).

27. Miscellaneity is a common feature of many genres of Middle English manuscript, not just those related to medicine or science; see Arthur Bahr, "Miscellaneity and Variance in Medieval Books," in *The Medieval Manuscript Book: Cultural Approaches*, ed. Michael Johnston and Michael Van Dussen (Cambridge: Cambridge University Press, 2015), 181–98; Margaret Connolly and Raluca Radulescu, eds., *Insular Books: Vernacular Manuscript Miscellanies in Late Medieval Britain* (Oxford: Oxford University Press, 2015).

28. On the emergence of the "mechanical arts" as an epistemological category, see George Ovitt, "The Status of the Mechanical Arts in Medieval Classifications of Learning," *Viator* 14 (January 1, 1983): 89–105.

29. On the practical manuscript as its own genre, see Melissa Reynolds, "'Here Is a Good Boke to Lerne': Practical Books, the Coming of the Press, and the Search for Knowledge, ca. 1400–1560," *Journal of British Studies* 58, no. 2 (April 2019): 259–88.

30. On the convergence of Latin and vernacular medical and scientific writing in English manuscripts, see Linda Ehrsam Voigts, "What's the Word? Bilingualism in Late-Medieval England," *Speculum* 71, no. 4 (1996): 813–26.

31. Many of these 182 manuscripts, originally created in the fifteenth century, have since been bound together as composite manuscripts and, as such, share the same shelfmark. All 182 once-separate fifteenth-century manuscripts are listed in the bibliography. In cases where multiple manuscripts are bound together into a single volume, the bibliography cites folio numbers indicating which portions of the composite volume are "practical manuscripts."

32. The term "epistemic object" derives from Hans-Jörg Rheinberger, *Toward a History of Epistemic Things: Synthesizing Proteins in the Test Tube* (Stanford, CA: Stanford

University Press, 1997). I adapt Rheinberger's influential concept of the "epistemic thing" as a locus for the production of scientific knowledge to include not just test tubes or laboratory specimens but also manuscripts.

33. This focus on the consequences of literacy within the lives of ordinary English people builds directly from the work of M. T. Clanchy in *From Memory to Written Record: England 1066–1307* (Oxford: Blackwell Press, 1993).

34. Information on the composition, contents, and reader marks in all 182 practical manuscripts studied for this book can be found in the Digital Appendices, accessible at https://readingpractice.github.io.

Chapter 1

1. Francis Blomefield, ed., "Clavering Hundred: Toft," in *An Essay Towards a Topographical History of the County of Norfolk: Volume 8*, British History Online (London: W. Miller, 1808), 61–64.

2. The makeup of BL Sloane 1764 is as follows: the plague treatise of John of Burgundy, ff. 5r–6v; various medical recipes, ff. 7r–11r; treatises on the virtues of herbs, ff. 11v–15v; *Thesaurus Pauperum*, ff. 16r–29v; directions for distilling waters, oils, and entretes, ff. 31r–46v; the *Circa instans*, ff. 47r–112v; the concordance of recipes, ff. 113r–114v.

3. Faye Getz, *Medicine in the English Middle Ages* (Princeton, NJ: Princeton University Press, 1998), 3–19.

4. BL Sloane 1764, ff. 3r–4v.

5. "A preciowss water for brenning or schaldyng or bytyng of hors or doge," BL Sloane 1764, f. 4v.

6. "fallyng of her," BL Sloane 1764, f. 114r.

7. Nicholas Orme, *Medieval Schools: From Roman Britain to Renaissance England* (New Haven, CT: Yale University Press, 2006), 58–66.

8. I draw these tuition rates from estimates in Orme, *Medieval Schools*, 132. For the average wages paid to day laborers in Oxford or Cambridge in 1450, see Jan Luiten van Zanden, "Wages and the Cost of Living in Southern England (London), 1450–1700," Institute of the Royal Netherlands Academy of Arts and Sciences, International Institute of Social History, n.d., http://www.iisg.nl/hpw/dover.php.

9. Orme, *Medieval Schools*, 225–54.

10. William Page, ed., "Alien Houses: The Priory of Toft Monks," in *A History of the County of Norfolk*, British History Online, vol. 2 (London: Victoria County History, 1906), 464–65. Cantele's name does not appear in the register of early Eton students; see Sir Wasey Sterry, ed., *The Eton College Register, 1441–1698, Alphabetically Arranged and Edited with Biographical Notes* (Eton: Spottiswoode, Ballantyne, & Co., Ltd., 1943).

11. Blomefield, "Clavering Hundred: Toft."

12. Orme, *Medieval Schools*, 229–40.

13. Barbara Hanawalt, *Growing up in Medieval London: The Experience of Childhood in History* (Oxford: Oxford University Press, 1993), 82.

14. The "Act for Pleading in English" is Edward III, "36 Edward III Stat. 1 c. 15: Item," *SOTR* vol. 1, 375–76. On English "firsts," see M. B. Parkes, *Scribes, Scripts, and Reader: Studies in the Communication, Presentation, and Dissemination of Medieval Texts* (London: Hambledon Press, 1991), 288.

15. On the *Liber uricrisiarum* and Daniel's other medical writings, see Sarah Star, ed., *Henry Daniel and the Rise of Middle English Medical Writing* (Toronto: University of Toronto Press, 2022).

16. These figures are drawn from David Cressy, *Literacy and the Social Order: Reading & Writing in Tudor & Stuart England* (Cambridge: Cambridge University Press, 1980), 112–67.

17. Margaret Spufford, "First Steps in Literacy: The Reading and Writing Experiences of the Humblest Seventeenth-Century Spiritual Autobiographers," *Social History* 4, no. 3 (1979): 407–35.

18. Clanchy, *From Memory to Written Record: England 1066–1307*, 246–47, 328–34; Parkes, *Scribes, Scripts, and Reader: Studies in the Communication, Presentation, and Dissemination of Medieval Texts*, 275, 280–83.

19. Adam Fox, *Oral and Literate Culture in England, 1500–1700* (Oxford: Clarendon Press, 2000), 37–50.

20. On the oral delivery of medieval literature, see Joyce Coleman, "Interactive Parchment: The Theory and Practice of Medieval English Aurality," *The Yearbook of English Studies* 25 (1995): 63–79; Ruth Crosby, "Oral Delivery in the Middle Ages," *Speculum* 11, no. 1 (1936): 88–110. On the public recitation of charters, see James Masschaele, "The Public Life of the Private Charter in Thirteenth-Century England," in *Commercial Activity, Markets and Entrepreneurs in the Middle Ages: Essays in Honour of Richard Britnell*, ed. Ben Dodds and Christian Drummond Liddy (Woodbridge, Suffolk: Boydell & Brewer Ltd, 2011), 205–9.

21. I use the term "middling sort" to describe the urban bourgeoisie and landed country farmers following Keith Wrightson, "Estates, Degrees, and Sorts: Changing Perceptions of Society in Tudor and Stuart England," in *Language, History, and Class*, ed. P. J. Corfield (Oxford: Blackwell Press, 1991), 30–52. On the now common characterization of the fifteenth century as a period of increasing book ownership, see Carol M. Meale, "Patrons, Buyers and Owners: Book Production and Social Status," in *Book Production and Publishing in Britain, 1375–1475* (Cambridge: Cambridge University Press, 1989), 201–38; Susan Hagen Cavanaugh, "A Study of Books Privately Owned in England: 1300–1450" (PhD diss., University of Pennsylvania, 1980).

22. Kathleen L. Scott, "Past Ownership: Evidence of Book Ownership by English Merchants in the Later Middle Ages," in *Makers and Users of Medieval Books: Essays in Honour of A. S. G. Edwards*, ed. Carol M. Meale and Derek Pearsall (Woodbridge, Suffolk: D. S. Brewer, 2014), 150–75; Caroline M. Barron, "What Did Medieval London Merchants Read?," in *Medieval Merchants and Money: Essays in Honour of James L. Bolton*, ed. Martin Allen and Matthew Davies (London: Institute for Historical Research, 2016), 43–70.

23. Christopher Dyer, *An Age of Transition? Economy and Society in England in the Later Middle Ages* (Oxford: Oxford University Press, 2005), 32–33. See also James Masschaele, "The Renaissance Depression Debate: The View from England," *The History Teacher* 27, no. 4 (August 1994): 405–16.

24. Christopher Dyer, *Standards of Living in the Later Middle Ages: Social Change in England c. 1200–1500*, rev. ed. (Cambridge: Cambridge University Press, 1998), 204–5.

25. On Thornton's literary and medical manuscripts, see Susanna Fein and Michael Johnston, eds., *Robert Thornton and His Books: Essays on the Lincoln and London Manuscripts* (New York: Boydell & Brewer, 2022). M. S. Ogden published an edition of Thornton's medical recipe collection: *The "Liber de Diversis Medicinis" in the Thornton*

Manuscript (MS Lincoln Cathedral A.5.2), Early English Text Society, Original Series 207 (London: Oxford University Press, 1938).

26. The Tollemache manuscript has been edited as Jeremy Griffiths, A. S. G. Edwards, and Nicolas Barker, eds., *The Tollemache Book of Secrets: a descriptive index and complete facsimile with an introduction and transcriptions together with Catherine Tollemache's Receipts of pastery, confectionary & c* (London: Roxburghe Club, 2001). The Paston manuscript was identified by A. I. Doyle as Boston, Harvard Medical School, Countway Library of Medicine, MS 19; see Norman Davis, ed., *The Paston Letters: A Selection in Modern Spelling*, Oxford World's Classics (Oxford: Oxford University Press, 1999), 168–69, no. 80.

27. M. A. Michael, "Urban Production of Manuscript Books and the Role of the University Towns," in *The Cambridge History of the Book in Britain, Vol. II: 1100–1400*, ed. Nigel Morgan and Rodney M. Thomson (Cambridge: Cambridge University Press, 2008), 189.

28. Michael, "Urban Production of Manuscript Books," 169, 175–83, 193.

29. For example, Linda Ehrsam Voigts identified several manuscripts of medicine and science created by the same scribe or workshop; see "The 'Sloane Group': Related Scientific and Medical Manuscripts from the Fifteenth Century in the Sloane Collection," *British Library Journal* 16, no. 1 (1990): 26–57. On the identification of scribes of well-known literary manuscripts, see, for example, Linne R. Mooney, "Vernacular Literary Manuscripts and Their Scribes," in *The Production of Books in England, 1350–1500*, ed. Alexandra Gillespie and Daniel Wakelin (Cambridge: Cambridge University Press, 2014), 192–211.

30. Cantele's manuscript is not featured in M. Benskin et al., eds., *eLALME: A Linguistic Atlas of Late Mediaeval Middle English* (Edinburgh: University of Edinburgh, 2013), http://www.lel.ed.ac.uk/ihd/elalme/elalme.html.

31. Meale, "Patrons, Buyers and Owners," 203.

32. C. Paul Christianson, "The Rise of London's Book Trade," in *The Cambridge History of the Book in Britain, Vol. III: 1400–1557*, ed. Lotte Hellinga and J. B. Trapp (Cambridge: Cambridge University Press, 1999), 129–34.

33. Parkes, *Scribes, Scripts, and Reader: Studies in the Communication, Presentation, and Dissemination of Medieval Texts*, 286.

34. On the earliest uses of paper in England, see Orietta Da Rold, *Paper in Medieval England: From Pulp to Fictions*, Cambridge Studies in Medieval Literature 112 (Cambridge: Cambridge University Press, 2020), 22–57.

35. The prices quoted here are drawn from a register recording parchment and paper purchases in 1359 published in Da Rold, *Paper in Medieval England*, 62.

36. R. J. Lyall, "Materials: The Paper Revolution," in *Book Production and Publishing in Britain, 1375–1475*, ed. Jeremy Griffiths and Derek Pearsall (Cambridge: Cambridge University Press, 1989), 11–13.

37. Da Rold finds that roughly the same percentage of manuscripts dateable from 1450 to 1500 in the Cambridge University Library were made from paper; see *Paper in Medieval England*, 155–56.

38. Da Rold, *Paper in Medieval England*, 167–70.

39. I was able to adduce the original size of the sheets of paper used for Cantele's manuscript using Will Noel's "Needham Calculator," Schoenberg Institute for Manuscript Studies, University of Pennsylvania Libraries, accessed 23 June 2023, http://www.needhamcalculator.net. The standard size of chancery paper was 31.5 x 46 cm;

see Will Noel and George Gordon, "The Needham Calculator (1.0) and the Flavors of Fifteenth-Century Paper," Schoenberg Institute for Manuscript Studies, University of Pennsylvania Libraries, 30 January 2017, https://schoenberginstitute.org/2017/01/30/the-needham-calculator-1-0-and-the-flavors-of-fifteenth-century-paper/. The price of paper cited here is drawn from the table in Da Rold, *Paper in Medieval England*, 62.

40. Parkes, *Scribes, Scripts, and Reader*, 284–86.

41. Joanne Filippone Overty, "The Cost of Doing Scribal Business: Prices of Manuscript Books in England, 1300–1483," *Book History* 11, no. 1 (September 12, 2008): 5–6.

42. On booklets, see P. R. Robinson, "The Format of Books—Books, Booklets and Rolls," in *The Cambridge History of the Book in Britain, Vol. II: 1100–1400*, ed. Nigel Morgan and Rodney M. Thomson (Cambridge: Cambridge University Press, 2008), 39–54.

43. On the prevalence of booklet construction in practical manuscripts, see Linda Ehrsam Voigts, "Scientific and Medical Books," in *Book Production and Publishing in Britain, 1375–1475*, ed. Derek Pearsall and Jeremy Griffiths (Cambridge: Cambridge University Press, 1989), 353.

44. Voigts, "The 'Sloane Group,'" 26–57.

45. On speculative manuscript production, see Andrew Taylor, "Manual to Miscellany: Stages in the Commercial Copying of Vernacular Literature in England," *The Yearbook of English Studies* 33 (2003): 1–17.

46. The pages of Cantele's manuscript have been cut from an earlier binding and remounted on paper stubs, making it impossible for me to determine the size or number of the quires.

47. In my corpus of practical manuscripts, original limp vellum or parchment bindings are still extant on Bibliothèque interuniversité de santé MS 3; Bodleian MSS Bodley 483 and Ashmole 1378; CUL MSS Additional 9308 and Additional 9309; MLM MS B.44; Glasgow MS Hunter 497; and Wellcome MS 404.

48. Davis, *The Paston Letters: A Selection in Modern Spelling*, 168–69, no. 80.

49. Bodleian MS Ashmole 1393, f. 18r.

50. CP 40/569, rot. 428d, *CCP*.

51. Overty, "The Cost of Doing Scribal Business," 5–7.

52. CP 40/576, rot. 112, *CCP*.

53. CP 40/569, rot. 212, *CCP*.

54. CP 40/664, rot. 454d, *CCP*.

55. Curt F. Bühler, *The Fifteenth-Century Book: The Scribes, the Printers, the Decorators*, The A. S. W. Rosenbach Fellowship in Bibliography (Philadelphia: University of Pennsylvania Press, 1960), 22–23.

56. Each of these social positions references a well-known fifteenth-century English amateur scribe: Robert Thornton (country gentleman), John Crophill (medical practitioner), and Robert Reynes (village reeve). On their manuscripts, see Ogden, *The "Liber de Diversis Medicinis" in the Thornton Manuscript (MS Lincoln Cathedral A.5.2)*; Lois Ayoub, "John Crophill's Books: An Edition of British Library MS Harley 1735" (PhD diss., University of Toronto, 1994); Robert Reynes, *The Commonplace Book of Robert Reynes of Acle: An Edition of Tanner MS 407*, ed. Cameron Louis (New York: Garland, 1980).

57. Neesbett's *Sururgia* is now Bodleian MS Ashmole 1438, pt. I, pp. 57–80. Aderston's recipe collection is Bodleian MS Ashmole 1389, ff. 1–133.

58. On the scattershot preservation of ancient medical texts in early medieval li-

braries, see Peregrine Horden, "What's Wrong with Early Medieval Medicine?," *Social History of Medicine* 24, no. 1 (April 1, 2011): 16–18.

59. On Old English medical writing, see M. L. Cameron, *Anglo-Saxon Medicine*, Cambridge Studies in Anglo-Saxon England 7 (New York: Cambridge University Press, 1993).

60. On early medieval prognostics, see Valerie Irene Jane Flint, *The Rise of Magic in Early Medieval Europe* (Princeton, NJ: Princeton University Press, 2020), 135–41. On the tradition in Old English, see Sándor Chardonnens, *Anglo-Saxon Prognostics, 900–1100: Study and Texts* (Leiden: Brill, 2007).

61. Charles Burnett, *The Introduction of Arabic Learning into England*, The Panizzi Lectures (London: The British Library, 1997).

62. Whereas only 160 medical manuscripts survive from the period before the year 1000, over 550 can be dated to the period between 1075 and 1225; see Monica H. Green, "Medical Books," in *The European Book in the Twelfth Century*, ed. Erik Kwakkel and Rodney M. Thomson, Cambridge Studies in Medieval Literature 101 (Cambridge: Cambridge University Press, 2018), 277–78.

63. The *Circa instans* herbal is ff. 47–112 in BL MS Sloane 1764. On the medical school at Salerno, see Paul Oskar Kristeller, "The School of Salerno," *Bulletin of the History of Medicine* 17 (January 1, 1945): 138–92.

64. On Constantine's translations of Arabic texts, see Charles Burnett and Danielle Jacquart, eds., *Constantine the African and 'Alī Ibn Al-'Abbās al-Magūsī: The Pantegni and Related Texts*, Studies in Ancient Medicine, 925–1421 10 (New York: E. J. Brill, 1994).

65. For a summary of the Galenic theory of therapeutic pharmacology, see Faith Wallis, "The Ghost in the *Articella*: A Twelfth-Century Commentary on the Constantinian *Liber Graduum*," in *Herbs and Healers from the Ancient Mediterranean through the Medieval West: Essays in Honor of John M. Riddle*, ed. Ann Van Arsdall and Timothy Graham (Aldershot: Ashgate, 2012), 112–14.

66. Many Salernitan medical authors, like Platearius, adapted theoretical medical texts into practical guides like the *Circa instans*; on this, see Green, "Medical Books," 287.

67. The Middle English translation has been edited as Macer, *A Middle English Translation of Macer Floridus De Viribus Herbarum*, ed. Gösta Frisk (Uppsala: Almqvist & Wiksells, 1949). The section of *De viribus herbarum* in Cantele's manuscript concerns the virtues of mugwort and wormwood; BL MS Sloane 1764, ff. 11v–15v.

68. Both have received modern editions: Gösta Brodin, ed., *Agnus castus: A Middle English Herbal Reconstructed from Various Manuscripts*, Essays and Studies on English Language and Literature 6 (Upsala: Lundequistska bokhandeln, 1950); and Pol Grymonprez, ed., *Here men may se the vertues of herbes: A Middle English herbal (MS. Bodley 483, ff. 57r–67v)*, Scripta: Medieval and Renaissance Texts 3 (Brussels: UFSAL, 1981).

69. Ovitt, "The Status of the Mechanical Arts in Medieval Classifications of Learning," 95.

70. Mark D. Jordan, "Medicine as Science in the Early Commentaries on 'Johannitius,'" *Traditio* 43 (1987): 121–45.

71. The division of the art of medicine into two parts is mapped out in the first few lines of the *Isagoge* of Johanittius, translated by Constantine the African; see *Articella*,

USTC 801793 (Pavia, Italy: Giacomo Pocatela, 1510), fol. 9b. The Middle English translation of the *Isagoge* appears in Bodleian MS Ashmole 1498, ff. 57r–62v, quoted at f. 57r.

72. Michael McVaugh, *The Rational Surgery of the Middle Ages*, Micrologus Library 15 (Firenze: SISMEL, 2006), 25–27, 38–61.

73. Lanfranc's *Chirurgia magna* has not yet received a modern edition and is thus best accessed in its early modern edition: Lanfranc of Milan, "Chirurgia Magna," in *Ars Chirurgica Guidonis Cauliaci* (Venice: Apud Iuntas, 1546), fols. 207–61. Guy de Chauliac's manual has been edited as *Inventarium Sive Chirurgia Magna, Volume One: Text*, ed. Michael McVaugh, Studies in Ancient Medicine 14 (Leiden: Brill, 1997). The Middle English translation of Chauliac's surgery has been edited as Guy de Chauliac, *The Cyrurgerie of Guy de Chauliac, Volume I: Text*, ed. Margaret S. Ogden, Early English Text Society 265 (London: Oxford University Press, 1971). Cantele's surgical recipes are BL Sloane MS 1764, ff. 31r–46v.

74. Other notable and popular all-purpose medical guidebooks by English practitioners were the *Rosa Anglica* of the physician John Gaddesden and the *Liber medicinarum* of the English surgeon John Arderne, both of whom were influenced by the *Compendium of medicine*; see Getz, *Medicine in the English Middle Ages*, 39–43; Peter Murray Jones, "University Books and the Sciences, c. 1250–1400," in *The Cambridge History of the Book in Britain, Vol. II: 1100–1400*, ed. Nigel Morgan and Rodney M. Thomson (Cambridge: Cambridge University Press, 2008), 458–61.

75. The recipe collection closes with "Explicit thesaurus pauperum," BL MS Sloane 1764, f. 29v. On the enormous popularity of this collection, see Lynn Thorndike, *A History of Magic and Experimental Science: During the First Thirteen Centuries of Our Era, Vol. II* (New York: Columbia University Press, 1923), 490–95.

76. Modern editions of these translations are Chauliac, *The Cyrurgerie of Guy de Chauliac, Volume I: Text*; Lanfranc of Milan, *Lanfranck's "Science of Cirurgie,"* ed. Robert von Fleischhacker, EETS, Original Series 102 (London: Oxford University Press, 1894); Faye M. Getz, ed., *Healing and Society in Medieval England: A Middle English Translation of the Pharmaceutical Writings of Gilbertus Anglicus* (Madison: University of Wisconsin Press, 2010).

77. On the difficulty of ascribing a Latin source for Middle English surgical collections, see Melissa Reynolds, "The *Sururgia* of Nicholas Neesbett: Writing Medical Authority in Later Medieval England," *Social History of Medicine* 35, no. 1 (February 2022): 144–69.

78. See Peter Murray Jones, "Medicine and Science," in *The Cambridge History of the Book in Britain, Vol. III: 1400–1557*, ed. Lotte Hellinga and J. B. Trapp (Cambridge: Cambridge University Press, 1999), 434–35.

79. "Here begynneth Medicynes that good leches hathe founden and drawen oute of bokys that is to sey Galyen and Achepeus and Ipocras for these wer the best leches in the worlde," Bodleian MS Bodley 483, f. 3v.

80. "Ipocras þis bok sende to þe emperor sesar wite þou wel þat þis bok ys leche to all þing þat hit doyþ teche," Bodleian MS Douce 84, f. 1r. On that treatise, see M. Teresa Tavormina, "The Middle English Letter of Ipocras," *English Studies* 88, no. 6 (December 2007): 632–52.

81. See table 1.2. For a breakdown of which manuscripts contain these practical texts, see the Digital Appendices at https://readingpractice.github.io.

82. "In þe moneth of may ryse up erly of þi bedde and erely eete and drynke and use hote metes," in TCC MS R.14.51, f. 27r.

83. Faith Wallis, "The Experience of the Book: Manuscripts, Texts, and the Role of Epistemology in Early Medieval Medicine," in *Knowledge and the Scholarly Medical Traditions*, ed. Don Bates (Cambridge: Cambridge University Press, 1995), 101–26.

84. See Alastair Minnis, *Medieval Theory of Authorship: Scholastic Literary Attitudes in the Later Middle Ages* (Philadelphia: University of Pennsylvania Press, 2012).

85. "Pueri enim sumus in collo gigantis, quia videre possumus quicquid gigas et aliquantulum plus," in Caulhiaco, *Inventarium sive chirurgia magna*, I, 1. On the trope that medieval philosophers stood on the shoulders of the ancients, see Brian Stock, "Antiqui and Moderni as 'Giants' and 'Dwarfs': A Reflection of Popular Culture?," *Modern Philology* 76, no. 4 (1979): 370–74.

86. The recipe to handle a serpent is found in BL Sloane MS 1315, ff. 97r. The recipes for the biting toad are found in TCC MS R.14.51, f. 4v.

87. Claire Jones, "Discourse Communities and Medical Texts," in *Medical and Scientific Writing in Late Medieval English*, ed. Irma Taavitsainen and Päivi Pahta (Cambridge: Cambridge University Press, 2004), 23–36; Irma Taavitsainen, "Scriptorial 'House-Styles' and Discourse Communities," in *Medical and Scientific Writing in Late Medieval English*, 209–40.

88. George R. Keiser, "Medicines for Horses: The Continuity From Script to Print," *Yale University Library Gazette* 69, no. 3–4 (1995): 111–28. Other copies of this text within my corpus of practical manuscripts are Bodleian MSS Wood empt. 18, ff. 61r–79v; Rawlinson C.506, ff. 287r–297v; BL MSS Sloane 686, ff. 49r–65v, and Sloane 372, ff. 113r–118v.

89. On these estate treatises, see Dorothea Oschinsky, "Medieval Treatises on Estate Management," *Economic History Review* 8, no. 3 (1956): 296–309.

90. These texts and their manuscript witnesses are listed in Ruth J. Dean and Maureen B. M. Boulton, *Anglo-Norman Literature: A Guide to Texts and Manuscripts*, Anglo-Norman Text Society, Occasional Series 3 (London: Anglo-Norman Text Society, 1999), nos. 362–64, 366–76, 378–80, 384–85, 387, 390–96, 398–404, 406, 408–11, 417–24, 428, 430–36, 439–40.

91. For example, the Middle English *Letter of Ipocras* uroscopy treatise was composed in Anglo-Norman French in the thirteenth century. On the collection in Anglo-Norman manuscripts, see Tony Hunt, *Popular Medicine in Thirteenth-Century England: Introduction and Texts* (Woodbridge, Suffolk: Boydell & Brewer Ltd., 1990), 100–141. See also Hunt's discussion of the Middle English translation of Roger Frugard's *Chirurgia* in BL Sloane MS 240 and its similarities to an earlier Anglo-Norman version; Tony Hunt, *Anglo-Norman Medicine I: Roger Frugard's Chirurgia, The Practica brevis of Platearius* (Woodbridge, Suffolk: Boydell & Brewer, 1994), 14–17.

92. On Anglo-Norman recipes for colors, see Tony Hunt, "Early Anglo-Norman Receipts for Colours," *Journal of the Warburg and Courtauld Institutes* 58 (1995): 203–9. On the development of Latin craft recipe collections in the twelfth century, see Lynn White, "Theophilus Redivivus," *Technology and Culture* 5, no. 2 (1964): 224–33.

93. "provyd for trowyth be one Johann Edward Brykyndynmaker the which maister geram the phizizion tawght hym in seynt margette in lothbury," Bodleian MS Ashmole 1443, f. 6r.

94. Master Geram's recipe (Bodleian MS Ashmole 1443, f. 6r.) for expelling bladder stones describes frying smallage, centory, and parsley in butter and then binding it to the patient's abdomen when still hot. A recipe for extracting the placenta after birth in

BL MS Harley 2320, f. 74r, similarly calls for frying ferns and then laying them while still hot on a woman's abdomen.

95. On the mutability of vernacular verse, see Pascale Bourgain, "The Circulation of Texts in Manuscript Culture," in *The Medieval Manuscript Book: Cultural Approaches* (Cambridge: Cambridge University Press, 2015), 140–59.

96. See table 1.2. For a breakdown of which manuscripts feature recipes, see the Digital Appendices at https://readingpractice.github.io.

97. BL MS Sloane 1764, ff. 5r–6v.

98. On the popularity of Burgundy's treatise and the variety of its forms in manuscript circulation, see Lori Jones, "Itineraries and Transformations: John of Burgundy's Plague Treatise," *Bulletin of the History of Medicine* 95, no. 3 (2021): 277–314.

99. Jacquart, "Theory, Everyday Practice, and Three Fifteenth-Century Physicians," 141–42, 159–60.

100. Justin Colson and Robert Ralley, "Medical Practice, Urban Politics and Patronage: The London 'Commonalty' of Physicians and Surgeons of the 1420s," *The English Historical Review* (October 29, 2015): 1102–31.

101. Carole Rawcliffe, *Medicine and Society in Later Medieval England* (Stroud, Gloucestershire: Alan Sutton, 1995), 108–12; Nancy Siraisi, "The Faculty of Medicine," in *A History of the University in Europe*, ed. Hilde de Ridder-Symoens (Cambridge: Cambridge University Press, 1991), 372–73.

102. Vivian Nutton makes this claim in "'A Diet for Barbarians': Introducing Renaissance Medicine to Tudor England," 275–77.

Chapter 2

1. On customs officials in fifteenth-century England, see H. S. Cobb, *The Overseas Trade of London: Exchequer Customs Accounts, 1480–1*, British History Online (London: London Record Society, 1990), xi–xlvii. Skires's customs records are TNA E 122/185/6, E 122/183/6, E 122/185/16, E 122/185/15, and E 122/96/43.

2. "hardeware vel haberdashery in quodam barell positus cuiusdam mercatoris alienquo," in TNA E 122/96/43, second membrane.

3. TNA E 122/96/43, second membrane.

4. P. R. Robinson was the first to make this connection in her chapter, "'Lewdecalendars' from Lynn," in *Tributes to Kathleen L. Scott, English Medieval Manuscripts: Readers, Makers, and Illuminators*, ed. Marlene Villalobos Hennessy (London: Harvey Miller Publishers, 2009), 221–22.

5. Duffy, *The Stripping of the Altars*, 11–46, quoted at 11.

6. On "labors of the month," see Colum Hourihane, *Time in the Medieval World: Occupations of the Months and Signs of the Zodiac in the Index of Christian Art* (State College: Pennsylvania State University Pess, 2007).

7. The other two identical lewdecalendars are Bodleian Library MS Douce 71 and Llubljana, National and University Library of Llubljana MS 160. All three feature the icon of St. Bavo for October first, a seventh-century Frankish noble whose cult was only celebrated in the Low Countries. Another two very similar folding calendars for German and Danish readers survive from the fourteenth and early sixteenth centuries, respectively: Berlin, Staatsbibliothek zu Berlin MS Libr. Pict. A 92 and Copenhagen, Royal Library of Denmark MS NKS 901.

8. See Robinson, "'Lewdecalendars' from Lynn," 221–22.

9. On the importance of anomalies for understanding premodern notions of natural order, see Lorraine J. Daston and Katharine Park, *Wonders and the Order of Nature, 1150–1750* (New York: Zone Books, 2001), 135–72.

10. Several minor saints celebrated only in Worcestershire and featured in the calendar are Saint Wulfstan (Jan. 19), Bishop of Worcester; Saint Oswald of Worcester (Feb. 28); and Saint Eadburga (June 15).

11. Bodleian MS Rawlinson D.939, section 1r (Harry and Talbat) and section 3v (Peris).

12. John B. Friedman, "Harry the Haywarde and Talbat His Dog: An Illustrated Girdle Book from Worcestershire," in *Art into Life: Collected Papers from the Kresge Art Museum Medieval Symposia*, ed. Carol Fisher and Kathleen L. Scott (East Lansing: Michigan State University Press, 1995), 115–53.

13. William Langland, *William Langland's The Vision of Piers Plowman* (London and New York: J. M. Dent and E. P. Dutton, 1978), Passus 19, ll. 327–34; John Wyclif, "De Apostasia Cleri," in *Select English Works of John Wyclif, Vol. 3*, ed. T. Arnold (Oxford: Clarendon Press, 1871), 430–40; John Wyclif, "De Officio Pastorali," in *The English Works of Wyclif Hitherto Unprinted*, ed. F. D. Matthew, Early English Text Society, Original Series 74 (London: Trübner & Co., 1880), 444. On the duties of the "haywarde" and "pinder," see Frances McSparran et al., eds., *Middle English Dictionary*, online ed. (Ann Arbor: University of Michigan Press, 2000–2018), http://quod.lib.umich.edu/m/middle-english-dictionary/, s.v. "hei-warde" and "pinder."

14. Geoffrey Chaucer, "The Nun's Priest's Prologue, Tale, and Epilogue," in *The Riverside Chaucer*, 3rd ed., ed. Larry D. Benson (Oxford: Oxford University Press, 2008), 260, ll. 3383.

15. The English translation of the *Legenda aurea* was printed four times before 1500, a marker of the work's considerable popularity among later medieval English readers. These editions are STC nos. 24873 (William Caxton, 1483), 24874 (William Caxton, 1483), 24875 (Wynkyn de Worde, 1493), and 24876 (Wynkyn de Worde, 1498).

16. Michael Baxandall famously coined the term "period eye" to describe the specific cultural expectations that conditioned a viewer's response to art in *Painting and Experience in Fifteenth-Century Italy: A Primer in the Social History of Pictorial Style*, 2nd ed. (Oxford: Oxford University Press, 1988), 31–40. See also Mary Carruthers's discussion of "reading pictures" in *The Book of Memory: A Study of Memory in Medieval Culture*, 2nd ed. (Cambridge: Cambridge University Press, 2008), 274–91.

17. "Nam quod legentibus scriptura, hoc idiotis praestat picture cernentibus, quia in ipsa ignorantes vident, quod sequi debeant, in ipsa legunt qui litteras nesciunt: unde praecipue gentibus pro lectione pictura est," in Gregory I, *Registrum Epistolarum Tomus II: Libri VIII–XIV Cum Indicibus et Praefatione*, ed. Paul Edwald and Ludovic Hartmann, Monumenta Germaniae Historica (Berlin: Weidmann, 1899), 270. See also Herbert L. Kessler, "Gregory the Great and Image Theory in Northern Europe during the Twelfth and Thirteenth Centuries," in *A Companion to Medieval Art: Romanesque and Gothic in Northern Europe*, ed. Conrad Rudolph (New York: John Wiley & Sons, Ltd., 2006), 151–72.

18. On the role of images in developing lay piety following Fourth Lateran, see Aden Kumler, *Translating Truth: Ambitious Images and Religious Knowledge in Late Medieval France and England* (New Haven, CT: Yale University Press, 2011), 4 and passim.

19. St. Thomas Aquinas, *Scriptura Super Sententiis Magistri Petri Lombardi, Tomus III*, ed. M. F. Moos (Paris: P. Lethielleux, 1933), 312.

20. On illustrated Bibles, see John Lowden, *The Making of the Bibles Moralisées*, vol. 1: The Manuscripts (University Park: Pennsylvania State University Press, 2000); on Books of Hours in England, see Eamon Duffy, *Marking the Hours* (New Haven, CT: Yale University Press, 2006); on one work of pictorial exegesis, see Adrian Wilson and Joyce Lancaster Wilson, *A Medieval Mirror: Speculum Humanae Salvationis 1324–1500* (Berkeley: University of California Press, 1985).

21. See, for example, the "labors of the month" and astrological illustrations in the "Bedford Hours," BL Additional MS 18850, ff. 1r–13v.

22. In both manuscripts, the illustration of the Annunciation of the Virgin employs the same vertical banderoles, and the Morgan *Speculum*'s illustration of Christ's harrowing of hell also follows the same composition as that in the Rawlinson calendar; see MLM MS M. 766 ff. 28v and 50v, Bodleian MS Rawlinson D.939, section 1r and section 5r. Wilson and Wilson contend that the illustrations in the Morgan *Speculum* are conspicuously unlike those in continental exempla, suggesting that the artisan responsible for that manuscript and whoever made the Rawlinson calendar shared a specifically *English* understanding of what those scenes should look like; see Wilson and Wilson, *A Medieval Mirror: Speculum Humanae Salvationis 1324–1500*, 47–49.

23. The prologue to the *Speculum* parroted Pope Gregory I's line that images "are the books of the lay people"; see Wilson and Wilson, *A Medieval Mirror: Speculum Humanae Salvationis 1324–1500*, 24.

24. Quoted in Margaret Aston, *England's Iconoclasts: Laws against Images* (Oxford: Clarendon Press, 1988), 99; see also Margaret Aston, *Lollards and Reformers: Images and Literacy in Late Medieval Religion*, History Series 22 (London: Hambledon Press, 1984), 130–31.

25. Aston, *England's Iconoclasts*, 96–159; K. B. McFarlane, *John Wycliffe and the Beginnings of English Nonconformity* (New York: Macmillan, 1953), 100–107.

26. *Dives & pauper*, STC 19213 (Westminster: Wynkyn de Worde, 1496), sig. a.vii *a–b*. On *Dives & pauper* in the context of Lollard iconoclasm, see Aston, *England's Iconoclasts*, 144–45.

27. Kathryn Kerby-Fulton argues that the English church was far more concerned with apocalypticism than Lollardy in the later fourteenth and fifteenth centuries; see *Books Under Suspicion: Censorship and Tolerance of Revelatory Writing in Late Medieval England* (Notre Dame, IN: University of Notre Dame Press, 2006), 6–14.

28. *KJV* 2 Thessalonians 2:3. On prophecy and its relation to the papal schism, see Laura Ackerman Smoller, *History, Prophecy, and the Stars: The Christian Astrology of Pierre d'Ailly, 1350–1420* (Princeton, NJ: Princeton University Press, 1994), 85–121.

29. On the apocalyptic writings of Pierre D'Ailly, see Smoller, *History, Prophecy, and the Stars*. On the apocalyptic writings of John of Rupescissa, see Leah DeVun, *Prophecy, Alchemy, and the End of Time: John of Rupescissa in the Late Middle Ages* (New York: Columbia University Press, 2009).

30. Bodleian MS Rawlinson D.939, f. section 1 recto.

31. Bodleian MS Rawlinson D.939, section 2a recto and 2c verso.

32. Bodleian MS Rawlinson D.939, section 4 recto (Adam and Eve and the Fall of Man); section 5 recto (the death of Adam and Eve and Christ's harrowing of hell); and section 6 recto (Christ's birth, the Adoration of the Magi, and crucifixion).

33. On the preponderance of medical materials within calendar manuscripts, see Faith Wallis, "Medicine in Medieval Calendar Manuscripts," in *Manuscript Sources of Medieval Medicine: A Book of Essays*, ed. M. R. Schleissner (New York: Garland, 1995), 105–43.

34. Hilary M. Carey, "What Is the Folded Almanac? The Form and Function of a Key Manuscript Source for Astro-Medical Practice in Later Medieval England," *Social History of Medicine* 16, no. 3 (December 1, 2003): 481–509. Physician's almanacs are not folded in the same style as the Rawlinson calendar, but rather take the form of "bat books"; see J. P. Gumbert, *Bat Books: A Catalogue of Folded Manuscripts Containing Almanacs or Other Texts*, Bibliologia 41 (Turnhout, Belgium: Brepols Publishers, 2016).

35. Hilary M. Carey, "Astrological Medicine and the Medieval English Folded Almanac," *Social History of Medicine* 17, no. 3 (December 1, 2004): 348. On the Arabic sources for astrological medicine, see Charles Burnett, "Astrology and Medicine in the Middle Ages," *The Bulletin of the Society for the Social History of Medicine* 37 (1985): 16–18.

36. One of the first Europeans to translate this theory for Latin readers was the Englishman Daniel of Morley; see Thorndike, *A History of Magic and Experimental Science: During the First Thirteen Centuries of Our Era, Vol. II*, 175–76. On the adoption of astrological medicine in England, see Roger French, "Foretelling the Future: Arabic Astrology and English Medicine in the Late Twelfth Century," *Isis* 87, no. 3 (1996): 455–57.

37. Roger Bacon, *Opus Majus, Volumes 1 and 2*, trans. Robert Belle Burke (Philadelphia: University of Pennsylvania Press, 1928), 214.

38. Frederick Hammond, "Odington, Walter (Fl. C. 1280–1301), Benedictine Monk and Scholar," in *ODNB*.

39. Cornelius O'Boyle, "Astrology and Medicine in Later Medieval England: The Calendars of John Somer and Nicholas of Lynn," *Sudhoffs Archiv* 89, no. 1 (2005): 1–2.

40. Modern editions of both *kalendaria* are Nicholas Lynn, *The Kalendarium of Nicholas of Lynn*, ed. Sigmund Eisner, The Chaucer Library (Athens: University of Georgia Press, 1980); John Somer, *The Kalendarium of John Somer*, ed. Linne R. Mooney (Athens: University of Georgia Press, 1998).

41. Geoffrey Chaucer, *A Treatise on the Astrolabe*, ed. W. W. Skeat (London: N. Trübner & Co., for the Chaucer Society, 1872), 3.

42. O'Boyle, "Astrology and Medicine in Later Medieval England: The Calendars of John Somer and Nicholas of Lynn," 2–3.

43. On the degrees of expertise assumed by differing genres of almanac, see Carey, "Astrological Medicine and the Medieval English Folded Almanac," 348–50.

44. O'Boyle, "Astrology and Medicine in Later Medieval England: The Calendars of John Somer and Nicholas of Lynn," 6–7. A handlist of surviving physician's almanacs from England is available in Carey, "What Is the Folded Almanac?," 505–9.

45. Astrological figures or tables using symbols appear on section 1v, section 3r, section 3v, section 4v, and section 5v of Bodleian MS Rawlinson D.939.

46. The full table extends from section 4v to section 5v of Bodleian MS Rawlinson D.939.

47. "whanne þe moone is in ony signe and abodi be lete blood or wounded . . . in þat signe þat haþ governaunce in ony place it is drede but þe body be soone deed or distried in þis world"; BL MS Sloane 100, f. 36r.

48. The pictorial history appears on section 5v of Bodleian MS Rawlinson D.939.

49. Middle English versions of the prognostication according to dominical letter on folio four of the Rawlinson calendar are found in BL MSS Harley 2320, ff. 73r; Harley 2252, ff. 141–42; and Sloane 393, ff. 73v–74v; Bodleian MSS Add. C.246, ff. 102v–103v and Tanner 407, ff. 53; Huntington MSS HM 505, f. 60v and HM 1336, f. 35r; and Wellcome MS 8004, ff. 68v–70r. Middle English versions of "Book of Thunder" prognostication on folio three of the Rawlinson calendar are found in Huntington MS HM 1336, ff. 34v–35r; Wellcome Library MS 8004, ff. 70r–70v; and BL Sloane MS 2584, ff. 33r–33v. These same prognostications can be found in Latin in BL Egerton MS 1995, ff. 61v–62v; CUL MS Dd.6.29, ff. 2v–16r; Bodleian MSS Douce 45, ff. 112v–114v, and Tanner 407, ff. 53; and Wellcome MS 537, ff. 12v–13v.

50. On early English prognostications, see Chardonnens, *Anglo-Saxon Prognostics, 900–1100*, 2. An early medieval Latin copy of the prognostication from dominical letter with Old English glosses has been edited by R. M. Liuzza in *Anglo-Saxon Prognostics: An Edition and Translation of Texts from London, British Library, MS Cotton Tiberius A.iii*, Anglo-Saxon Texts 8 (Cambridge: D. S. Brewer, 2011), 154–57.

51. Bede, *Bede: The Reckoning of Time*, ed. Faith Wallis, Translated Texts for Historians 29 (Liverpool: Liverpool University Press, 1999).

52. On the relationship between time-reckoning and predictive meteorology in early medieval Europe, and especially in Bede's writings, see Anne Lawrence-Mathers, *Medieval Meteorology: Forecasting the Weather from Aristotle to the Almanac* (Cambridge: Cambridge University Press, 2019), 21–39.

53. Bodleian MS Rawlinson D.939, section 3r.

54. Bodleian MS Rawlinson D.939, section 6v.

55. See my discussion of the verse *storia lunae* in BL Harley MS 1735, below. See also Irma Taavitsainen, "The Identification of Middle English Lunary MSS," *Neuphilologische Mitteilungen* 88, no. 1 (1987): 18–26; Irma Taavitsainen, "Storia Lune and Its Paraphrase in Prose: Two Versions of a Middle English Lunary," in *Neophilologica Fennica*, ed. Leena Kahlas-Tarkka, Mémoires de La Société Néophilologique de Helsinki 45 (Helsinki: Société Philologique, 1987), 521–55.

56. Keith Thomas gives a number of examples of similar tallying systems in "The Meaning of Literacy in Early Modern England," in *The Written Word: Literacy in Transition*, ed. Gerd Baumann (Oxford: Clarendon Press, 1986), 108–9.

57. Bodleian MS Ashmole 8 recto.

58. The same sword symbol appears on day eighteen of the Ashmole version, with the caption "a manqueler" above it; Bodleian MS Ashmole 8 recto.

59. Sonja Drimmer, *The Art of Allusion: Illuminators and the Making of English Literature, 1403–1476* (Philadelphia: University of Pennsylvania Press, 2019), 5 and passim.

60. Wilson and Wilson, *A Medieval Mirror: Speculum Humanae Salvationis 1324–1500*, 25–27.

61. On prophecy as a reading of divine providence, see Marjorie Reeves, *The Influence of Prophecy in the Later Middle Ages: A Study in Joachimism*, ACLS Humanities E-Book (Oxford: Clarendon Press, 1969), ix.

62. This point is argued by Lawrence Duggan as a criticism of scholarly acceptance that pictures really were books of the illiterate. See Lawrence G. Duggan, "Was Art Really the 'Book of the Illiterate'?," *Word and Image* 5, no. 3 (January 1989): 227–51; Lawrence G. Duggan, "Reflections on 'Was Art Really the "Book of the Illiterate"?,'"

in *Reading Images and Texts*, ed. Mariëlle Hageman and Marco Mostert, vol. 8 (Turnhout: Brepols Publishers, 2005), 109–19.

63. The six pictorial practical manuscripts created in the fifteenth century are Bodleian MS Ashmole 8 (c. 1400); BL MSS Egerton 2724 (c. 1430); Harley 2332 (c. 1411); Royal 17 A.xvi (c. 1420); Oslo, Schøyen MS 1581 (c. 1425); and MLM MS M.941 (c. 1433). BL MS Additional MS 17367 (1535) is the only known exemplar from the sixteenth century.

64. It is notoriously difficult to ascertain the identities of English manuscript illustrators and to attribute manuscripts to their workshops; see Drimmer, *The Art of Allusion*, 27–37.

65. BL Egerton MS 2724. Pictorial prognostications by dominical letter: f. 4r–5r; *homo signorum*: 6r; prognostications from thunder: 4v, 5v, 6v, 7v, 8v, and 9v. Though the catalogue entry from the British Library dates this calendar to 1404, I arrived at a date of 1430 based on images in the pictorial history on f. 10v: one depicts bodies in shrouds, representing the Black Death, captioned with the number 82, indicating that eighty-two years had passed since the 1348 epidemic, and the second shows a young king with a scepter, captioned with the number 8, indicating that eight years had passed since the coronation of Henry VI, on August 31, 1422.

66. BL MS Egerton 2724, f. 2r.

67. B. Dodwell, "The Foundation of Norwich Cathedral," *Transactions of the Royal Historical Society* 7 (1957): 1–18.

68. MLM MS M.941.

69. There are a number of transcription errors in the Latin abbreviations scattered throughout MLM MS M.941. For example, the icon for "All Saints' Day" on November first is labeled "Oimi Seor" for *Omnium sanctorum*, f. 1v. In the pictorial history, *ab origine* is abbreviated "aboreuicui" and *Nativitas domini* "Nativitasdb," f. 3r.

70. The pictorial liturgical calendar is BL MS Harley 2332, ff. 1v–13v; the *homo signorum* is f. 18r; the pictorial table of astrological influences is ff. 18v–19r; and the pictorial prognostication by dominical letter is ff. 19v–20r.

71. The manuscript's calendar also features the feast of St. Maurus on January 15, only celebrated in Benedictine houses, and the feast of St. Wilfrid, Bishop of York; BL MS Additional 46143, f. 5v and 14v. The tables of lunar and solar positions (ff. 5v–17r) are also for the latitude of York.

72. The dates and times of the eclipse tables in both manuscripts (BL MSS Harley 2332, ff. 14v–17r and Additional 46143, ff. 18v–22r) are an exact match and both are derived from Nicholas of Lynn's *kalendarium*. The monthly tables of lunar and solar positions in the Harley almanac are *not* a match to those in Additional MS 46143, but instead are very close to those in surviving copies of John Somer's *kalendaria*.

73. Wellcome Library MS 8004, p. 6.

74. O'Boyle, "Astrology and Medicine in Later Medieval England: The Calendars of John Somer and Nicholas of Lynn."

75. The two manuscripts are BL MS Royal 17 A.xvi (1420) and Schøyen Collection MS 1581 (1425).

76. The reader mark of Thomas Rowland, who described himself as "priour of the monastery," is on BL MS Royal 17 A.xvi, f. 31r. On Luffield and Rowland's appointment as prior, see William Page, ed., "Houses of Benedictine Monks: The Priory of Luffield," in *A History of the County of Buckingham: Volume 1*, British History Online (London: Victoria County History, 1905), 347–50.

77. Other practical manuscripts feature illustrated prognostications, though none are pictographic in the same way. For example, the copy of the Middle English prognostic treatise known as the *Wise Book of Astronomy and Philosophy* in Bodleian MS Rawlinson D.1220 features illustrated roundels of astrological signs. On *The Wise Book of Astronomy and Philosophy*, see Carrie Griffin, ed., *The Middle English "Wise Book of Philosophy and Astronomy": A Parallel-Text Edition. Edited from London, British Library, MS Sloane 2453 with a Parallel Text from New York, Columbia University, MA Plimpton 260*, Middle English Texts 47 (Heidelberg: Winter, 2013).

78. Harvard, Houghton Library MS Richardson 35, f. 1v. On the tremendous popularity of the prose *Brut* in later medieval England, see Lister M. Matheson, *The Prose* Brut*: The Development of a Middle English Chronicle*, Medieval & Renaissance Texts & Studies 180 (Tempe, AZ: Medieval & Renaissance Texts & Studies, 1998).

79. Houghton MS Richardson 35, f. 25r. On the relationship between the illustrations in Bodleian MS Rawlinson D.939 and the Houghton *Brut*, see Sonja Drimmer, "The Shapes of History: Houghton Library, MS Richardson 35 and Chronicles of England in Codex and Roll," in *Beyond Words: New Research on Manuscripts in Boston Collections*, ed. Jeffrey F. Hamburger et al., Studies and Texts 221; Text Image Context: Studies in Medieval Manuscript Illumination 8 (Toronto: PIMS, 2021), 253–68.

80. The heraldry illustrations in Houghton MS Richardson 35 are ff. 95r–101v.

81. BL Harley MS 1735, ff. 1–13v.

82. Peter Murray Jones, "Crophill, John (d. in or after 1485)," in *ODNB*.

83. BL Harley MS 1735, f. 2r.

84. BL MS Harley 1735, f. 4r.

Chapter 3

1. The final leaf of the manuscript contains an *ex libris* mark: "Iste liber constat Roberto Taylour de Boxforde," written by Taylor's scribe; Huntington MS HM 1336, f. 36r.

2. The manuscript composed for Taylor is Huntington MS HM 1336, ff. 1–18, 30–35. These medical recipes are found (in order) on ff. 4r, 4v, and 5v.

3. Huntington MS HM 1336, ff. 34v–35r and 35v–36r.

4. Huntington MS HM 1336, ff. 3v, 12r, and 15r.

5. "Nunc scripsi totum pro Christo da michi potum"; Huntington MS HM 1336, f. 36v. Wysbech identified himself as "Symon Wysbech studens in jure canonico" and "Symon Wysbech scolaris cantabrigiensis" on the same page.

6. Jenny Stratford, "The Manuscripts of John, Duke of Bedford: Library and Chapel," in *England in the Fifteenth Century: Proceedings of the 1986 Harlaxton Symprium*, ed. Daniel Williams (Woodbridge, Suffolk: Boydell & Brewer, 1987), 329–51; M. J. Barber, "The Books and Patronage of Learning of a Fifteenth-Century Prince," *The Book Collector* 12, no. 3 (1963): 312.

7. That manuscript has been edited as Chauliac, *The Cyrurgerie of Guy de Chauliac, Volume I: Text*.

8. Keiser, *MWME*, 3596.

9. Faye Marie Getz, "Gilbertus Anglicus Anglicized," *Medical History* 26, no. 4 (October 1982): 437.

10. Bodleian MS Ashmole 1396, ff. 2v–269v.

11. Bodleian MS Ashmole 1505. On the identity of the manuscript's owner, see

Linda Ehrsam Voigts, "The Master of the King's Stillatories," in *The Lancastrian Court: Proceedings of the 2001 Harlaxton Symposium*, ed. Jenny Stratford, Harlaxton Medieval Studies 13 (Donington, Lincolnshire: Shaun Tyas, 2003), 232–55.

12. Parma's treatise is ff. 1–59v and Arderne's treatise is on ff. 60–75v of BL Sloane MS 277, with the diagrams on f. 64r. A modern edition of the treatise is John Arderne, *Treatises of Fistula in Ano, Haemorrhoids, and Clysters by John Arderne*, ed. D'Arcy Power, EETS, Original Series 139 (London: Oxford University Press, 1919). See also Peter Murray Jones, "Four Middle English Translations of John of Arderne," in *Latin and Vernacular: Studies in Late Medieval Manuscripts*, ed. Alastair Minnis (Woodbridge, Suffolk: Boydell & Brewer, 1989), 61–89.

13. Portions of the *Chirurgia* appear on ff. 34v–77v of University of Glasgow, MS Hunter 95. BL Sloane MS 1 contains Chauliac's "Treatise on Fractures" and "Treatise on Wounds." These manuscripts have both been edited by Björn Wallner: Guy de Chauliac, *An Interpolated Middle English Version of The Anatomy of Guy de Chauliac*, ed. Björn Wallner (Lund: Lund University Press, 1995); Guy de Chauliac, *The Middle English Translation of Guy de Chauliac's Treatise on Wounds, Book III of the Great Surgery*, ed. Björn Wallner (Lund: CWK Gleerup, 1976).

14. New York Academy of Medicine MS 12.

15. BL MS Sloane 2584, ff. 2r, 3v, and 6r (ink- and color-making); f. 8r (catching game); ff. 8v, 13r–29v (medical recipes; the second, more substantial collection begins with the incipit "Man that will of lechecraft lere"); ff. 30r–31r (treatise on perilous days); ff. 33r–33v (prognostications from thunder); ff. 35r–v (prognostication according to the position of the sun in the zodiac); ff. 35v–36v (prognostications from Christmas Day); ff. 90r–91v (treatise on weights and measures); f. 92r (table of multiplication factors), ff. 102r–117v (treatise on equine medicine). On the recipe group, "The man that wyll of lechecreft lere," see Keiser, *MWME*, 3840.

16. Wellcome MS 406, ff. 1v–34r (directions for bloodletting); ff. 4v–13v (medical recipes); ff. 14r20v (Virtues of Rosemary).

17. BL MS Additional 34210 has a large ornamented initial "M" on f. 5r., an ornamented "A" on f. 24v, and an ornamented "H" on f. 25r.

18. Every text in this manuscript opens with a large blue initial, decorated with red penwork that runs down the margins.

19. When one quire of parchment (usually around 8–12 leaves) ended in the middle of a text, the scribe would write the first word of the next quire, the "catchword," at the bottom of the last page of the previous quire. The beasts drawn around the catchwords in Wellcome MS 5262 are found on ff. 12v, 17v, and 24v.

20. BL Harley MS 2320, ff. 1–4 (calendar), 5–30 (Book of Nativities), 31–52r (*storia lunae* prognostication), and 52r–70v (silk-braiding directions). On these silk-braiding instructions, see E. G. Stanley, "Directions for Making Many Sorts of Laces," in *Chaucer and Middle English Studies in Honour of Rossell Hope Robbins*, ed. Beryl Rowland (Kent, OH: Kent State University Press, 1974), 89–103.

21. The silkwomen of London described their craft as the "craftes of wymmen . . . of tyme that noo mynde runneth unto the contrarie"; TNA SC 8/29/1411, 1455. On London's silkwomen, see Marian K. Dale, "The London Silkwomen of the Fifteenth Century," *Economic History Review* 4, no. 3 (October 1933): 324–35; Maryanne Kowaleski and Judith M. Bennett, "Crafts, Guilds, and Women in the Middle Ages: Fifty Years after Marian K. Dale," *Signs* 14, no. 2 (Winter 1989): 474–501.

22. For records of London silkwomen, see Kay Lacey, "The Production of 'Narrow

Ware' by Silkwomen in Fourteenth and Fifteenth Century England," *Textile History* 18, no. 2 (November 1987): 200–204. For records of silkwomen outside London, see Marjorie K. McIntosh, *Working Women in English Society 1300–1620* (Cambridge: Cambridge University Press, 2005), 223.

23. The historiated initials are on ff. 5r, 31r, and 52r, with the portrait of the woman spinning on f. 52r. Historiated initials are not common in English practical manuscripts, though they are found in other vernacular genres of medical writing like the French *régime du corps*; see Jennifer Borland, *Visualizing Household Health: Medieval Women, Art, and Knowledge in the* Régime Du Corps (State College: Pennsylvania State University Press, 2022).

24. For example, the De Brailes Hours, one of the earliest Books of Hours created for English use, features a portrait of its female patron; see Duffy, *Marking the Hours*, 30–34.

25. On Books of Hours and lay book ownership, see Andrew Taylor, "Into His Secret Chamber: Reading and Privacy in Late Medieval England," in *The Practice and Representation of Reading in England*, ed. James Raven, Helen Small, and Naomi Tadmor (Cambridge: Cambridge University Press, 2007), 41–61; Paul Saenger, "Books of Hours and the Reading Habits of the Later Middle Ages," in *The Culture of Print: Power and the Uses of Print in Early Modern Europe*, ed. Roger Chartier, trans. Lydia G. Cochrane (Princeton, NJ: Princeton University Press, 1987), 141–73.

26. Kathryn M. Rudy, "Dirty Books: Quantifying Patters of Use in Medieval Manuscripts Using a Densitometer," *Journal of Historians of Netherlandish Art* 2, no. 1–2 (Summer 2010).

27. Rudy has documented this exact behavior in "Touching the Book Again: The Passional of Abbess Kunigunde of Bohemia," in *Codex und Material*, Wolfenbütteler Mittelalter-Studien 34 (Wiesbaden: Harrassowitz Verlag in Kommission, 2018), 247–58.

28. On the scarcity of experiential reader marks in late medieval English recipe books, see Reynolds, "'Here Is a Good Boke to Lerne': Practical Books, the Coming of the Press, and the Search for Knowledge, ca. 1400–1560," 270–72.

29. Pamela H. Smith, "In the Workshop of History: Making, Writing, and Meaning," *West 86th* 19, no. 1 (2012): 4–31.

30. On the social value of recipe collections, see Melissa Reynolds, "How to Cure a Horse, or, the Experience of Knowledge and the Knowledge of Experience," *Historical Studies in the Natural Sciences* 52, no. 4 (August 2022): 546–52.

31. A great deal of recent scholarship has focused on reconstructing early recipes. One notable example is the Making and Knowing Project et al., eds., *Secrets of Craft and Nature in Renaissance France: A Digital Critical Edition of BnF Ms. Fr. 640* (New York: Making and Knowing Project, 2020), https://edition640.makingandknowing.org.

32. "To þre þinges god ȝeveþ virtu: to worde to herbis & to stonis. Deo gracias," Wellcome MS 537, f. 310v.

33. "In the beginning there was the Word, and the Word was with God, and the Word was God," *KJV* John 1:1.

34. St. Augustine provided the most influential description of how divine power was manifest in the created world in Book XXI of *De civitate dei*, discussed in Daston and Park, *Wonders and the Order of Nature, 1150–1750*, 39–41.

35. "Deus qui creasti hominem & mirabiliter reformasti & qui dedisti medicinam ad recuperandam sanitatem humanorum corporum da benedictionem tuam super

hoc antedotum ut in quorum corporum intervenerit sanitatem accipere mereant Per dominum nostrum Jehum cristum et cetera." This prayer appears amid Middle English medical recipes in CUL MS Dd.5.76, f. 6v–7r.

36. "Here men may see the virtus of diverse herbes which ben hoote and which ben colde and to how many thynges thr arn gode," BL Sloane 393, f. 13r. Thirteen other manuscripts in my corpus contain this incipit, followed by some combination of herbal texts: Bodleian MSS Ashmole 1477 II, ff. 36r–39v; Ashmole 1477 III, ff. 2v–6r; Ashmole 1438 I, ff. 1–6; Bodley 483, ff. 57r–80v; Selden Supra 73, ff. 116–123r; BL MSS Additional 12056, ff. 3r–12v; Additional 34210, ff. 25r–41v; Lansdowne 680, ff. 2r–21v; Royal 17 A.xxxii, ff. 8r–20v; Sloane 540A, ff. 1r–2v; Sloane 3160, ff. 102r–108v; and CUL MS Additional 9309, ff. 29r–33v. The herbal appears in TCC MS O.1.13, ff. 1–14, without the incipit. This herbal has been edited as Grymonprez, *Here Men May Se the Vertues of Herbes.*

37. Peter Cantele's manuscript, discussed in chapter 1, contains a copy of the Middle English translation of the *Circa instans* herbal. See BL Sloane MS 1764, ff. 47r–112v, with this entry on Aloe at f. 47r.

38. "Þus seyth ypocras þe good surgyon / And socrates & galyen," BL MS Sloane 468, f. 7r. This recipe collection has been edited from the copy in BL MS Additional 33996 as Fritz Heinrich, *Ein Mittelenglisches Medizinbuch* (Halle: Niemeyer, 1896).

39. CUL MS Additional MS 9308, f. 49r, transcribed in Lea T. Olsan, "The Corpus of Charms in the Middle English Leechcraft Remedy Books," in *Charms, Charmers and Charming: International Research on Verbal Magic*, ed. Jonathan Roper, Palgrave Historical Studies in Witchcraft and Magic (New York: Palgrave Macmillan, 2009), 219–20. In addition to invoking "holy mothers," this charm also incorporates the *SATOR AREPO* motif, which dates to early Latin manuscripts of the eighth century; see Lea Olsan, "The Marginality of Charms in Medieval England," *The Power of Words: Studies on Charms and Charming in Europe*, ed. James Kapalo, Éva Pócs, and William Francis Ryan (Budapest and New York: Central European University Press, 2013), 143n29.

40. CUL MS Additional 9308, f. 14v, transcribed in Olsan, "The Corpus of Charms in the Middle English Leechcraft Remedy Books," 217.

41. CUL MS Additional 9308, f. 78r, transcribed in Olsan, 224.

42. On this established set of sixteen charms in the "leechcraft" collection, see Olsan, "The Corpus of Charms in the Middle English Leechcraft Remedy Books."

43. Don C. Skemer, *Binding Words: Textual Amulets in the Middle Ages*, Magic in History (State College: Pennsylvania State University Press, 2006), 32, 58–64. See also Catherine Rider, "Common Magic," in *The Cambridge History of Magic and Witchcraft in the West*, ed. S. J. Collins (Cambridge: Cambridge University Press, 2015), 305.

44. See Lea T. Olsan, "Charms and Prayers in Medieval Medical Theory and Practice," *Social History of Medicine* 16, no. 3 (December 1, 2003): 352–53.

45. "In tribus enim dicunt phisici precipuam vim nature esse constituam: in verbis, et herbis et in lapidus. De virtute autem herbarum et lapidum aliquid scimus, de virtute verborum parvum vel nihil novimus," in Thomas of Cobham, *Summa Confessorum*, ed. F. Broomfield, Analecta Mediaevalia Namurcensia 25 (Louvain: Editions Nauwelaerts, 1968), 478; cited in Rider, "Medical Magic and the Church in Thirteenth-Century England," 10.

46. On these medical writers' use of charms and their labeling them as *experimenta*, see Olsan, "Charms and Prayers in Medieval Medical Theory and Practice." On the

incompatibility of *experimenta* within natural philosophical theory, see Daston and Park, *Wonders and the Order of Nature, 1150–1750*, 128–29.

47. "Empirica. Quamis ego delicto ad has res par[andum?], tantum est bonum scribere in libro nostro ut non remaneat tractatus sine eis quae dixerunt antique," Gilbertus Anglicus, *Compendium medicine*, USTC 143565 (Lyons: Jacques Sacon, 1510), f. 327r; quoted in Olsan, 351n43.

48. CUL MS Dd.4.44, ff. 1r and 2r. "The Boke of Marchalsi" is found in another eight manuscript witnesses; see Keiser, "Medicines for Horses: The Continuity from Script to Print," 115, 128n10.

49. CUL MS Dd.4.44, f. 19v and 19v–25v.

50. CUL MS Dd.4.44, ff. 5v–7r.

51. CUL MS Dd.4.44, f. 19v.

52. CUL MS Dd.4.44, f. 25v–26v.

53. Slips or even rolls or parchment were sometimes used as amulets in this manner. A popular Middle English charm in BL Royal MS 18 A.vi, f.103, instructed a reader to wear a textual amulet to protect the wearer in battle. Wellcome MS 410, a single roll of parchment containing sixteen recipes and charms, may have been intended as an instructional manual *and* an amulet. On amulets, see Skemer, *Binding Words*.

54. CUL MS Dd.4.44, ff. 27v–28v (four charms for "farcine"); f. 28r (charm to stanch blood); f. 29r (sage leaf charm); f. 31r (a charm for worms); f. 32r (*longes miles* charm); f. 33r (charm to make a horse stand still and charm for trenches); f. 34v (charm for worms in a horse); and f. 37r (a charm for the toothache).

55. "Take a sawge lef þat is not perced and wryte þis on with a penne with ynke: *In principio principio erat verbum angelus nunciat* and þanne gif hyt þe seke to ete and let þe seke seye first v. *pater nosters* in þe wershyppe of þe v. woundes of oure lord jehus xiste criste and v. aveys in þe wership of þe v. ioyes of oure lady and þanne in þe secunde day take a noþer lef and write þis on *Et verbum erat apud deum Johannes Johannes predicat* and seye þe prayers for seyde and þe þrydde day take a noþer lef and write þis on *Et deus erat verbum Cristus tonat* and gif hit þe seke and let hym seye þe prayers forseyde and by goddis grace he shal be hele"; CUL MS Dd.4.44, f. 29r. On this charm, see Keiser, *MWME*, 3869–70; Olsan, "The Corpus of Charms in the Middle English Leechcraft Remedy Books," 230–31. A Latin version of the charm is discussed in Olsan, "Charms and Prayers in Medieval Medical Theory and Practice," 358. An Anglo-Norman version of the charm utilizing a non-descript "herbive" is edited by Hunt in *Popular Medicine in Thirteenth-Century England: Introduction and Texts*, 84.

56. *KJV* John 1:1.

57. Duffy, *The Stripping of the Altars: Traditional Religion in England, 1400–1580*, 215.

58. These seventeen extant copies of the charm in addition to CUL MS Dd.4.44 are: Beinecke Library MS Takamiya 46, f. 36r; Bodleian MSS Ashmole 1443, f. 175r; Ashmole 1477, f. 15r; Laud. Misc. 553, f. 26r; BL MSS Arundel 272, f. 7v, 15v; Harley 1600, f. 9v, 34v; Lansdowne 680, f. 66v; Sloane 140, f. 47v–48r; Sloane MS 382, ff. 222v–223r; Sloane 468, ff. 70v–71r; Sloane 2457/58, ff. 29r; Sloane 2584, f. 25v; Sloane 3160, f. 116v; CUL MSS Additional 9308, f. 85r–v; Dd.5.76, f. 24r; Ee.1.15, f. 16v; TCC MS O.1.13, f. 172r–v.

59. CUL MS Dd.4.44, ff. 29r–v.

60. BL MS Sloane 3160, f. 102r. The recipe for the sage leaf charm is f. 116v.

61. CUL MS Additional 9308, ff. 55r–56r.

62. BL MS Harley 1600, f. 9v.

63. "+ ay + loy + sadoloy + demicaloy + liberator deum," BL MS Sloane 372, f. 25r.

64. Huntington MS HM 19079, f. 1r.

65. "In al þys bok þer as þou seest vyolet i don in electewarys oþer in syrypes þou shalt take þe flour of vyolet & nogt þe lefes. Ofer al in þys bok þer as þou fyndest in electewaryes þys word cytre þus ywryte oþer þus cytr. it schulde be safren. Roust of yren is y don in electewaryes bote þat ne is nogt of old rousted yron bote it is þat strof þat is y beten awey at þe onfold & me clepeþ þat in him me places smyþes colm. Bote er þan me do it in medycynes it schal be furst smale ypoudres & seþþe be soden in vynegre þryes oþer fouresyþes," Huntington MS HM 19079, f. 237v.

66. "her beþ medycynes þat beþ ydrawen out of goode leches bokes," Huntington MS HM 19079, f. 1r.

67. Neesbett's name does not appear in Francis Collins, ed., *Register of the Freemen of the City of York, from the City Records. Vol. 1: 1272–1558* (Durham: Andrews and Co., 1896); nor is he listed in Charles H. Talbot and Eugene Ashby Hammond, *The Medical Practitioners in Medieval England: A Biographical Register* (Wellcome Historical Medical Library, 1965); nor is he mentioned in the supplement to this register, Faye Getz, "Medical Practitioners in Medieval England," *Social History of Medicine* 3, no. 2 (August 1, 1990): 245–83.

68. Bodleian MS Ashmole 1438 I, pp. 57–80. On p. 61, Neesbett mentions two towns, Selby and Nun Appleton, both in the Northern Riding of Yorkshire.

69. Bodleian MS Ashmole 1438, I, p. 57, features a single boundary line to mark the upper and right-hand margin, in dry-point; pp. 64–65 feature boundary lines delineating all of the margins, in ink.

70. "þe white of þe egges be rynnyng as watter & þat is cald glare þat wryters make rede ynke wit," Bodleian MS Ashmole MS 1438, I, p. 61.

71. Like Neesbett's *Surugia*, the Middle English translation of Guy de Chauliac's *Chirurgia magna* retained the prologue with his authorial claim, as did the Middle English translation of John of Arderne's *De arte phisicali et de cirurgia*; see Chauliac, *The Cyrurgerie of Guy de Chauliac, Volume I: Text*, v–vi; John Arderne, *De Arte Phisicali et de Cirurgia of Master John Arderne, Surgeon of Newark. Dated 1412*, ed. D'Arcy Power and Eric Miller (New York: W. Wood, 1922), 1.

72. "þis is þe mervellest wyrkyr þat I know for al sores þat er corryp," Bodleian MS Ashmole 1438, I, p. 64.

73. "I culd have wryttyn ȝow oyntmentes þat had ben more costly . . . bott þise schall wyrke als wele as it," Bodleian MS Ashmole 1438, I, pp. 64–65.

74. "dram of camfore þat poticarys hase to sell it is lyke on fro farom to set as it wer allum bott it is soft," Bodleian MS Ashmole 1438, I, p. 70.

75. Bodleian MS Ashmole 1438, I, pp. 66–67. Juhani Norri, *Dictionary of Medical Vocabulary in English, 1375–1550: Body Parts, Sicknesses, Instruments, and Medicinal Preparations* (New York: Routledge, 2016), s.v. unguentum, B63. unguentum judeorum. Norri gives a transcription of a similar recipe from BL MS Sloane 73, f. 177v.

76. "þis is þe oyntment þat I told yow of & aman war prikyd wit a thorn or prike nost bot take a blak wowle lok & wete it in þis oyntment & ly þer to & he schal fele no werk nor no dysse[ase] after þat," Bodleian MS Ashmole 1438, I, p. 66.

77. "þis is a preciose thyng I lat yow wytt . . . kepe þam privay to your selfe," Bodleian MS Ashmole 1438, I, p. 65.

78. "Whatt mysters me to wryte yow mo when þese er sufficient," Bodleian MS Ashmole 1438, I, p. 67.

79. A "William Aderston, of London, surgeon" is listed as the plaintiff in a trespassing case against the sheriffs of London, TNA C 1/140/13. Aderston is not listed in either Talbot and Hammond, *The Medical Practitioners in Medieval England*, or Getz's supplement, "Medical Practitioners in Medieval England." The collection has to postdate the arrival of the sweating sickness in England, first seen in 1485, because it contains a recipe for a "goode remedy for þe new dyse yese [disease] of þe hede & stomake & swete *with* payne in þe bake"; Bodleian MS Ashmole 1389, f. 37v.

80. "This I have moche used and lovyd, for *with* it I helyd þe scheryff of Brystowe," Bodleian MS Ashmole 1389, f. 61r.

81. "probatum est per me W Aderston," Bodleian MS Ashmole 1389, f. 14r.

82. On the phenomenon of amateur scribal work in the fifteenth century, see Bühler, *The Fifteenth-Century Book*, 22–24.

83. George R. Keiser, "Robert Thornton's *Liber de Diversis Medicinis*: Text, Vocabulary, and Scribal Confusion," in *Rethinking Middle English: Linguistic and Literary Approaches*, ed. Nikolaus Ritt and Herbert Schendl (Frankfurt: Peter Lang, 2005), 30–41.

84. Misspelled Latin words are found in Griffiths, Edwards, and Barker, *The Tollemache Book of Secrets*, 48, 124.

85. These three once-separate manuscripts are Bodleian MS Ashmole 1481 1B (ff. 4–12), 1E (ff. 25–35), and 1F (ff. 36–41).

86. Bodleian MS Ashmole 1481 E, ff. 34v–35r.

87. On John Tyryngham and the notes he left in this manuscript, see Caleb Prus, "John Tyryngham: An Ordinary Medical Practitioner in Ashmole 1481," (master's thesis, University of Oxford, 2022).

88. The silkwomen of London described their craft as the "craftes of wymmen . . . of tyme that noo mynde runneth unto the contrarie"; see TNA SC 8/29/1411, 1455.

89. Davis, *The Paston Letters*, 257, no. 139.

90. For other examples of women's recipe knowledge operating as a form of political power, see Sharon T. Strocchia, *Forgotten Healers: Women and the Pursuit of Health in Late Renaissance Italy*, I Tatti Studies in Italian Renaissance History (Cambridge, MA: Harvard University Press, 2019), 14–49; Sheila Barker and Sharon Strocchia, "Household Medicine for a Renaissance Court: Caterina Sforza's Ricettario Reconsidered," in *Gender, Health, and Healing, 1250–1550*, ed. Sara Ritchey and Sharon Strocchia (Amsterdam: Amsterdam University Press, 2020), 139–65; Alisha Rankin, *Panaceia's Daughters: Noblewomen as Healers in Early Modern Germany* (Chicago: University of Chicago Press, 2013).

Chapter 4

1. John L. Flood, "'Safer on the Battlefield Than in the City': England, the 'Sweating Sickness,' and the Continent," *Renaissance Studies* 17, no. 2 (2003): 148–50.

2. Jacobi Joannes, *Here begynneth a litil boke the whiche traytied and reherced many gode thinges necessaries for the infirmite & grete skenesse called pestilence*, trans. Bengt Knutsson, STC 4589 (London: W. de Machlinia, 1485).

3. Jacobi, *Here begynneth a litill boke necessarye & behouefull agenst the pestilence*, STC 4590 (London: W. de Machlinia, 1485), fols. 1b–2a. On the theory that Henry VII was patron of this publication, see George R. Keiser, "Two Medieval Plague Treatises and Their Afterlife in Early Modern England," *Journal of the History of Medicine* 58 (July 2003): 318–19.

4. Eight of the eleven vernacular Italian medical books published before 1480, and three of the five vernacular German medical texts published before 1475, were recipes or regimens against the plague. Italy: USTC 996074 (1475?); USTC 994937 (1475); USTC 994936 (1476?); USTC 999899 (1477); USTC 997047 (1478); and USTC 993281 (1478). Germany: USTC 742826 (1472?); USTC 749219 (1473); and 749220 (1474).

5. Joannes Jacobi, *Regime de l'epidimie*, USTC 71138 (Lyon: Guillaume le Roy, 1476?).

6. The other two editions are Joannes, *Here begynneth a litill boke necessarye & behouefull agenst the pestilence*, STC 4590; Jacobi Joannes, *A passing gode lityll boke necessarye & behouefull agenst the pestilence*, trans. Bengt Knutsson, STC 4591 (London: W. de Machlinia, 1485).

7. Edward Gordon Duff, *The Printers, Stationers, and Bookbinders of Westminster and London from 1476 to 1535* (Cambridge: Cambridge University Press, 1906), 53. The printer Peter Schöffer of Mainz was the first to use a printed title page in 1463; see Garold Cole, "The Historical Development of the Title Page," *Journal of Library History (1966–1972)* 6, no. 4 (1971): 305.

8. On early printers' rivalries and alliances, see Atkin and Edwards, "Printers, Publishers and Promoters to 1558," 33–34.

9. Katherine Pantzer, Alfred W. Pollard, and G. R. Regrave, eds., *A Short-Title Catalogue of Books Printed in England, Scotland, and Ireland, and of English Books Printed Abroad, 1475–1640*, rev. and enlarged 2nd ed., 3 vols. (London: Bibliographical Society, 1976).

10. Roger Chartier was the first to describe how printers modeled their books "on the expectations and abilities attributed to the public at which they [were] aimed"; see *The Order of Books: Readers, Authors, and Libraries in Europe between the Fourteenth and Eighteenth Centuries*, 14.

11. Duff, *The Printers, Stationers, and Bookbinders of Westminster and London from 1476 to 1535*, 6.

12. Juliana Berners, *Here in thys boke afore ar contenyt the bokys of haukyng and huntyng*, STC 3308 (Saint Albans: s.n., 1486); *In this tretyse that is cleped gouernayle of helthe*, STC 12138 (Westminster: William Caxton, 1490).

13. N. F. Blake, "Worde, Wynkyn de (d. 1534/5)," in *ODNB*.

14. Juliana Berners, *This present boke shewyth the manere of hawkynge & huntynge*, STC 3309 (Westminster: Wyknyn de Worde, 1496); Juliana Berners, *The boke of hawkynge, and huntynge, and fysshynge*, STC 3309.5 (London: Wynkyn de Worde, 1518); John of Burgundy, *Here begynneth a lytell treatyse called the gouernall of helthe*, STC 12139 (London: Wynkyn de worde, 1506); Jacobi Joannes, *Here begynneth a treatyse agaynst pestelence & of ye infirmits*, STC 4592 and 4592.5 (London: Wynkyn de Worde, 1509 and 1511); *Proprytees & medicynes of hors*, STC 20439.5 and 20439.3 (Westminster and London: Wynkyn de Worde, 1497 and 1502); Walter de Henley, *Boke of husbandry*, STC 25007 (London: Wynkyn de Worde, 1508); *The crafte of graffynge & plantynge of trees*, STC 5953 (London: Wynkyn de Worde, 1518).

15. This text has been edited as David Scott-Macnab, *The Middle English Text of "The Art of Hunting" by William Twiti*, Middle English Texts 40 (Heidelberg: Winter, 2009).

16. A handlist of the fifty-nine manuscript witnesses to Lydgate's "Dietary" can be found in Linne R. Mooney et al., eds., *The DIMEV: An Open-Access, Digital Edition of the Index of Middle English Verse*, DIMEV no. 1356, accessed 24 June 2023, www

.dimev.net. The Middle English translation of John of Burgundy's dietary is found in Bodleian MS Ashmole 1498, ff. 51–56v, and BL MS Sloane 3215, ff. 2–17v.

17. That Middle English translation is in BL Sloane MS 404. The text has been edited by Joseph P. Pickett, "A Translation of the 'Canutus' Plague Treatise," in *Popular and Practical Science of Medieval England*, ed. Lister M. Matheson, Medieval Texts and Studies, No. 11 (East Lansing, MI: Colleagues Press, 1994), 263–82.

18. On the origins of the *Proprytees & medicynes of hors*, see Keiser, "Medicines for Horses: The Continuity from Script to Print." De Worde's treatise contains recipe texts similar to those in BL Sloane MS 686, ff. 49r–65v; Bodleian MSS Wood emptor 18, ff. 61r–79v; and Wood D.8, ff. 114r–128r.

19. De Henley's treatise was preserved in at least seven fifteenth-century manuscripts; see Oschinsky, "Medieval Treatises on Estate Management," 296–99.

20. David G. Cylkowski, "A Middle English Treatise on Horticulture: Godfridus Super Palladium," in *Popular and Practical Science of Medieval England*, ed. Lister M. Matheson, Medieval Texts and Studies, No. 11 (East Lansing, MI: Colleagues Press, 1994), 301–29.

21. Julia Boffey has shown that early printers routinely worked from manuscripts that circulated among the merchant networks of fifteenth-century London; see *Manuscript and Print in London, c. 1475–1530* (London: The British Library, 2012), 54–55.

22. Though we lack precise figures for most early print runs, 500 editions per run in the sixteenth century is a good estimation based on Eric Marshall White, "A Census of Print Runs for Fifteenth-Century Books," *Consortium for European Research Libraries* (blog), 2012, https://www.cerl.org/_media/resources/links_to_other_resources/printruns_intro.pdf.

23. Peter W. M. Blayney, *The Stationers' Company and the Printers of London, 1501–1537* (Cambridge: Cambridge University Press, 2013), 31–32.

24. Atkin and Edwards, "Printers, Publishers and Promoters to 1558," 30.

25. Blayney, *The Stationers' Company and the Printers of London, 1501–1537*, 29.

26. The longer and larger edition is the 1496 "Book of Saint Albans," STC 3309.

27. Dorne sold two copies of the *Medicine for hors* and one copy of the *Boke of husbandry*; see F. Madan, "Day-Book of John Dorne, Bookseller in Oxford, A.D. 1520," in *Collecteana*, ed. Charles Robert Leslie Fletcher, First Series 5 (Oxford: Oxford Historical Society at the Clarendon Press, 1885), 78–139. On wages in early sixteenth-century England, see Van Zanden, "Wages and the Cost of Living in Southern England (London), 1450–1700," http://www.iisg.nl/hpw/dover.php.

28. The payment is listed in Lady Margaret's account book, University of Cambridge, St. John's College Archives, SJLM/1/1/3/3, and quoted in Keiser, "Medicines for Horses: The Continuity From Script to Print," 116.

29. Martha W. Driver, "Ideas of Order: Wykyn de Worde and the Title Page," in *Texts and Their Contexts: Papers from the Early Book Society*, ed. V. J. Scattergood and Julia Boffey (Dublin: Four Courts Press, 1997), 87–149.

30. The Oxford bookseller Dorne was careful to mark which of his books were sold "*ligatum*," or bound, but none of the practical works he sold are marked as such; see Madan, "Day-Book of John Dorne, Bookseller in Oxford, A.D. 1520," 72–177.

31. John of Burgundy, *In this tretyse that is cleped gouernayle of helthe*, STC 12138, sig. A.i *b*; Joannes, *A passing gode lityll boke necessarye & behouefull agenst the pestilence*, STC 4591, unsigned fol. 3v; de Henley, *Boke of husbandry*, STC 25007, sig. A.i *b*.

32. Peter Blayney notes that under Henry VII, "the idea of exclusive rights had

apparently not yet infected the English book trade." See *The Stationers' Company and the Printers of London, 1501–1537*, 114.

33. Pamela Neville-Sington, "Pynson, Richard (c. 1449–1529/30)," in *ODNB*.

34. On the establishment of royal privilege, see Blayney, *The Stationers' Company and the Printers of London, 1501–1537*, 160–73.

35. Blayney, *The Stationers' Company and the Printers of London, 1501–1537*, 170.

36. John Fitzherbert, *Here begynneth a newe tracte or treatyse moost profitable for all husbandmen*, STC 10994 (London: Richard Pynson, 1523). Pynson had published at least two short almanacs with royal privilege prior to this publication.

37. Fitzherbert, *Here begynneth a newe tracte or treatyse moost profitable for all husbandmen*, fol. lx.

38. De Henley, *Boke of husbandry*, sigs. A.iv–A.iv *b*; Fitzherbert, *Here begynneth a newe tracte or treatyse moost profitable for all husbandmen*, sig. A.v.

39. Blayney, *The Stationers' Company and the Printers of London, 1501–1537*, 234. *The vertues & proprytes of herbes*, STC 13175.1 (London: Richard Banckes, 1525); *The seynge of urynes*, STC 22153 (London: [J. Rastell for] Richarde Banckes, 1525); *The treasure of pore men*, STC 24199 (London: Richard Bankes, 1526).

40. *The seynge of urynes*, sig. H.iii *b*. Sarah Neville also notes that Banckes included the abbreviated titles for each work in each book's signature line, which would have aided binders in combining all three texts in one compositive volume; see Sarah Neville, *Early Modern Herbals and the Book Trade: English Stationers and the Commodification of Botany* (Cambridge: Cambridge University Press, 2022), 139–40.

41. CUL Sel.5.175.

42. The contents of the book combined two Middle English uroscopy treatises and may have been based on British Library MS Sloane 382; see M. Teresa Tavormina, "Uroscopy in Middle English: A Guide to the Texts and Manuscripts," *Studies in Medieval and Renaissance History* 11 (2014): 19–22.

43. *The treasure of pore men*, fol. 10a; BL Sloane MS 372, f. 21r.

44. *The treasure of pore men*, fol. 10b; BL Arundel MS 272, f. 1v. The Arundel manuscript contains the recipe collection known by its incipit, "The man that wyll of lechecraft lere," and numerous recipes in Banckes's *Treasure of pore men* were drawn from that collection. See, Keiser, *MWME*, 3840.

45. *The treasure of pore men*, fol. 11b; BL Additional MS 12056 f. 12b.

46. *The treasure of pore men*, fol. 11a; Bodleian MS Ashmole 1438 I, p. 58.

47. On the emergence of organizational paratexts in the Latin Middle Ages, see Ann M. Blair, *Too Much to Know: Managing Scholarly Information before the Modern Age* (New Haven, CT: Yale University Press, 2010), 33–46.

48. BL Sloane MS 372, ff. 2r–14v.

49. BL MS Sloane 1315, f. 87v; BL MS Royal 18 A.vi, f. 64r; Wellcome Library MS 409, f. 144v.

50. *The treasure of pore men*, fol. 13a.

51. See Wear, *Knowledge and Practice in Early Modern English Medicine, 1550–1680*, 35. For an excellent overview of humanist printers and correctors, see Anthony Grafton, *Inky Fingers: The Making of Books in Early Modern Europe* (Cambridge, MA: Harvard University Press, 2020), 29–55.

52. Manutius did not always interpret ancient Greek correctly, either; see Anthony Grafton, *Defenders of the Text: The Traditions of Scholarship in an Age of Science, 1450–1800* (Cambridge, MA: Harvard University Press, 1991), 27–28.

53. For the locations of English print shops, see Appendices G and H in Blayney, *The Stationers' Company and the Printers of London, 1501–1537*, 968–1000. See also Peter W. M. Blayney, *The Bookshops in Paul's Cross Churchyard*, Occasional Papers of the Bibliographical Society 5 (London: The Bibliographical Society, 1990), 11.

54. Banckes continued to work as a publisher in the interim; see Blayney, *The Stationers' Company and the Printers of London, 1501–1537*, 180–81.

55. *A boke of the propertyes of herbes the which is called an Herbal*, STC nos. 13175.4 (John Scot, 1537); 13175.5 (Robert Redman, 1539); 13175.7 (Elisabeth Redman, 1541); 13175.8 (Thomas Petyt, 1541). On the publication history of Banckes's herbal, see Neville, *Early Modern Herbals and the Book Trade*, 134–40.

56. *The grete herbal*, STC 13176 (Southwark: Peter Treveris, 1526). Later editions are STC nos. 13177 (Peter Treveris, 1529); 13177.5 (Peter Treveris, 1529); and 13178 (Thomas Gibson, 1539).

57. Identical recipes involving "lye of vervayne" and one requiring its user to chew the "rote of pelater or Spain" are both in *This is the myrour or glasse of helth*, STC 18214a (London: Robert Wyer, 1531), sig. E.ii–E.ii *b*, as well as in *The treasure of pore men*, fol. 10–10 *b*. On the medieval sources of the plague tract, see Keiser, "Two Medieval Plague Treatises and Their Afterlife in Early Modern England."

58. Editions of Moulton's *This is the myrour or glasse of helthe* are STC nos. 18214a (Robert Wyer, 1531); 18214a.3 (Robert Wyer, 1536); 18214a.5 (Robert Wyer, 1536); 18214a.7 (Robert Wyer, 1536); 18216 (Robert Redman, 1540); 18225.2 (Robert Wyer, 1540); 18219 (Elisabeth Redman, 1541); 18220 (William Middleton, 1545); 18225.4 (Thomas Petyt, 1545). Editions of *The antidotharius* are STC nos. 675.3 (Wyer, 1535); 675.7 (Wyer, 1535?).

59. On just one page of *The antidotharius*, Avicenna, William of Saliceto, Lanfranc of Milan, Henry de Mondeville, and Guy de Chauliac are all cited as authorities; see *The antidotharius*, STC 675.7 (London: Robert Wyer, 1535), sigs. B.iii *a–b*.

60. These editions are STC nos. 24200 (Robert Redman, 1539); 24201 (Thomas Petyt, 1539); 24202 (Robert Redman, 1540); 24202.5 (Thomas Petyt, 1540).

61. See the Digital Appendices for a database of practical books published in England between 1485 and 1600, available at https://readingpractice.github.io.

62. Blayney, *The Stationers' Company and the Printers of London, 1501–1537*, 100–101.

63. On the breakdown of De Worde's total output of around 850 editions, see Atkin and Edwards, "Printers, Publishers and Promoters to 1558," 30, 32.

64. Blayney, *The Stationers' Company and the Printers of London, 1501–1537*, 257.

65. Paul Slack first noted the derivative quality of English vernacular medical print in "Mirrors of Health and Treasures of Poor Men: The Uses of the Vernacular Medical Literature of Tudor England," in *Health, Medicine and Mortality in the Sixteenth Century*, ed. Charles Webster (Cambridge: Cambridge University Press, 1979), 240.

66. For a list of these titles and a network analysis showing how printers were connected to one another through their publication of the same practical texts, see the Digital Appendices, https://readingpractice.github.io.

67. Neville, *Early Modern Herbals and the Book Trade*, 101, 138.

68. Quoted from the original proclamation, transcribed in Blayney, *The Stationers' Company and the Printers of London, 1501–1537*, 482.

69. Geoffrey Eatough, "Paynell, Thomas (d. 1564?)," in *ODNB*.

70. *Regimen sanitatis Salerni*, STC 21596 (London: Thomas Berthelet, 1528). Subsequent editions are STC nos. 21597 (1530), 21598 (1535), and 21599 (1541).

71. Correspondence between Berthelet and Paynell is discussed and quoted in Cavallo and Storey, "Regimens, Authors and Readers," 26.

72. Ullrich von Hutten, *De morbo gallico*, trans. Thomas Paynell, STC 14024 (London: Thomas Berthelet, 1533). Later editions were published under the title *Of the wood called guaiacum* and are STC nos. 14025 (1536), 14026.5 (1539), 14026 (1539), and 14027 (1540).

73. On his translation of the Galenic corpus, see Vivian Nutton, "Linacre, Thomas (c. 1460–1524), Humanist Scholar and Physician," in *ODNB*. Berthelet's editions of *The castell of health* are STC nos. 7642.5 (1537), 7643 (1539), 7642.7 (1539), 7644 (1541), 7645 (1541), 7646 (1544), 7646.5 (1547), 7647 (1550).

74. By contrast, Berthelet published just thirty-four practical books over a career that spanned from 1528 to 1557. For the titles of these editions, see the Digital Appendices, https://readingpractice.github.io.

75. Wyer reissued this same altered version again in 1545 (STC 18225.4), 1547 (STC 18225.6), and 1555 (STC 18225.8). On the publication history of the Moulton plague treatise, see Keiser, "Two Medieval Plague Treatises and Their Afterlife in Early Modern England," 309–10.

76. Sarah Neville has suggested that the four printers who produced those editions may well have joined together to share the printing expenses of a run of several hundred books, which they then each fitted with their own title pages and sold at their own shops; see Neville, *Early Modern Herbals and the Book Trade*, 138. Their editions are STC nos. 13175.4 (John Skot, 1537), 13175.5 (Robert Redman, 1539), 13175.7 (Elizabeth Redman, 1541), and 13175.8 (Thomas Petyt).

77. Richard Banckes and Robert Wyer, *Hereafter foloweth the knowledge, properties, and the vertues of herbes*, STC 13175.6 (London: Robert Wyer, 1540).

78. For criticism of Wyer's edition, see Francis R. Johnson, "'A Newe Herball of Macer' and Banckes's 'Herball': Notes on Robert Wyer and the Printing of Cheap Handbooks of Science in the Sixteenth Century," *Bulletin of the History of Medicine* 15, no. 3 (March 1944): 246–60. For the rebuttal, see Neville, *Early Modern Herbals and the Book Trade*, 143–46.

79. Banckes, *The vertues & proprytes of herbes*, sig. A.i *b*.

80. Banckes and Wyer, *Hereafter foloweth the knowledge, properties, and the vertues of herbes*, sig. A.i *b*, and BL MS Sloane 2460, f. 2r. I do not mean to suggest that this manuscript was necessarily Wyer's source, but rather that the existence of this manuscript witness suggests Wyer was working from a manuscript within this family of the *Agnus castus*.

81. *A newe herball of Macer, translated out of Laten in to Englysshe*, STC 13175.8c (London: Robert Wyer, 1543).

82. That prologue to the Middle English *De viribus herbarum* is found in Bodleian MS Rawlinson C.81, f. 18r, while the Middle English version in CUL MS Ee.1.15 is simply titled "Macer."

83. Neville makes this suggestion in *Early Modern Herbals and the Book Trade*, 149.

84. *A lytel herball of the properties of herbes newly amended and corrected*, STC 13175.13 (London: William Powell, 1550).

85. None of the surviving copies of Powell's edition contain the "certain additions at the back of the boke," but Askham apparently authored a treatise on herbs and astrology in the same year, which may have been what Powell intended to add to his edition; see Neville, *Early Modern Herbals and the Book Trade*, 153–54.

86. *Macers herbal: Practysyd by Doctor Lynacro*, STC 13175.13c (London: Robert Wyer, 1552).

87. Spurious authorial attributions using famous names were common in early print; see Ann Blair, "Authorship in the Popular 'Problemata Aristotelis,'" *ESM Early Science and Medicine* 4, no. 3 (1999): 189–227.

88. *Hereafter foloweth the iudgement of all vrynes*, STC 14834 (London: Robert Wyer, 1555).

89. Data collected from the English Short Title Catalogue, http://estc.bl.uk, accessed 30 September 2021.

90. William Turner, *A new herbal*, STC 14365 (London: Steven Mierdman, 1551), sig. A.iii. On Turner's use of Fuchs's images from *De historia stirpium*, see Brent Elliott, "The World of the Renaissance Herbal," *Renaissance Studies* 25, no. 1 (2011): 34.

91. Andreas Vesalius and Thomas Gemini, *Compendiosa totius anatomie delineatio, ære exarata: per Thomam Geminum*, trans. Nicholas Udall, STC 11716 (London: Nicholas Hill, 1553). Gemini's English edition draws its text from Thomas Vicary's *A profitable Treatise of the Anatomie of Man's body* (1548), which was itself based on the anatomy of thirteenth-century author Henry de Mondeville; see Sanford V. Larkey, "The Vesalian Compendium of Geminus and Nicholas Udall's Translation: Their Relation to Vesalius, Caius, Vicary, and De Mondeville," *The Library* s4-XIII, no. 4 (1933): 374–78.

92. On Vesalius's and Fuchs's illustrations in print, see Kusukawa, *Picturing the Book of Nature*.

93. "Anglia nunc ostendit, ubi figuras Epitomes meae, adeo obscure & citra picturae artificium, non tamen absque sumptu, quisquis tandem illum tulerit in aere, sunt imitavi: ut pudeat aliquem arbitrari, a me ita exiisse," in Andreas Vesalius, *Radicis chynae Usus*, USTC 149871 (Lyon: excudebat Jean Frellon, 1547), 283; translated and quoted in Larkey, "The Vesalian Compendium of Geminus and Nicholas Udall's Translation: Their Relation to Vesalius, Caius, Vicary, and De Mondeville," 372.

94. Johns, *The Nature of the Book*, 6–11, 34–40.

95. Edward Arrer, ed., *A Transcript of the Registers of the Company of Stationers of London. Electronic Reproduction*, vol. 1: 1554–1640: (New York: Columbia University Press, 2007), xxiv.

96. On the workings of the Stationers' guild, see Johns, *The Nature of the Book*, 200–230; Adrian Johns, *Piracy: The Intellectual Property Wars from Gutenberg to Gates* (Chicago: University of Chicago Press, 2009), 24–27.

97. Johns, *The Nature of the Book*, 54–56.

98. Arrer, *A Transcript of the Registers of the Company of Stationers of London*, ff. 24, 61.

99. King's edition was not a reprint of De Worde's earlier editions but was rather drawn from a manuscript source containing the version of this text closest to Bodleian MS Rawlinson C.506, ff. 287r–297v, and CUL MS Ll.1.18, ff. 64r–79v; on this, see Keiser, "Medicines for Horses: The Continuity From Script to Print," 120–21.

100. *A treatyse: contaynynge the orygynall causes, and occasions of the diseases, growynge on horses*, STC 24237.5 (London: John King, 1560).

101. *A litle herball of the properties of herbes, newly amended & corrected*, STC 13175.19 (London: John King, 1561); *The greate herball*, STC 13179 (London: John King, 1561).

102. *Medicines for horses*, STC 20439.7 (London: W. Copeland, 1565). This was in fact a reedition of De Worde's version and not King's.

103. *A booke of the properties of herbs. Also a general rule . . . by W.C.*, STC 13175.19c (London: John Awdely?, 1567). This edition was the revised text edited by William Copeland in 1552, and not King's version.

104. More's response to the sweat in a letter to Erasmus is excerpted in Flood, "'Safer on the Battlefield Than in the City': England, the 'Sweating Sickness,' and the Continent," 147–51, quoted at 147.

105. John Caius, *A boke, or counseill against the disease commonly called the sweate, or sweatyng sicknesse*, STC 4343 (London: Richard Grafton, 1552), fol. 39a.

106. Caius, *A boke, or counseill against the disease commonly called the sweate, or sweatyng sicknesse*, fol. 4b, 8b.

Chapter 5

1. On Barton and her attainder for treason, see Diane Watt, "Reconstructing the Word: The Political Prophecies of Elizabeth Barton (1506–1534)," *Renaissance Quarterly* 50, no. 1 (1997): 136–63.

2. TNA SP 1/58, f. 89; transcribed and abridged in J. S. Brewer, ed., "6652: Prophecies," in *Letters and Papers, Foreign and Domestic, of the Reign of Henry VIII*, vol. 4: Part III: 1529–30 (London: Longman, 1876), 2997.

3. An excellent overview of these events is Haigh, *English Reformations: Religion, Politics, and Society under the Tudors*, 88–102.

4. On the origins of the prognostication in the early Middle Ages, see Chardonnens, *Anglo-Saxon Prognostics, 900–1100*, 43–50.

5. Though the almanac, *Planeten-Tafel, sive Ephemerides 1448*, USTC 741136 (Mainz: Type of the 36-line Bible, 1458), is dated on its title page at least five years before the first Gutenberg Bible, bibliographers date its publication to around 1458.

6. Regiomontanus, *Ephemerides, 1475–1506*, USTC 748421 (Nuremberg: Johann Müller of Königsberg, 1474).

7. Regiomontanus, *Kalendarium*, USTC 748412 (Nuremberg: Johann Müller of Königsberg, 1474). On "physician's almanacs," see Carey, "What Is the Folded Almanac?"

8. The earliest surviving prognostications published in England are two fragments of an English translation of William Parron's prognostication for 1498 (STC nos. 385.3 and 385.7), which Bernard Capp suggests may have been translated by Parron himself; see *English Almanacs, 1500–1800: Astrology and the Popular Press* (Ithaca, NY: Cornell University Press, 1979), 26. On English court astrology in the fourteenth and earlier fifteenth centuries, see Carey, *Courting Disaster: Astrology at the English Court and University in the Later Middle Ages*.

9. Hillary Carey gives an excellent summary of the range of practices encompassed within medical astrology, and the sources required for that range, in "Astrological Medicine and the Medieval English Folded Almanac."

10. On the rediscovery of Ptolemy in Greek, see Ornella Faracovi, "The Return to Ptolemy," in Brendan Dooley, ed., *A Companion to Astrology in the Renaissance*, Brill's Companions to the Christian Tradition 49 (Leiden: Brill, 2014), 87–98. The literature on the Hermetic corpus is vast, but a good starting point is Frances Amelia Yates, *Giordano Bruno and the Hermetic Tradition* (London: Routledge and Kegan Paul, 1964), and Brian P. Copenhaver, ed., *Hermetica: The Greek Corpus Hermeticum and the*

Latin Asclepius in a New English Translation, with Notes and Introduction (Cambridge: Cambridge University Press, 1992).

11. Most famous of these critics of judicial astrology was the scholar Giovanni Pico della Mirandola; see Robert Westman, *The Copernican Question: Prognostication, Skepticism, and Celestial Order* (Berkeley: University of California Press, 2011), 82–87.

12. On astrology at the Milanese court, see Monica Azzolini, *The Duke and the Stars Astrology and Politics in Renaissance Milan*, I Tatti Studies in Italian Renaissance History (Boston: Harvard University Press, 2013).

13. Parron commissioned at least two lavish manuscripts of astrological material as gifts for Henry: Bodleian Library MS Selden Supra 77 and British Library MS Royal 12 B.vi; see Hilary M. Carey, "Henry VII's Book of Astrology and the Tudor Renaissance," *Renaissance Quarterly* 65, no. 3 (2012): 661–710.

14. William Parron, *A prognostication for 1498*, STC 385.3 (Westminster: Wynkyn de Worde, 1498), unsigned first page.

15. Martha Carlin, "Parron, William (b. before 1461, d. 1518)," in *ODNB*.

16. Jasper Laet, *The pnostication of Maister Jasp Laet, practised in the towne of Anterpe, for the yere of Our Lorde, M.D.XX*, STC 470.6 (London: Richard Pynson, 1520), unsigned.

17. William Rede, *Almanach ephemerides in anno domini m. d. vii.*, STC 504 (London: R. Pynson and Rouen, R. Mace, 1507).

18. *Almanacke for xii. yere*, STC 387 (London: Wynkyn de Worde, 1508).

19. The sixteenth-century almanac in Bodleian MS Ashmole 340 is ff. 1–24. Additional early sixteenth-century vernacular manuscripts created as guidebooks to judicial astrology are Bodleian MS Ashmole 210, ff. 16–43, and Bodleian MS Ashmole 349.

20. Madan, "Day-Book of John Dorne, Bookseller in Oxford, A.D. 1520," 78–139.

21. On printed almanacs and readers' shifting sense of time, see Anne Lawrence-Mathers, "Domesticating the Calendar: The Hours and the Almanac in Tudor England," in *Women and Writing, c. 1340–1650: The Domestication of Print Culture*, ed. Anne Lawrence-Mathers and Phillipa Hardman (Woodbridge, Suffolk: Boydell & Brewer, 2010), 34–61; Allison A. Chapman, "Marking Time: Astrology, Almanacs, and English Protestantism," *Renaissance Quarterly* 60, no. 4: 1257–90; Linne R. Mooney, "English Almanacs from Script to Print," in *Texts and Their Contexts: Papers from the Early Book Society*, ed. V. J. Scattergood and Julia Boffey (Dublin: Four Courts Press, 1997), 11–25.

22. Eisenstein, *The Printing Revolution in Early Modern Europe*, 220.

23. On woodcut illustration in early English print, see Martha W. Driver, *The Image in Print: Book Illustration in Late Medieval England and Its Sources* (London: The British Library, 2004).

24. Perhaps the most well-known examples of European blockbooks are *Biblia Pauperum*, examples of which are Library of Congress Incun. X .B562 (Netherlands or Germany, 1470); *Biblia Pauperum*, Princeton University Scheide Library S2.9 (Nuremburg: Hans Sporer, 1471). East Asian printing, which predated European printing by several centuries, was always done with woodblocks; on this, see Kai-Wing Chow, "Reinventing Gutenberg: Woodblock and Moveable-Type Printing in Europe and China," in *Agent of Change: Print Culture Studies after Elizabeth L. Eisenstein*, ed. Eric N. Lindquist, Eleanor F. Shevlin, and Sabrina Alcorn Baron (Amherst: University of Massachusetts Press, 2007), 169–92.

25. Joannes Regiomontanus and Hans Biefftruck, *Calendarium*, Library of Congress Incun. 1474.M82 (Nuremberg: Sold by Hans Biefftruck?, 1474).

26. Edward Hodnett, *English Woodcuts, 1480–1535* (Oxford: Oxford University Press, 1973), 75–76; Edward Hodnett, *English Woodcuts, 1480–1535: Additions & Corrections* (London: The Bibliographical Society, 1973), 1.

27. *Here begynneth the kalender of shepherdes*, STC 22408 (London: Richard Pynson, 1506), sig. K.ii.

28. *Le compost et kalendrier des bergiers* was published in seven different editions in France before the year 1500: USTC nos. 70063–67, 95900 (Paris: Guy Marchant, 1491–1497); 89882 (Lyon: Guillaume Balsarin, 1499).

29. *The kalendayr of the shyppars*, trans. Alexander Barclay, STC 22407 (Paris: Antoine Vérard, 1503).

30. *Here begynneth the kalender of shepherdes*, sig. A.ii.

31. Hodnett, *English Woodcuts, 1480–1535*, 39–40. For a description of several of the English-printed volumes with woodcuts drawn from Vérard's originals, see Martha W. Driver, "Woodcuts and Decorative Techniques," in *A Companion to the Early Printed Book in Britain, 1476–1558*, ed. Vincent Gillespie and Susan Powell (Woodbridge, Suffolk: D. S. Brewer, 2014), 99–106.

32. *The kalender of shepeherdes*, ed. Wynkyn de Worde and Robert Copeland, STC 22409 (London: Wynkyn de Worde, 1516). Blayney speculates that the surviving 1516 edition is actually a reprint of a lost 1508 edition, given that 1508 appears in the colophon; see Blayney, *The Stationers' Company and the Printers of London, 1501–1537*, 125.

33. *The kalender of shepeherdes*, STC 22409, sigs. A.iv *b*–C.iii, K.vii *b*, M.viii *b*, N.ii *b*.

34. Pynson reprinted his folio edition in 1510 and 1517 (STC nos. 22409.3 and 22409.7); De Worde reprinted his quarto edition in 1516 and 1528 (STC nos. 22409.5 and 22411).

35. *Here begynneth the kalender of shepherdes*, sig. A.iii.

36. Natalie Zemon Davis notes a similarly elite readership for French editions of *Le compost et kalendrier des bergiers*; see *Society and Culture in Early Modern France* (Stanford, CA: Stanford University Press, 1975), 191.

37. Bodleian Library Douce A 632 (dateable to 1522); British Library C.36.a.5 (dateable to 1537); British Library C.41.a.28 (dateable to 1538); and British Library C.29.c.6 (dateable to 1542).

38. The woodblocks were likely produced in England, as each calendar contains English saints and a pictorial history with events only meaningful in England, like the martyrdom of Thomas Becket.

39. *To them that before this image of pity devoutly say v. pater noster, v. aves & a credo piteously* survives in twenty-seven copies (STC nos. 14077c.6–.23B), dating from 1487 to 1534, which may in fact represent hundreds, if not thousands, of printed copies that are now lost. See Tessa Watt's discussion of these icons in Watt, *Cheap Print and Popular Piety, 1550–1640*, 131–33; see also Hodnett, *English Woodcuts, 1480–1535*, nos. 350, 381, 390, 454, 459, 568, 1374, 2016, 2024, 2380, 2039, 2062, and 2498; Hodnett, *English Woodcuts, 1480–1535: Additions & Corrections*, nos. 2507, 2508, and 2513.

40. *Dives & Pauper*, sig. a.vii–a. vii *b*.

41. Aston, *England's Iconoclasts: Laws Against Images*, 160–73.

42. Sir Thomas More, *A dyaloge of syr Thomas More knyghte: one of the counsayll of our souerayne lorde the kyng and chauncelloure of hys duchy of Lancaster*, STC 18085 (London: Printed by William Rastell, 1530).

43. See, for example, William Tyndale's condemnation of the Gregorian dictum regarding images in *An answere vnto Sir Thomas Mores dialoge made by Willyam Tindale*, STC 24437 (Antwerp: S. Cock, 1531), sig. C.xiiii.

44. Duffy, *The Stripping of the Altars: Traditional Religion in England, 1400–1580*, 381–83.

45. Martin Bucer, *A treatise declaryng & shewig dyuers causes taken out of the holy scriptures of the sentences of holy faders & of the decrees of deuout emperours*, trans. William Marshal, STC 24239 (London: T. Godfray for William Marshal, 1535).

46. David Wilkins, ed., *Concilia Magnae Britanniae et Hiberniae, ab Anno MCCCL ad Annum MDXLV. Volumen Tertium* (London: R. Gosling, F. Gyles, T. Woodward, & C. Davis, 1737), 821–24; Duffy, *The Stripping of the Altars: Traditional Religion in England, 1400–1580*, 391–95.

47. Church of England, *The institvtion of a Christen man*, STC 5163 (London: Thomas Berthelet, 1537).

48. Wilkins, *Concilia Magnae Britanniae et Hiberniae*, 816; Duffy, *The Stripping of the Altars: Traditional Religion in England, 1400–1580*, 401–7.

49. The Six Articles of 1539 did not expressly treat the use of images, but one of the manuscript copies of the *Rationale of Ceremonial* (1540) includes a passage on "the Right use of Images" that describes them as "unlearned men's books." See Cyril S. Cobb, ed., *The Rationale of Ceremonial, 1540–1543, with Notes and Appendices and an Essay on the Regulation of Ceremonial during the Reign of King Henry VIII* (London: Longmans, Green & Co., 1910), 44–45; "31 Henry VIII c. 14: An Acte abolishing diversity in Opynions," in *SOTR vol.* 3, 739–43; Duffy, *The Stripping of the Altars: Traditional Religion in England, 1400–1580*, 427–29.

50. BL C.36.aa.5 (dateable to 1537); BL C.41.a.28 (dateable to 1538); and BL C.29.c.6 (dateable to 1542).

51. Thanks to the dates added to its pictorial history, BL MS Additional 17367, sec. 10v, we know that the manuscript was created in 1535.

52. See BL Additional MS 17367, sec. 10v.

53. W. H. Frere and W. P. M. Kennedy, eds., *Visitation Articles and Injunctions of the Period of the Reformation, Vol. II: 1536–1558* (London: Longmans, Green & Co., 1910), 126, no. 28; see also Aston, *England's Iconoclasts*, 254–63.

54. Frere and Kennedy, *Visitation Articles and Injunctions of the Period of the Reformation, Vol. II: 1536–1558*, 107, no. 34.

55. Duffy, *The Stripping of the Altars: Traditional Religion in England, 1400–1580*, 450–70. See "3 & 4 Edward VI c. 10: An Acte for the abolishinge and puttinge awaye of diverse Bookes and Images," in *SOTR vol.* 4, 110–11.

56. "5 & 6 Edward VI c. 3: An Acte for the Keping of Hollie Daies and Fastinge Dayes," in *SOTR vol.* 4, 132–33.

57. The English church did replace some of the old feast days with new Protestant festivals; see David Cressy, *Bonfires and Bells: National Memory and the Protestant Calendar in Elizabethan and Stuart England* (London: Weidenfeld and Nicolson, 1989).

58. Andrew Boorde, *The pryncyples of astronamye*, STC 3386 (London: Robert Coplande, 1547), unsigned second page.

59. Boorde, *The pryncyples of astronamye*, unsigned second page.

60. On Melancthon's astrology, see Westman, *The Copernican Question*, 110–13; Robin B. Barnes, *Astrology and Reformation* (New York: Oxford University Press, 2015), 131–37.

61. Barnes, *Astrology and Reformation*, 49–55.

62. Quoted and translated in Aby Warburg, *The Renewal of Pagan Antiquity: Contributions to the Cultural History of the European Renaissance*, trans. David Britt, Getty Research Publications (Los Angeles: Getty Research Institute, 1999), 655.

63. On the Lutheran doctrine of the "priesthood of all believers," see B. A. Gerrish, "Priesthood and Ministry in the Theology of Luther," *Church History* 34, no. 4 (December 1965): 404–22.

64. See the "Bishop's Book," or *The institvtion of a Christen man*, sig. D. iv.

65. The original German work is Otto Brunfels, *Almanach ewig werend, Teütszch vnd Christlich Practick*, USTC 610939 (Straßburg: Prüß d.J., 1526).

66. Otto Brunfels, *A very true pronosticacion, with a kalender*, trans. John Ryckes, STC 421.17 (London: John Byddell, 1536), sig. A.i *b*, discussed in Blayney, *The Stationers' Company and the Printers of London, 1501–1537*, 490.

67. Brunfels, *A very true pronosticacion, with a kalender*, sig. A. i *b*–A.ii.

68. Brunfels, *A very true pronosticacion, with a kalender*, sig. B.iii *b*.

69. *A faythfull and true pronostication vpon the yere .M.CCCCC.xlviii*, trans. Miles Coverdale, STC 20423 (London: [J. Hereford] for Richard Kele, 1547?), sig. A.ii *b*–A.iii.

70. Boorde, *The pryncyples of astronamye*, unsigned second page.

71. *Calendar of the Patent Rolls Preserved in the Public Record Office. Philip and Mary, Vol. I: 1553–1554* (London: His Majesty's Stationery Office, 1937), 261, 266.

72. John Proctor, *The historie of Wyates rebellion with the order and maner of resisting the same*, STC 20407 (London: Robert Calye, 1554).

73. *Calendar of the Patent Rolls Preserved in the Public Record Office. Philip and Mary, Vol. II: 1554–1555* (London: His Majesty's Stationery Office, 1936), 270.

74. Leonard Digges, *A prognostication of right good effect*, STC 435.35 (London: Thomas Gemini, 1555).

75. Digges, *A prognostication of right good effect*, sig. *.iii *a*.

76. Capp, *English Almanacs, 1500–1800: Astrology and the Popular Press*, 28–29.

77. Andrew Boorde, *A pronostycacyon or an almanacke for the yere of our lorde, M. CCCCC. xlv*, STC 416.5 (London: s.n., 1545). Though now lost, Boorde mentions other prognostications published by Richard Copeland in *The pryncyples of astronamye*, unsigned second page.

78. Askham's guide to astrological medicine is *A litell treatyse of astrouomy*, STC 857a.5 (London: William Powell, 1550). His almanacs are STC nos. 410 (William Powell, 1548); 410.1 (William Powell, 1549); 410.2 and 410.3 (William Powell, 1551); 410.4 (William Powell for Richard Kele, 1552); 410.5 (William Powell, 1553); 410.6 (William Powell, 1554); 410.7 (Thomas Marshe, 1555); 410. 8 (William Powell, 1555); 410.9, 410.10, and 410.11 (Thomas Marshe, 1556); 410.12 (Thomas Marshe, 1557).

79. *The pronostycacyon for euer of Erra Pater: A jewe borne in jewery, a doctour in astronomye, and physycke*, STC 439.3 (London: Robert Wyer, 1540), sig. B.iv *b*–B.v.

80. The other eight sixteenth-century editions of *The pronostycacyon for euer of Erra Pater* are STC 339.5 (Wyer, 1545), 439.7 (Wyer, 1550), 439.9 (Wyer, 1552), 439.11 (Wyer, 1554), 439.13 (Wyer, 1555), 439.15 (Thomas Colwell, 1562), 439.17 (Thomas East, 1581), and 439.18 (Thomas East, 1598).

81. Digges, *A prognostication of right good effect*, sig. *.iv *a*–*b*.

82. Digges, *A prognostication of right good effect*, sig. *.iii *b*.

83. Anthony Askham, *An almanacke and prognostication, for the yere of our Lorde God M. D.L.V.*, STC 410.8 (London: William Powell, 1555), sig. A.v *b*.

84. Askham, *An almanacke and prognostication, for the yere of our Lorde God M. D.L.V.*, sig. A.iii *b*.

85. Digges, *A prognostication of right good effect*, sig. *.iii *a*–*b*.

86. Digges, *A prognostication of right good effect*, sig. *.iv *b*.

87. Leonard Digges, *A prognostication euerlasting of ryght good effecte*, STC 435.37 (London: Thomas Gemini, 1556).

88. Gemini advertised his instrument-making abilities in another one of Digges's titles, *A book named Tectonicon*, published in 1562; see Peter Murray Jones, "Gemini [Geminus, Lambrit], Thomas (fl. 1540–1562), Engraver, Printer, and Instrument Maker," in *ODNB*.

89. Digges, *A prognostication of right good effect*, sig. B.ii *b*.

90. Digges, *A prognostication of right good effect*, sig. D.i.

91. Digges, *A prognostication of right good effect*, sig. C.iii *b*–C.iv.

92. J.A., *A Perfyte pronostycacion perpetuall*, STC 406.3 (London: Robert Wyer, 1556). The annual predictions in J.A.'s prognostication do not match those in Wyer's editions of the *Erra pater*, but are rather much closer to those in Digges's *A prognostication of right good effect*.

93. J.A., *A Perfyte pronostycacion perpetuall*, sig. A.i and A.ii *b*.

94. Brewer, "6652: Prophecies."

95. Aston, *England's Iconoclasts*, 254–63; Duffy, *The Stripping of the Altars: Traditional Religion in England, 1400–1580*, 450–81.

96. Eisenstein, *The Printing Revolution in Early Modern Europe*, 24–25.

97. Patrick Collinson, *From Iconoclasm to Iconophobia: The Cultural Impact of the Second English Reformation* (Reading: University of Reading Press, 1986).

98. Digges, *A prognostication of right good effect*, sig. *.iii.

99. Leonard Digges, Nicolaus Copernicus, and Thomas Digges, *A prognostication everlasting of right good effecte: fruitfully augmented by the auctour*, STC 435.47 (London: Thomas Marshe, 1576), sig. M. 1–O.3.

100. Digges, Copernicus, and Digges, *A prognostication euerlastinge of right good effecte*, sig. M. 1.

Chapter 6

1. *The gospelles of dystaves*, STC 12091 (London: Wynkyn de Worde, 1510).

2. *Le traittie intitule les evangiles des quenouilles faittes a l'onneur et exaucement des dames*, USTC 70278 (Bruges: Colard Mansion, 1479), based on Paris, Bibliothèque nationale de France fr. 2151. On Mansion's career as a printer, see Paul Saenger, "Colard Mansion and the Evolution of the Printed Book," *The Library Quarterly* 45, no. 4 (October 1975): 409–10. On the manuscript tradition of the "distaff gospels," see Madeleine Jeay and Kathleen Garay, *The Distaff Gospels: A First Modern English Edition of Les Évangiles Des Quenouilles* (Peterborough, ON: Broadview Press, 2006), 19–20, 23–25.

3. The quarto editions from the press of Matthias Huss in Lyon are USTC nos. 765438 (1482–83); 70279 (1482); 765608 (1485–87); 765609 (1485–87); and 767229 (1498). The sedecimo edition from Jean Mareschal (1493) is USTC 63527. On later

translations of the "distaff gospels," see Kathleen Loysen, *Conversation and Storytelling in Fifteenth- and Sixteenth-Century French Nouvelles* (New York: Peter Lang, 2004), 19.

4. Suzanne Hull determined that only five titles printed in the first century of English print (1475–1575) even claimed to be for women readers, and all of them—including *The gospelles of dystaves*—were authored by men; see *Chaste, Silent, & Obedient: English Books for Women, 1475–1640* (San Marino, CA: Huntington Library Press, 1982), 7.

5. Cressy, *Literacy and the Social Order: Reading & Writing in Tudor & Stuart England,* 115–19.

6. *The gospelles of dystaves,* sig. A.ii–iii, A.iv, A.v *b*, A.vi *b*, B.ii *b*, C.i, C.viii *b*, D.iv, E.iv–iv *b*.

7. Huntington MS HM 58, f. 52r, contains a recipe "For to knowe wether a child in a womans womb be a knave or a mayden child," as does Griffiths, Edwards, and Barker, *The Tollemache Book of Secrets,* 48.

8. Digges, *A prognostication of right good effect,* sig. B.ii *b*.

9. John Partridge, *The treasurie of commodious conceits, & hidden secrets and may be called, the huswiues closet, of healthfull prouision,* STC 19425.5 (London: Richard Jones, 1573).

10. W. H. Frere and W. P. M. Kennedy, eds., *Visitation Articles and Injunctions of the Period of the Reformation, Vol. III: 1559–1575* (London: Longmans, Green & Co., 1910), 5, 383.

11. *The boke of secretes of Albertus Magnus,* STC 258.5 (London: J. King, 1560).

12. For example, BL MS Sloane 2584 contains the "holy mothers" charm (ff. 25v–26r) and BL MS Sloane 1315 contains charms to help a woman conceive (f. 93v, 94v) and a charm "To cawse a man to love hys wyffe and the woman to love her howsbande" (f. 94r). For more examples, see Peter Murray Jones and Lea T. Olsan, "Performative Rituals for Conception and Childbirth in England, 900–1500," *Bulletin of the History of Medicine* 89, no. 3 (October 27, 2015): 406–33.

13. The eleventh-century codification of church law known as the *Decretum* characterized simple magic like charms and rituals as "illas vanitates aut consensisti quas stultae mulieres facere solent"; see Burchard of Worms, *Burchardi Wormanciensis Ecclesiae Episcopi Decretorum Libri* (Paris: Joannem Foucherium, 1549), Liber XIX, Cap. 5, cols. 964A–965C.

14. On the scarcity and leniency of prosecutions for common magic in fifteenth-century England, see Karen Jones and Michael Zell, "'The Divels Speciall Instruments': Women and Witchcraft before the 'Great Witch-Hunt,'" *Social History* 30, no. 1 (2005): 51–52.

15. "Prohibitum est exercitium ab ecclesia catholica," Huntington MS HM 58, f. 84r. The two charms are much abridged versions of the popular *Maria peperit* charm identified by Lea Olsan in eleven other fifteenth-century medical collections; see Olsan, "The Corpus of Charms in the Middle English Leechcraft Remedy Books," 219–20, 226.

16. Hans Peter Broedel, *The Malleus Maleficarum and the Construction of Witchcraft: Theology and Popular Belief* (Manchester: Manchester University Press, 2003), 177–79.

17. Michael David Bailey, "The Feminization of Magic and the Emerging Idea of the Female Witch in the Late Middle Ages," *Essays in Medieval Studies* 19, no. 1 (2002): 120–34.

18. Stuart Clark, "The 'Gendering' of Witchcraft in French Demonology: Misogyny or Polarity?," *French History* 5, no. 4 (December 1, 1991): 426–37.

19. For example, the *Decretum* recommends fasting as penance for performing charms and healing rituals; see Burchard of Worms, *Burchardi Wormanciensis Ecclesiae Episcopi Decretorum Libri*, Liber XIX, Cap. 5, cols. 964A–965C.

20. The writ, dated January 2, 1406, is summarized in *Calendar of the Patent Rolls Preserved in the Public Record Office: Henry IV, Vol. III, 1405–1408* (London: Mackie and Co, 1907), 112. The original Latin is quoted in Jones and Zell, "'The Divels Speciall Instruments,'" 48.

21. William Hale, ed., *A Series of Precedents and Proceedings in Criminal Causes Extending from the Year 1475 to 1640: Extracted from Act-books of Ecclesiastical Courts in the Diosese of London, Illustrative of the Discipline of the Church of England* (London: Francis & John Rivington, 1847), 3.

22. On the origins of the fifteenth-century "witch craze," see Richard Kieckhefer, *European Witch Trials: Their Foundations in Popular and Learned Culture, 1300–1500* (Berkeley: University of California Press, 1976).

23. BL MS Additional 34210, f. 45r, 47r.

24. BL MS Sloane 382, ff. 164v.

25. "Mugworth and modywort as is one Ipocras seiþ it is hoot & drye in þre degres . . . Also ȝif a childe be ded in his moder wombe take mugwort [illegible] it smale & make a plaster and leye to þe wombe al colde an be þe grace off god sche schal have delyveraunce"; BL MS Additional 12056, f 3r.

26. Montserrat Cabré and Fernando Salmón, "Blood, Milk, and Breastbleeding: The Humoral Economy of Women's Bodies in Medieval Medicine," in *Gender, Health, and Healing, 1250–1550*, ed. Sara Ritchey and Sharon Strocchia (Amsterdam: Amsterdam University Press, 2020), 93–117.

27. On these Hippocratic treatises, and the origins of "hysteria" and "green sickness" in ancient Greek medicine, see Helen King, *Hippocrates' Woman: Reading the Female Body in Ancient Greece* (London: Routledge, 2002), 188–246.

28. The treatise has been edited by Monica H. Green and Linne R. Mooney as "The Sickness of Women," in *Sex, Aging, and Death in a Medieval Medical Compendium: Trinity College Cambridge MS R.14.52, Its Texts, Language, and Scribe, Vol. II*, ed. M. Teresa Tavormina, Medieval and Renaissance Texts and Studies 292 (Tempe, AZ: ACMRS, 2006), 455–568. For other Middle English copies of the text, see Monica H. Green, "Obstetrical and Gynecological Texts in Middle English," *Studies in the Age of Chaucer* 14, no. 1 (1992): 53–88.

29. See Katharine Park, *Secrets of Women: Gender, Generation, and the Origins of Human Dissection* (New York: Zone Books, 2006), 132–41; Monica H. Green and Daniel Lord Smail, "The Trial of Floreta d'Ays (1403): Jews, Christians, and Obstetrics in Later Medieval Marseille," *Journal of Medieval History* 34, no. 2 (June 2008): 185–211.

30. Monica H. Green, *Making Women's Medicine Masculine: The Rise of Male Authority in Pre-Modern Gynaecology* (Oxford: Oxford University Press, 2008). On Trota's collection of reproductive recipes, see Monica H. Green, *The Trotula: A Medieval Compendium of Women's Medicine*, The Middle Ages Series (Philadelphia: University of Pennsylvania Press, 2001).

31. The treatise is discussed at length in Monica H. Green, "From 'Diseases of Women' to 'Secrets of Women': The Transformation of Gynecological Literature in the Later Middle Ages," *Journal of Medieval and Early Modern Studies* 30, no. 1 (Janu-

ary 1, 2000): 5–40; Green, *Making Women's Medicine Masculine*, 209–20; Park, *Secrets of Women*, 82–85.

32. Translated and quoted in Park, *Secrets of Women*, 84.

33. Ps.-Albertus Magnus, *Women's Secrets: A Translation of Pseudo-Albertus Magnus's De Secretis Mulierum with Commentaries*, SUNY Series in Medieval Studies (Albany: State University of New York Press, 1992), 1.

34. Green, "From 'Diseases of Women' to 'Secrets of Women.'"

35. Park, *Secrets of Women*, 77–120.

36. "every woman lettrid rede it to other unlettrid and helpe hem and conseyle hem in here maladyes with owte shewynge here dishese to man"; CUL MS Ii.6.33 II, f. 1r–v.

37. This treatise was identified by Monica Green as a compilation of the *Trotula Maior*, the *Trotula minor*, and elements of Muscio's *Gynaecia*; see Green, "Obstetrical and Gynecological Texts in Middle English," 64–68.

38. *The gospelles of dystaves*, sig. A.ii–iv.

39. *The gospelles of dystaves*, sigs. A.ii.

40. *The gospelles of dystaves*, sig. E.v *b*.

41. *The gospelles of dystaves*, sig. A.iv.

42. That manuscript is Chantilly, Musée Condé MS 654. See also Jeay and Garay, *The Distaff Gospels*, 23–25.

43. William Eamon, *Science and the Secrets of Nature: Books of Secrets in Medieval and Early Modern Culture* (Princeton, NJ: Princeton University Press, 1994), 139–47.

44. Girolamo Ruscelli, *The Secretes of the reuerende Maister Alexis of Piemount*, trans. William Ward, STC 293 (London: John Kingstone for Nicolas Inglande, 1558). On English books of secrets, see Allison Kavey, *Books of Secrets: Natural Philosophy in England, 1550–1600* (Urbana and Chicago: University of Illinois Press, 2007).

45. Tessa Storey, *Italian Book of Secrets Database* (Leicester: University of Leicester, 2008), dataset, https://hdl.handle.net/2381/4335.

46. Partridge, *The treasurie of commodious conceits, & hidden secrets and may be called, the huswiues closet, of healthfull prouision*.

47. Ruscelli, *The secretes of the reverende Maister Alexis of Piemount*, sig. *.ii.

48. Ruscelli, *The secretes of the reverende Maister Alexis of Piemount*, sig. *.ii.

49. Eamon, *Science and the Secrets of Nature*, 11, 93–133, 234–66, 353.

50. Ruscelli, *The secretes of the reverende Maister Alexis of Piemount*, sig. *.ii.

51. "non solamente da grandi huomini per dottrina, & da gran Signori, ma ancora da povere feminelle," in Girolamo Ruscelli, *Secreti del reverendo donno Alessio piemontese* (Venice: Iuntas, 1555), sig. A.ii *b*; Ruscelli, *The secrets of the reverende Maister Alexis of Piemount*, sig. *.ii.

52. Ruscelli, *The secrets of the reuerande Maister Alexis*, sig. +.iv *b*.

53. On this dynamic between orality and literacy, see Gretchen V. Angelo, "Author and Authority in the *Evangiles Des Quenouilles*," *Fifteenth-Century Studies* 26 (2001): 21–22.

54. Monica H. Green has demonstrated that the German text by Eucharius Rösslin was based on a fifteenth-century German manuscript, which itself was based on the *Practica* of the northern Italian physician Michele Savonarola, composed between 1440 and 1466; see "The Sources of Eucharius Rösslin's 'Rosegarden for Pregnant Women and Midwives' (1513)," *Medical History* 53, no. 2 (April 2009): 167–92.

55. Eucharius Rösslin, *The byrth of mankynde, newly translated out of Laten into*

Englysshe, trans. Richard Jonas, STC 21153 (London: Thomas Raynald, 1540), fols. LXXXIII *b*–LXXXVII.

56. Rösslin, *The byrth of mankynde, newly translated out of Laten into Englysshe*, fol. VII *b*. On the political and religious stakes of publishing a childbirth manual in Henry VIII's England, see Mary E. Fissell, *Vernacular Bodies: The Politics of Reproduction in Early Modern England* (Oxford: Oxford University Press, 2004), 29–35.

57. Rösslin, *The byrth of mankynde, newly translated out of Laten into Englysshe*, fol. VIII *a*.

58. Rösslin, *The byrth of mankynde, newly translated out of Laten into Englysshe*, fol. VI *b*.

59. Peter M. Blayney notes that Raynald may have owned a roller-press himself, as Raynald also had in his possession the engraved plates for an anatomical pamphlet, STC 564.2, published for the first time in 1540; see *The Stationers' Company and the Printers of London, 1501–1537*, 439–43.

60. Thomas Raynald and Eucharius Rösslin, *The byrth of mankynde, otherwyse named the womans booke*, STC 21154 (London: Tho. Raynald, 1545), sig. B.ii *b*–B.iii, B. v *b*. On the differences between Raynald's revised edition and the original English version, see Fissell, *Vernacular Bodies*, 29–35.

61. Subsequent editions are STC nos. 21155, 21156, 21157, 21157.5, 21158, 21159, 21160, 21161, 21162, 21163, 21164; Wing R1782C.

62. King's licenses for other practical books are discussed at more length in chapter 4 and listed in Arrer, *A Transcript of the Registers of the Company of Stationers of London*, fol. 59b.

63. *Liber aggregationis seu liber secretorum Alberti magni de virtutibus herbarum lapidum et animalium*, STC 258 (London: William Machlinia, 1483?); *Secreta mulierum et virorum*, STC 273 (London: William Machlinia, 1483?). The publication dates are suggestions of the STC, but both books share the same black-letter type associated with Machlinia's earlier years printing at "Flete-Bridge"; see Henry Robert Plomer, *A Short History of English Printing, 1476–1898*, ed. Alfred W Pollard, The English Bookman's Library (London: Kegan Paul, Trench, Trübner, and Company, 1900), 27.

64. *The boke of secretes of Albertus Magnus*, sigs. A.v *b*–A.vi, C.i, E.ii *b*–C.iii.

65. *The boke of secretes of Albertus Magnus*, sig. A.i *b*.

66. The subsequent editions are STC nos. 259, 260, 261, 262, 263, 264, 265, 266, 266.5, 267; Wing nos. A875I, A875J.

67. Thomas Lupton, *A thousand notable things, of sundry sortes*, STC 16955 (London: John Charlewood for Hugh Spooner, 1579), sig. A.ii.

68. Lupton, *A thousand notable things, of sundry sortes*, sig. A.iii.

69. The establishment of the Stationers' Register is discussed in chapter 4. On the regulatory powers of the Stationers' guild and register, see Johns, *The Nature of the Book*, 200–230; Johns, *Piracy*, 24–27.

70. Lupton, *A thousand notable things, of sundry sortes*, sigs. A.iii–A.iii *b*.

71. Lupton, *A thousand notable things, of sundry sortes*, sig. A.iii.

72. For transcriptions of these early modern reader marks, see the Digital Appendices: https://readingpractice.github.io.

73. *The Institution of a Christen Man*, sig. O.iv–O.iv *b*.

74. The earliest mention of witchcraft or sorcery in visitation articles dates from the 1538 visitation of Salisbury by Nicholas Shaxton; see Frere and Kennedy, *Visitation Articles and Injunctions of the Period of the Reformation, Vol. II: 1536–1558*, 58.

75. Frere and Kennedy, II, 111, no. 60.

76. Though the publication date in the title of the play is 1538, the English Short Title Catalogue gives a publication date of 1548. John Bale, *A comedy concernynge thre lawes, of nature Moses, & Christ, corrupted by the Sodomytes, Pharysees and Papystes*, STC 1287 (Wesel: Nicolaus Bamburgenses, 1538), sig. B.ii *b*–B.iii.

77. Though the midwife-witch was a powerful literary figure, David Harley has shown that midwives were only very rarely prosecuted for witchcraft; see David Harley, "Historians as Demonologists: The Myth of the Midwife-Witch," *Social History of Medicine* 3, no. 1 (1990): 1–26.

78. Questions concerning the practice of midwives appear in earlier Visitation Articles, but these were not assigned to their own section, nor did they associate midwifery with witchcraft; Frere and Kennedy, *Visitation Articles and Injunctions of the Period of the Reformation, Vol. II: 1536–1558*, II, 356, no. 116.

79. Frere and Kennedy, *Visitation Articles and Injunctions of the Period of the Reformation, Vol. III: 1559–1575*, 5, 383.

80. Fissell, *Vernacular Bodies*, 15–52.

81. On later sixteenth-century witchcraft prosecutions in England, see Alan Macfarlane in *Witchcraft in Tudor and Stuart England* (New York: Harper & Row, 1970), 28, 70, 255–306.

82. On the disappearance of traditional institutions of charity as a factor in the English witch craze, see Keith Thomas, *Religion and the Decline of Magic*, reprint (London: Penguin Books, 1991), 638–80; Macfarlane, *Witchcraft in Tudor and Stuart England*, 195–98, 200–211.

83. On this pattern of accusations in England, see Fissell, *Vernacular Bodies*, 75–89. Lyndal Roper has shown a similar pattern for witchcraft cases in seventeenth-century Augsburg; see "Witchcraft and Fantasy in Early Modern Germany," in *Oedipus and the Devil: Witchcraft, Religion, and Sexuality in Early Modern Europe* (New York: Taylor & Francis, 1994), 200–227.

84. *The Examination and Confession of Certaine Wytches at Chensforde in the countie of Essex*, STC 19869.5 (London: William Powell, 1566), sig. A.vii. The Frauncis case is discussed in further detail in Fissell, *Vernacular Bodies*, 74–88.

85. BL MS Additional 34210, f. 5r.

86. BL Additional MS 34210, ff. 18v–19r.

87. BL Sloane MS 962, f. 50v.

88. "and sche schall have child if þat it be so þat her tyme be comyn," BL MS Sloane 393, f. 56v.

89. Wellcome MS 409, ff. 11v, 12v, 13r, 79v, 80r. Starys's ownership mark, "Who sowever on me dothe loke I am John Starys boke," is on f. 146r.

90. Hannah Marcus has shown that a far more systematic program of censorship developed in Italy following the establishment of the papal index of prohibited books. See *Forbidden Knowledge: Medicine, Science, and Censorship in Early Modern Italy* (Chicago: University of Chicago Press, 2020).

91. For a list of manuscripts with censored charms or recipes as well as descriptions of these recipes' contents and the style of cancellation, see the Digital Appendices, https://readingpractice.github.io.

92. BL Sloane MS 1315, ff. 98v, 103v–104r.

93. Charms in Bodleian MS Ashmole 1477 2 that are nearly entirely crossed out

are ff. 7v–8r, 9r, 11v–12v, 15v, 17v–19v, 22r–23r. Charms that remain legible and are only crossed out with a single X are on ff. 16r, 27v–28r, 33v, 39v, 44r.

94. CUL MS Ee.1.15, ff. 14v, 15r, 6v, and 94v.

95. BL Lansdowne MS 680, ff. 48v and 66v.

96. Wellcome MS 5262, f.38v; Bodleian MS Rawlinson C.506, f. 95r.

97. Wellcome MS 404, f. 6v and 8v.

98. Wellcome MS 404, f. 32r.

99. Tollemache owned the fifteenth-century collection known as *The Tollemache Book of Secrets* and composed her own recipe collection, known as *Recepts of pastery, confectionary, & c.* Both have been edited and transcribed in Griffiths, Edwards, and Barker, *The Tollemache Book of Secrets.*

100. Leong has published widely on early modern women's recipe collections, but especially important works are "Making Medicines in the Early Modern Household," *Bulletin of the History of Medicine* 82, no. 1 (Spring 2008): 145–68; "Collecting Knowledge for the Family: Recipes, Gender and Practical Knowledge in the Early Modern English Household," *Centaurus: International Magazine of the History of Science and Medicine* 55, no. 2 (May 2013): 81–103; and *Recipes and Everyday Knowledge.*

101. On Woolley's writing, see Margaret J. M. Ezell, "Cooking the Books, or, the Three Faces of Hannah Woolley," in *Reading and Writing Recipe Books, 1550–1800*, ed. Sara Pennell and Michelle DiMeo (Manchester: Manchester University Press, 2013), 159–78.

102. This dynamic was first recognized by Harold Love in *Scribal Publication in Seventeenth-Century England* (Oxford: Clarendon Press, 1993).

103. On the binary worldview that structured early modern attitudes toward women and witchcraft, see Clark, "The 'Gendering' of Witchcraft in French Demonology: Misogyny or Polarity?"

Chapter 7

1. Buttus's "book of medicines" is Bodleian MS Rawlinson C.816.

2. Charles Trice Martin and Rachel Davies, "Sir William Butts (c. 1485–1545)," in *ODNB*.

3. Bodleian MS Rawlinson C.816, f. 2r.

4. Buttus was one of the men who traveled to Newfoundland in 1527, and his account of this journey was recorded by Richard Hakluyt in his *The principall navigations, voiages and discoveries of the English nation made by sea or over land*, STC 12625 (London: By George Bishop and Ralph Newberie, 1589), 517–19.

5. Bodleian MS Rawlinson C.816, f. 10r.

6. The history of early modern recipe collection has been most capably documented by Elaine Leong in multiple publications, most notable of them *Recipes and Everyday Knowledge*. See also Sara Pennell and Michelle DiMeo, eds., *Reading and Writing Recipe Books, 1550–1800* (Manchester: Manchester University Press, 2013).

7. Walker's manuscript is Bodleian MS Additional C.246, and the sketch of the prognostication by dominical letter (incomplete, ending with the year C) is ff. 73r–v.

8. As William Sherman has shown, early modern English readers were inveterate annotators of books, whether printed or in manuscript; see *Used Books: Marking Readers in Renaissance England* (Philadelphia: University of Pennsylvania Press, 2009).

9. See the Digital Appendices for a database of all manuscripts with reader marks with transcriptions of those with names or dates: https://readingpractice.github.io.

10. See Margaret Connolly, *Sixteenth-Century Readers, Fifteenth-Century Books*, Cambridge Studies in Paleography and Codicology 16 (Cambridge: Cambridge University Press, 2019).

11. He valued the book so much that he specifically mentioned it in his will, bequeathing *The secretes of the reverende maister Alexis* to Mr. Frannces Anger; see BL MS Additional 39227, f. 121.

12. Noah Millstone makes exactly this argument in *Manuscript Circulation and the Invention of Politics in Early Stuart England*, Cambridge Studies in Early Modern British History (Cambridge: Cambridge University Press, 2016), 29–54.

13. Buttus's recipe collection, discussed above, is Bodleian MS Rawlinson C.816. Tollemache's recipe book has been edited along with the fifteenth-century manuscript she owned as Griffiths, Edwards, and Barker, *The Tollemache Book of Secrets*.

14. On these recipe collections, see Leong, *Recipes and Everyday Knowledge*.

15. A number of fifteenth-century manuscripts feature tables of contents or indices added by later early modern readers: BL MSS Additional 4698, ff. 5–6; Sloane 357, f. 61v; Sloane 393, ff. 1–12; and Bodleian MS Ashmole 1468, p. 176.

16. "Wrytyn & fynyshyd þe ere of owr lord M. CCCCC & xi þe rayin of king hary þe viij[th] þe iii yere // þe xvii day of januer [17 January 1512]," Wellcome MS 406, f. 24r. "G. Martyn" dated his marginal notations with the year 1568 in Bodleian MS Ashmole 1477, ff. 20r–v.

17. See Elaine Leong, "'Herbals She Peruseth': Reading Medicine in Early Modern England," *Renaissance Studies* 28, no. 4 (September 2014): 556–78.

18. Huntington MS HM 58, f. 1r. Thomas Cartwright, *An hospitall for the diseased*, STC 4303.5 (London: by [R. Tottell? for] Thomas Man and William Hoskins, 1578).

19. Huntington MS HM 58, ff. 23r and 27v. On Banckes's herbal, the first printed version of the Middle English *Agnus castus* herbal, see chapter 4.

20. Dib's manuscripts are BL MSS Sloane 1764 and Sloane 121 and Wellcome MS 404, the latter of which bears his signature "Mr Thomas dib borne 1561 the first of May" on the front flyleaf.

21. Say's signature, which looks to be late fifteenth century or early sixteenth century in its style, appears on BL MS Sloane 1764 at f. 19v and BL MS Sloane 121 at f. 110v. Intriguingly, a Thomas Say from Abingdon, Oxfordshire, bequeathed "English bookes" to his daughter, Ann Say, in his will of 1501; see TNA PRO 11/287/12.

22. BL MS Sloane 1764, f. 1r.

23. *This lytell practyce of Iohannes de Vigo in medycyne*, STC 24725 (London: Robert Wyer, 1550).

24. A search of the Voigts-Kurts database returns fourteen manuscripts with the "Sphere of life and death" attributed to Pythagoras, including two manuscripts analyzed for this book: British Library MSS Additional 4698, f. 2, and Royal 17 A.xxxii, f. 3.

25. BL MS Sloane 2584 features color-making recipes on ff. 1–7 and medical recipes on ff. 13–29.

26. BL Sloane MS 121, f. 2r–2v. I have searched the Universal Short Title Catalogue and Leon Voet and Jenny Voet-Grisolle, *The Plantin Press (1555–1589): A Bibliography of the Works Printed and Published by Christopher Plantin at Antwerp and Leiden,*

Vols. I–V (Amsterdam: Van Hoeve, 1980), and have been unable to identify the titles of these books.

27. BL Sloane MS 121, f. 127v.

28. Dib's prognostications and horoscopes are BL Sloane MS 121, ff. 118–126. On Simon Forman and his notebooks, see Lauren Kassell, *Medicine and Magic in Elizabethan London: Simon Forman, Astrologer, Alchemist, and Physician* (Oxford: Oxford University Press, 2005).

29. BL Sloane MS 121, f. 114.

30. BL Sloane MS 1764, f. 29v.

31. Dib's ownership mark appears in Wellcome MS 404, front flyleaf.

32. Wellcome MS 404, f. 23v.

33. Wellcome MS 404, f. 24r.

34. Rampling, *The Experimental Fire: Inventing English Alchemy, 1400–1700*, 97–98 and passim.

35. Thorowgood's annotation of his daughter's birth is on CUL MS Dd.10.44, f. 102v and the signatures of John Young and John Tight are on 149v. Thorowgood was sworn in as an alderman of the City of London for Cheap Ward in May 1589; see Alfred P. Beaven, ed., "Alderman of the City of London: Cheap Ward," in *The Alderman of the City of London Temp. Henry III–1912*, British History Online (London: Corporation of the City of London, 1908), 99–106.

36. Smerthwaite's name appears on the list of barber-surgeons given in the modern edition of Thomas Vicary, *Anatomie of the Bodie of Man*, ed. Frederick J. Furnivall and Percy Furnivall, Early English Text Society, Extra Series 53 (Berlin & New York: C. Scriber & Co, 1888), 244.

37. "wrytyn & fynyshed þe ere of our lord M CCCCC & xi . . . þe xvii day of januer," Wellcome MS 406, f. 24r.

38. "surgeon to þe kynge Henry þe eaygth," BL MS Sloane 1, f. 314v.

39. "anotte of an acte made the therd yere of the rayne of king henry the viij for the stabling of physions and surgants"; Bodleian MS Bodley 483, f. 117v.

40. Bodleian MS Rawlinson C.816, f. 10r.

41. "This is John Rice is boke, the whiche cost him xxv d.," BL MS Royal 17 A.xxxii, f. 2r. The records of John Dorne, an Oxford bookseller, suggest that early sixteenth-century volumes on these topics sold for around two pence each; see F. Madan, "Day-Book of John Dorne, Bookseller in Oxford, A.D. 1520."

42. Bodleian MS Ashmole 1396, front flyleaf verso. The same page features another mark in the same hand, dated 1598. Forman's association with the manuscript is stated in William Henry Black, *A Descriptive, Analytical, and Critical Catalogue of the Manuscripts Bequeathed unto the University of Oxford by Elias Ashmole* (Oxford: Oxford University Press, 1845), cols. 1089–90.

43. See chapter 4 for a discussion of these marketing tactics.

44. Eisenstein, *The Printing Revolution in Early Modern Europe*, 209–50.

45. Huntington MS HM 58, ff. 23r and 27v.

46. Digges, *A prognostication of right good effect*, sig. *.iii *a–b*. On the "stigma of print" owing to its association with money and the trades, see Love, *Scribal Publication in Seventeenth-Century England*.

47. Leong, *Recipes and Everyday Knowledge*, 27–40.

48. These sources were collected in the *Collectanea satis copiosa*, presented to

Henry by a team of advisors in September 1530. See Haigh, *English Reformations: Religion, Politics, and Society Under the Tudors*, 102. The *Collectanea* is now held in the British Library as MS Cotton Cleopatra E. VI.

49. N. R. Ker estimates that just a small fraction of the manuscripts that once existed in medieval England survived the dissolution of the monasteries and the subsequent crackdown on Catholic books that characterized Edward VI's reign; see Jennifer Summit, *Memory's Library: Medieval Books in Early Modern England* (Chicago: University of Chicago Press, 2008), 103–35.

50. On this, see Alexandra Walsham, *The Reformation of the Landscape: Religion, Identity, and Memory in Early Modern Britain and Ireland* (Oxford: Oxford University Press, 2011).

51. Margaret Aston, "English Ruins and English History: The Dissolution and the Sense of the Past," *Journal of the Warburg and Courtauld Institutes* 36 (1973): 231–55.

52. Gregory B. Lyon, "Baudouin, Flacius, and the Plan for the Magdeburg Centuries," *Journal of the History of Ideas* 64, no. 2 (2003): 253–72.

53. On Parker's scholarly practice, see Madeline McMahon, "Matthew Parker and the Practice of Church History," in *Confessionalisation and Erudition in Early Modern Europe: An Episode in the History of the Humanities*, ed. Nicholas Hardy and Dmitri Levitin, Proceedings of the British Academy (Oxford and New York: Oxford University Press, 2020), 116–53. On Parker's network of influence as a collector, see Benedict Scott Robinson, "'Darke Speech': Matthew Parker and the Reforming of History," *The Sixteenth Century Journal* 29, no. 4 (1998): 1061–83.

54. Stow's collection has been catalogued in Janet Wilson, "A Catalogue of the 'Unlawful' Books Found in John Stow's Study on 21 February 1568/9," *British Catholic History* 20, no. 1 (May 1990): 1–30. Dee's is catalogued in John Dee, *The Private Diary of Dr. John Dee, and the Catalog of His Library of Manuscripts*, ed. J. O. (James Orchard) Halliwell-Phillipps (London: The Camden Society, 1892).

55. Huntington MS HM 19079, f. 1r.

56. Huntington MS HM 19079, f. 15v.

57. On this collection, see Getz, *Healing and Society in Medieval England*.

58. On this, see Anthony Grafton, *Forgers and Critics: Creativity and Duplicity in Western Scholarship*, new ed. (Princeton, NJ: Princeton University Press, 2019), 99–123.

59. Barrett L. Beer, "Stow, John (1524/5–1605)," in *ODNB*; R. Julian Roberts, "Dee, John (1527–1609)," in *ODNB*; Wyman H. Herendeen, "Camden, William (1551–1623)," in *ODNB*.

60. "Parishes: Cropthorne," in *A History of the County of Worcester*, British History Online, vol. 3 (London: Victoria County History, 1913), 322–29.

61. I thank especially Teresa M. Tavormina, who shared with me via personal correspondence notes on Henry's manuscripts compiled by the late Lister Matheson. Practical manuscripts that can be confirmed as having belonged to Dyngley are Bodleian MS Rawlinson C.506; Wellcome MS 5262; TCC MSS O.8.35 and R.14.52; and BL MS Royal 17 A.xxxii.

62. The manuscript contains a prayer to St. Kenelm, patron saint of Winchcombe Abbey, on f. 61v. The relationship between St. Kenelm and Winchcombe Abbey is noted in the Wellcome's catalogue entry for the manuscript. On Winchcombe's dissolution, see William Page, ed., "Houses of Benedictine Monks: The Abbey of Winch-

combe," in *A History of the County of Gloucester: Volume 2*, British History Online (London: Victoria County History, 1907), 66–72.

63. Charms are partially obscured by smeared ink in Bodleian MS Rawlinson C.506, ff. 95r and 119, and illustrations of saints partially obliterated in Wellcome MS 5262, ff. 1r–3r.

64. Humfrey Harrison's signature is on f. 345v, Bodleian MS Rawlinson C.506. The record of his affiliation with Alstonefield is Cheshire Archives, DCH/O/1, a copy deed of a land gift made on 24 March 1474.

65. A. P. Baggs et al., "Houses of Cistercian Monks: The Abbey of Combermere," in *A History of the County of Chester: Volume 3*, ed. C. R. Elrington and B. E. Harris (London: Victoria County History, 1980), 150–56.

66. "for megrenes," Bodleian MS Rawlinson C.506, f. 123r.

67. TCC MS O.8.35, f. 126r–127r. Henry's signature and the date 1554 are on the back of the first flyleaf of the manuscript.

68. Dyngley's plea against Worcester Cathedral is not dated, but it is addressed to Nicholas Bacon, Keeper of the Great Seal (1558–1579), and the Crown's response looks to be dated "anno regne E tertio," which would place the proceeding sometime in 1560–1561; TNA UK, C 3/52/26.

69. BL Royal MS 17 A.xxxii, ff. 5r, 89r, and 119r. Mary's christening on 12 April 1560 is recorded in Frederick Arthur Crisp, ed., *The Parish Register of Cropthorne, Worcestershire: 1557–1717* (London: Printed at the Private Press of F. A. Crisp, 1896), 3.

70. Barbara's christening on 25 July 1561 and Alice's on 11 December 1562 are both recorded in Crisp, *The Parish Register of Cropthorne, Worcestershire*, 3–4. His other six children (and their ages) are recorded in the Worcestershire Visitation of 1569; see W. P. W. Phillimore, ed., *The Visitation of the County of Worcester Made in the Year 1569: With Other Pedigrees Relating to That County from Richard Mundy's Collection* (London: Harleian Society, 1888), 50.

71. Dyngley's copy of the poem is on ff. 125v–126r, BL MS Royal 17 A.xxxii. The full poem appears in Foxe, *Actes and monuments of these latter and perillous dayes touching matters of the church*, 1263.

72. BL MS Royal 17 A.xxxii, f. 126r.

73. Wellcome MS 244, ff. 1v–3v.

74. Wellcome MS 244, p. 356.

75. Wellcome MS 244, f. 4r.

76. TCC MS R.14.52, f. 28r and 269r.

77. Crisp, *The Parish Register of Cropthorne, Worcestershire*, 21.

78. It should be noted that Dyngley's Protestantism, his roles as a county bureaucrat, and his pursuit of intellectual advancement make him a model of the kind of social advancement Keith Wrightson and David Levine have identified as central to the social stratification of local English communities in the later sixteenth century; see *Poverty and Piety in an English Village: Terling, 1525–1700*, 2nd ed. (Oxford: Clarendon Press, 1995), 175–84, 200–208.

79. BL MS Royal 17 A.xxxii, f. 126r.

80. The younger Henry attended St. Edmund Hall, Oxford, from 1594 to 1595; see Joseph Foster, *Alumni Oxonienses: The Members of the University of Oxford, 1500–1714* (Oxford, 1891), 44.

81. The emplastrum recipe is in Wellcome MS 244, pp. 51 and 63, and the note about the water is on p. 335.

82. See the Appendix in Matthew Phillpott, *The Reformation of England's Past: John Foxe and the Revision of History in the Late Sixteenth Century*, Routledge Research in Early Modern History (New York: Routledge, 2018), 215–26. On the incredible range of manuscript and printed sources consulted by Foxe, see Elizabeth Evenden and Thomas S. Freeman, *Religion and the Book in Early Modern England: The Making of Foxe's "Book of Martyrs,"* Cambridge Studies in Early Modern British History (Cambridge: Cambridge University Press, 2011).

83. Foxe, *Actes and monuments of these latter and perillous dayes touching matters of the church*, 7–16.

84. See, for example, Foxe's translation of Gregory the Great's letter to Augustine outlining the foundation of the English church in *Actes and monuments of these latter and perillous dayes touching matters of the church*, 16–17.

85. BL MS Royal 17 A.xxxii, ff. 125v–126r.

86. Buttus bequeathed part two of Foxe's *Actes and monuments* to Mr. John Payne and his "pictor of Mr John Fox that made the booke of actes & monuments" to Sir Drew Drewry; Norfolk Record Office, NCC Will Register Apleyard 373, Butts, Thomas, esquire, of Great Ryburgh, 1592, f. 375, 383.

87. The earliest entries in Wellcome MS 244 and Bodleian MS Rawlinson C.816 are dated 1564.

88. Alexandra Walsham, *Providence in Early Modern England* (Oxford University Press, 1999), 1–2, 5.

89. Foxe, *Actes and monuments of these latter and perillous dayes touching matters of the church*, 347.

90. On Foxe's vision of the English church as "elect," see William Haller, *The Elect Nation: The Meaning and Relevance of Foxe's Book of Martyrs* (New York: Harper & Row, 1963). For a summary and criticism of Haller's argument, see Patrick Collinson, *This England: Essays on the English Nation and Commonwealth in the Sixteenth Century*, Politics, Culture and Society in Early Modern Britain (Manchester: Manchester University Press, 2011), 193–94.

91. On English nationalism, antiquarianism, and nostalgia for the past, see Collinson's collected essays in *This England: Essays on the English Nation and Commonwealth in the Sixteenth Century*, 193–215, 245–308.

92. Thomas Gale, *Certaine vvorkes of chirurgerie, nevvly compiled and published by Thomas Gale, maister in chirurgerie*, STC 11529 (London: Rouland Hall, 1563). On Thomas Gale's gift of his books to William Gale, see Erin Connelly, "'My Written Books of Surgery in the Englishe Tonge': The London Company of Barber-Surgeons and the Lylye of Medicynes," *Manuscript Studies: A Journal of the Schoenberg Institute for Manuscript Studies* 2, no. 2 (2017): 369–91.

93. Brooke's ownership mark is on f. 245r, and the *ex libris* mark of the rector of St. Peter's church is on f. 246r, Bodleian MS Ashmole 1505. On Brooke's identity as Henry VI's distiller, see Voigts, "The Master of the King's Stillatories."

94. Bodleian MS Ashmole 1505, f. 3r.

95. Bodleian MS Ashmole 1505, f. 2r.

96. The date of the work's composition appears at the end of the Latin prologue to the text: "Inchoatus fuit liber iste cuius auxilio magni dei in preclari studio motis pesulani post annum lecture nostre xx. Anno domini M. CCC. III mense Iulii," Ber-

nardus de Gordonio, *Practica, seu Lilium medicinae*, USTC 996910 (Naples: Fransesco del Tuppo, for Berdardinus Gerardinus, 1480), unsigned first page.

97. Bodleian MS Ashmole 1505, f. 246r.

98. Frampton first published in 1577 a translation of just Monardes's first volume under the title *Three Bookes written in the Spanish tonge* (STC 18005), but the title was changed early in the print run to *Joyfull newes out of the newe founde worlde* (STC 18005a).

99. Nicolás Monardes, *Joyfull newes out of the newe founde worlde*, trans. John Frampton, STC 18006.5 (London: Thomas Dawson for William Norton, 1580), sig. *.3. On Frampton's translation, see Alisha Rankin, "New World Drugs and the Archive of Practice: Translating Nicolás Monardes in Early Modern Europe," *Osiris* 37 (June 1, 2022): 76.

100. Monardes, *Joyfull newes out of the newe founde worlde*, trans. Frampton, fol. 1–1v.

101. In Thomas Lupton's *A thousand notable things, of sundry sortes*, he urged his readers to buy the first of Frampton's editions, because "whosoever doth buy it therefore doth not pay the hundredth part that it is worth"; see Lupton, *A thousand notable things*, 280.

102. Page Life, "Bright, Timothy (1549/50–1615)," in *ODNB*. Timothy Bright, *A treatise: wherein is declared the sufficiencie of English medicines, for cure of all diseases, cured with medicine*, STC 3750 (London: Henrie Middleton for Thomas Man, 1580).

103. Bright, *A treatise: wherein is declared the sufficiencie of English medicines*, 5.

104. Bright, *A treatise: wherein is declared the sufficiencie of English medicines*, 7.

105. Bright, *A treatise: wherein is declared the sufficiencie of English medicines*, 27–28. Pliny the Elder, *Naturalis Historia*, ed. Karl Friedrich Theodor Mayhoff (Leipzig: Teubner, 1906), bk. 22, ch. 56, and bk. 24, ch. 2.

106. Bright, *A treatise: wherein is declared the sufficiencie of English medicines*, 33–34.

107. Cooper discusses similar rhetoric in publications by Symphorien Champier, Bartholomäus Carrichter, and Johan van Beverwyck in *Inventing the Indigenous: Local Knowledge and Natural History in Early Modern Europe* (Cambridge: Cambridge University Press, 2007), 32–45.

108. Bruce T. Moran, "The 'Herbarius' of Paracelsus," *Pharmacy in History* 35, no. 3 (1993): 104. Paracelsus's text is discussed in Cooper, *Inventing the Indigenous: Local Knowledge and Natural History in Early Modern Europe*, 22–32.

109. John Foxe and Timothy Bright, *An abridgement of the booke of acts and monumentes of the Church*, STC 11229 (London: J. Windet, at the assignment of Master Tim Bright, 1589), sig. *.ii *b*.

110. Bright, *A treatise: wherein is declared the sufficiencie of English medicines*, 9–11.

111. Bright, *A treatise: wherein is declared the sufficiencie of English medicines*, 11–12.

112. Banckes, *The vertues & proprytes of herbes*, sig. E.ii *b*–E.iii.

113. Bright, *A treatise: wherein is declared the sufficiencie of English medicines*, 30.

114. Bright, *A treatise: wherein is declared the sufficiencie of English medicines*, 27.

115. The very earliest framework elaborating environmental influences on human health, and the foundation for such thinking in Western medicine through the early modern period, was the Hippocratic "Airs, Waters, Places," in *Hippocrates, Volume I*, trans. Paul Potter, Loeb Classical Library 147 (Cambridge, MA: Harvard University Press, 2022), 73–75.

116. Bright, *A treatise: wherein is declared the sufficiencie of English medicines*, 29.

117. Thomas Man, Bright's publisher, seems to have realized that the recipes in Bright's treatise were deficient. When he reissued the treatise in 1615, he appended to it "A collection of medicines growing (for the most part) within our English climat" and added a table of contents similar to those in other herbals; see Timothy Bright, *A treatise, wherein is declared the sufficiencie of English medicines, for cure of all diseases, cured with medicines. Whereunto is added A collection of medicines growing (for the most part) within our English climat, approoved and experimented against the jaundise, dropsie, stone, falling-sicknesse, pestilence,* STC 3752 (London: H. Lownes for Thomas Man, 1615), 70–127.

118. Walsham, *Providence in Early Modern England*, 2.

119. John Gerard's signature and the date 1577 are on f. 118v, BL MS Sloane 7. Gerard's herbal is *The herball or Generall historie of plantes: Gathered by John Gerarde of London master in Chirurgerie*, STC 11750 (London: by John Norton, 1597).

120. In the preface to the 1633 revised edition of Gerard's text, Thomas Johnson famously accused Gerard of plagiarizing from a translation of Dodoens's *Stirpium historiae pemptades sex*, and of then disguising the theft by reorganizing the work using Matthias L'Obel's new classification system; see John Gerard and Thomas Johnson, *The herball or Generall historie of plantes*, STC 11751 (London: Adam Islip, Joice Norton, and Richard Whitakers, 1633), sig. ¶¶¶.1 *b*.

121. Harkness, *The Jewel House*, 44–56.

122. On Gerard's compilation strategies and his use of earlier herbal texts, see Neville, *Early Modern Herbals and the Book Trade*, 237–62.

123. See, especially, Steven Shapin, "The House of Experiment in Seventeenth-Century England," *Isis* 79, no. 3 (September 1988): 373–404; Steven Shapin, *The Social History of Truth: Civility and Science in Seventeenth-Century England* (Chicago and London: University of Chicago Press, 1994). Peter Dear's influential essay comparing Descartes's method with Bacon's philosophy is "Miracles, Experiments, and the Ordinary Course of Nature," *Isis* 81, no. 4 (December 1990): 663–83.

Conclusion

1. Bacon, *The tvvoo bookes of Francis Bacon, of the proficience and aduancement of learning*, sigs. G.ii–G.iii *b*.

2. Bacon, The *tvvoo bookes of Francis Bacon, of the proficience and aduancement of learning*, sig. F.iii. Important works that center "ordinary" practitioners in the history of science are Harkness, *The Jewel House*; Long, *Artisan/Practitioners and the Rise of the New Sciences, 1400–1600*; and Smith, *The Body of the Artisan*.

3. Bacon, *The tvvoo bookes of Francis Bacon, of the proficience and aduancement of learning*, sigs. C.i, F.iii.

4. Thomas, *Religion and the Decline of Magic*, 769–99, quoted at 793, 799.

5. Steven Shapin most famously demonstrated how texts could stand in for the in-person witness of experimentation in his landmark essay, "Pump and Circumstance: Robert Boyle's Literary Technology," first published in 1984 and reprinted in *Never Pure*, 89–116.

6. These old texts were Gemini, Udall, Vesalius, and de Mondeville, *Compendiosa totius anatomie delineatio, ære exarate*, printed in 1553, featuring Vesalius's anatomical diagrams and a text adapted from Henry de Mondeville's thirteenth-century surgical treatise, and Thomas Digges's reedition of his father's *A prognostication euerlastinge*

of right good effecte, published in a fifth edition in 1576 with Copernicus's heliocentric model of the universe appended to it.

7. This opposition has, traditionally, dominated histories of science that center on the contributions of ordinary people over learned philosophers; see, for example, Long, *Artisan/Practitioners and the Rise of the New Sciences, 1400–1600*; Smith, *The Body of the Artisan*.

Bibliography

Pre-1500 "Practical Manuscripts" Consulted

BEINECKE LIBRARY, NEW HAVEN, CT

Beinecke 171
Takamiya 46
Takamiya 59

BIBLIOTHÈQUE INTERUNIVERSITÉ DE SANTÉ, PARIS

MS 3

BODLEIAN LIBRARY, OXFORD

Add. B. 60
Ashmole 8
Ashmole 189 1 (ff. 1–69)
Ashmole 340 3 (ff. 41–57)
Ashmole 1378
Ashmole 1389 1 (pp. 1–266)
Ashmole 1389 2 (pp. 267–286)
Ashmole 1393 1 (ff. 1–6, 55–57, 70)
Ashmole 1393 2 (ff. 7–18, 27–42)
Ashmole 1393 3 (ff. 19–26)
Ashmole 1396
Ashmole 1413 2 (pp. 25–154)
Ashmole 1432 I 1 (pp. 3–24)
Ashmole 1432 I 2 (pp. 25–70, 85–100)
Ashmole 1432 I 4 (pp. 71–84)
Ashmole 1432 II 5 (pp. 81–144)
Ashmole 1438 I 3 (pp. 57–80)
Ashmole 1438 I 4 (pp. 81–92)
Ashmole 1438 I 5 (pp. 94–130)
Ashmole 1438 I 6 (pp. 131–146)
Ashmole 1438 II 1 (pp. 1–64)
Ashmole 1443
Ashmole 1444 2 (pp. 87–110)
Ashmole 1444 3 (pp. 113–192)
Ashmole 1444 4 (pp. 193–304)
Ashmole 1447 1 (pp. 1–18)
Ashmole 1447 2 (pp. 19–74)
Ashmole 1447 3 (pp. 75–104)
Ashmole 1447 4 (pp. 105–164)
Ashmole 1447 5 (pp. 166–202)
Ashmole 1468
Ashmole 1477 1 (pp. 1–97)
Ashmole 1477 2 (ff. 1–47)
Ashmole 1477 3 (ff. 1–12)
Ashmole 1481 1B (ff. 4–12)
Ashmole 1481 1E (ff. 25–35)
Ashmole 1481 1F (ff. 36–41)
Ashmole 1498
Ashmole 1505
Bodley 178
Bodley 483
Bodley 1031

Douce 84 1 (ff. i–vi, 1–24)
Douce 84 2 (ff. 25–32)
Douce 84 4 (ff. 46–53)
Douce 290 (ff. 134–157)
Douce 304
Hatton 29 1 (ff. 1–59)
Hatton 29 2 (ff. 60–81)
Hatton 29 3 (ff. 82–96)
James 43
Laud Misc. 553
Lyell 37
Radcliffe Trust e.4
Rawlinson C.81
Rawlinson C.211
Rawlinson C.299
Rawlinson C.506
Rawlinson D.939
Rawlinson D.1220
Rawlinson D.1222
Selden Supra 73 1 (ff. 1–26)
Selden Supra 73 2 (ff. 28–36)
Selden Supra 73 3 (ff. 37–75)
Selden Supra 73 5 (ff. 27, 116–129)
Wood empt. 18 1 (ff. 1–56)

BRITISH LIBRARY, LONDON

Additional 4698
Additional 12056
Additional 19674 2
Additional 34210
Additional 37786
Arundel 272 1 (ff. 1–32)
Arundel 272 2 (ff. 32–64)
Egerton 827
Egerton 2724
Harley 937
Harley 1600
Harley 1735
Harley 2320
Harley 2332
Harley 2340
Harley 2381
Harley 3840 (ff. 139–178)
Harley 5086
Lansdowne 680
Royal 17 A.xvi
Royal 17 A.xxxii
Royal 17 C.xv
Royal 18 A.vi 1 (ff. 1–21)
Royal 18 A.vi 2 (ff. 22–34)
Royal 18 A.vi 3 (ff. 35–55)
Royal 18 A.vi 4 (ff. 57–58)
Royal 18 A.vi 5 (ff. 59–63)
Royal 18 A.vi 6 (ff. 64–87)
Royal 18 A.vi 7 (ff. 88–103)
Sloane 1
Sloane 7
Sloane 73 3 (ff. 28–50)
Sloane 73 6 (ff. 148–170)
Sloane 73 8 (ff. 176–194)
Sloane 76
Sloane 100
Sloane 120 1 (ff. 1–61)
Sloane 120 2 (ff. 62–95)
Sloane 121
Sloane 140
Sloane 240
Sloane 277
Sloane 357
Sloane 372
Sloane 382 3 (ff. 206–264)
Sloane 393 1 (ff. 1–75)
Sloane 393 2 (ff. 76–149)
Sloane 468
Sloane 521 (ff. 198–275)
Sloane 540A
Sloane 610
Sloane 686
Sloane 1314
Sloane 1315 2 (ff. 16–162)
Sloane 1764
Sloane 2457/2458
Sloane 2460
Sloane 2584 (ff. 1–92, 102–116)
Sloane 3153 (ff. 2–97)
Sloane 3160 (ff. 87–172)
Sloane 3449
Sloane 3489
Stowe 982

CAMBRIDGE UNIVERSITY LIBRARY, CAMBRIDGE

Additional 9308
Additional 9309
Dd.4.44
Dd.5.76 1 (ff. 1–29)
Dd.5.76 2 (ff. 30–75)
Dd.6.29
Dd.10.44
Ee.1.13 1 (ff. 1–96)
Ee.1.13 2 (ff. 97–139)
Ee.1.15
Ii.6.33 2
Kk.6.33

HENRY E. HUNTINGTON LIBRARY, SAN MARINO, CA

HM 58
HM 505
HM 1336
HM 19079

THE MORGAN MUSEUM AND LIBRARY, NEW YORK

B.21
B.44
M.941

NATIONAL LIBRARY OF MEDICINE, BETHESDA, MD

MS 4
MS E 30

ROYAL SOCIETY OF LONDON ARCHIVES, LONDON

MS 45

THE SCHOYEN COLLECTION, LONDON AND OSLO

MS 1581

TRINITY COLLEGE, CAMBRIDGE

O.1.13 1 (ff. 1–82)
O.1.13 3 (ff. 142–165)
O.1.13 4 (ff. 166–172)
O.8.35
O.9.32
O.9.37
O.9.39 1 (ff. 1–18)
O.9.39 2 (ff. 19–29)
R.14.51
R.14.52

UNIVERSITY OF GLASGOW LIBRARY, GLASGOW

Hunter 95
Hunter 117
Hunter 307
Hunter 497
Hunter 509

WELLCOME LIBRARY, LONDON

Medical Society of London 136
Wellcome 397
Wellcome 404 1 (ff. 1–34, 41–46)
Wellcome 404 2 (ff. 35–40)
Wellcome 406 1 (ff. 1–22)
Wellcome 409
Wellcome 410
Wellcome 5262
Wellcome 5650
Wellcome 8004

Other Manuscripts Cited

Berlin, Staatsbibliothek zu Berlin MS Libr. Pict. A 92
Cambridge, MA, Harvard University, Houghton Library MS Richardson 35
Cambridge, MA, Harvard Medical School, Countway Library of Medicine MS 19
Cambridge, Cambridge University Library MS Ll.1.18
Cambridge, St. John's College Archives SJLM/1/1/3
Chantilly, Musée Condé MS 654
Chester, Cheshire Archives DCH/O/1
Copenhagen, Royal Library of Denmark MS NKS 901
Llubljana, Slovenia, National and University Library MS 160
London, Welcome Library, MS 244
London, Wellcome Library, MS 537
New York, Morgan Library and Museum MS M.766
New York, Morgan Library and Museum MS M.1117
New York, New York Academy of Medicine MS 12
Norwich, Norfolk Record Office NCC Will Register Apleyard 373
Paris, Bibliothèque nationale de France MS fr. 2151
Washington, DC, Folger Shakespeare Library MS V.a.438

BODLEIAN LIBRARY, OXFORD

Additional C. 246
Ashmole 210
Ashmole 349
Ashmole 1378
Douce 45
Douce 71
Rawlinson C.816
Tanner 407
Selden Supra 77
Wood D.8

THE BRITISH LIBRARY, LONDON

Additional 17367
Additional 18850
Additional 39227
Additional 46143
Additional 70517
Cotton Cleopatra E. VI
Egerton MS 1995
Harley 950
Harley 2252
Royal 12 B.vi
Sloane 962
Sloane 2507
Sloane 3215

THE NATIONAL ARCHIVES OF THE UNITED KINGDOM, KEW

C 1/140/13
C 3/252/66
E 122/183/6
E 122/185/15
PRO 11/287/12
SC 8/29/1411
C 3/52/26
E 122/96/43
E 122/185/6
E 122/185/16
SC 8/29/1410
SP 1/58

Early Printed Books

Publisher's location is London unless otherwise indicated.

Almanacke for xii. yere. EEBO. STC 387. Wynkyn de Worde, 1508.

Anglicus, Gilbertus. *Compendium medicinae tam morborum universalium quam particularium, nondum medicis sed et cyrurgicis utilissimum.* USTC 143565. Lyon: Jacques Sacon, 1510.

The antidotharius. EEBO. STC 675.3. Robert Wyer, [1535?]. Reissued in the same year as STC 675.7.

Articella. USTC 801793. Pavia, Italy: Giacomo Pocatela, 1510.

Askham, Anthony. *A litell treatyse of astrouomy: very necessary for physyke and surgerye planetes, sygnes and constellacyons.* EEBO. STC 857a.5. William Powell, 1550.

———. *A prognosticacion made for the yere of oure Lord god a thousande fyue hundreth xlviii.* EEBO. STC 410. William Powell, 1548. Other annual almanacs produced by Askham were published by William Powell in 1549 as STC 410.1; in 1551 as STC 410.2 and STC 410.3; in 1552 as STC 410.4; in 1553 as STC 410.5; and in 1554 as 410.6. Askham's almanacs were published by Thomas Marshe in 1555 as STC 410.7; in 1556 as STC 410.9, 410.10, and 410.11; and in 1557 as STC 410.12.

———. *An almanacke and prognostication, for the yere of our Lorde God M. D. L.V.* EEBO. STC 410.8. William Powell, 1555.

Bacon, Francis. *The tvvoo bookes of Francis Bacon, of the proficience and aduancement of learning, diuine and humane: to the king.* Princeton University Scheide Library. For Henrie Tomes, 1605.

Bale, John. *A comedy concernynge thre lawes, of nature Moses, & Christ, corrupted by the Sodomytes. Pharysees and Papystes: Compyled by Iohan Bale. Anno M. D.XXXVIII.* EEBO. STC 1287. Wesel: Nicolaum Bamburgensem [i.e., Derick van der Straten], 1548.

Barclay, Alexander, trans. *The kalendayr of the shyppars.* STC 22407. Paris: Antoine Verard, 1503.

Bernard of Gordon. *Practica, seu Lilium medicinae.* National Library of Medicine 2211018r. USTC 996910. Naples: Fransesco del Tuppo, for Berdardinus Gerardinus, 1480.

Berners, Juliana. *Here in thys boke afore ar contenyt the bokys of haukyng and huntyng.* EEBO. STC 3308. Saint Albans: s.n., 1486. Reissued by Wynkyn de Worde in 1496 as STC 3309.

Berners, Juliana, and Wynkyn de Worde. *The boke of hawkynge, and huntynge, and fysshynge.* EEBO. STC 3309.5. Wynkyn de Worde, 1518.

Biblia pauperum. Library of Congress Incun. X.B562. Netherlands or Germany: s.n., 1470.

Biblia pauperum. Princeton University Scheide Library S2.9. Nuremburg: Hans Sporer, 1471.

The boke of secretes of Albertus Magnus of the vertues of herbes, stones, and certayne beasts. EEBO. STC 258.5. J. King, 1560.

A boke of the propertyes of herbes the which is called an Herbal. EEBO. STC 13175.4. John Scot, 1537. Reissued in 1539 by Robert Redman as STC 13175.5; in 1539 by Elizabeth Redman as STC 13175.7; in 1541 by Thomas Petyt as STC 13175.8; in 1546 by William Middleton as STC 13175.10; in 1547 by Robert Copeland as STC 13175.11; and in 1548 by John Waley as STC 13175.12.

Boorde, Andrew. *A pronostycacyon or an almanacke for the yere of our lorde, M. CCCCC*. EEBO. STC 416.5. s.n., 1545.

———. *The pryncyples of astronamye the whiche diligently perscrutyd is in maners pronosticacyon to the worldes end*. EEBO. STC 3386. Robert Copland, 1547.

Bright, Timothy. *A treatise: wherein is declared the sufficiencie of English medicines, for cure of all diseases, cured with medicine*. Huntington Library 59341. STC 3750. Henry Middleton for Thomas Man, 1580.

———. *A treatise, wherein is declared the sufficiencie of English medicines, for cure of all diseases, cured with medicines. Whereunto is added A collection of medicines growing (for the most part) within our English climat, approoved and experimented against the jaundise, dropsie, stone, falling-sicknesse, pestilence*. STC 3752. National Library of Medicine, Bethesda, MD. H. Lownes for Thomas Man, 1615.

Brunfels, Otto. *Almanach ewig werend, Teütszch vnd Christlich Practick*. Straßburg. USTC 610939. Prüß d.J., 1526.

———. *A very true pronosticacion, with a kalender, gathered out of the moost auncyent bokes of ryght holy astronomers: for the yere of our lorde M. CCCCC. xxxvj*. Translated by John Ryckes. EEBO. STC 421.17. John Byddell, 1536.

Bucer, Martin. *A treatise declaryng & shewig dyuers causes taken out of the holy scriptures of the sentences of holy faders & of the decrees of deuout emperours*. Translated by William Marshal. EEBO. STC 24239 T. Godfray for William Marshal, 1535.

Burchard of Worms. *Burchardi Wormanciensis Ecclesiae Episcopi Decretorum Libri Viginti*. Paris: Joannem Foucherium, 1549.

Caius, John. *A boke, or counseill against the disease commonly called the sweate, or sweatyng sicknesse*. EEBO. STC 4343. Richard Grafton, 1552.

Cartwright, Thomas. *An hospitall for the diseased*. EEBO. STC 4303.5. [R. Tottell? for] Thomas Man and William Hoskins, 1578.

Church of England. *The institvtion of a Christen man*. EEBO. STC 5163. Thomas Berthelet, 1537.

Copeland, William, and Richard Banckes, eds. *A boke of the propreties of herbes called an herbal . . . by W.C*. EEBO. STC 13175.15. William Copland for John Wight, 1552. Reissued by William Copland for Richard Kele in 1552 as STC 13175.15a; by William Copland in 1559 as STC 13175.18; and by J. Awdely in 1567 as STC 13175.19c.

Coverdale, Miles, ed. *A faythfull and true pronostication vpon the yere .M.CCCCC. xlviii and parpetually after to the worldes ende*. EEBO. STC 20423. [J. Herford for] Richard Kele, 1547.

The crafte of graffynge & plantynge of trees. EEBO. STC 5953. Wynkyn de Worde, 1518.

Digges, Leonard. *A prognostication of right good effect, fructfully augmented*. EEBO. STC 435.35. Thomas Gemini, 1555. Reissued in 1556 by Thomas Gemini as STC 435.37.

Digges, Leonard, Nicolaus Copernicus, and Thomas Digges. *A prognostication euerlastinge of right good effecte . . . Lately corrected and augmented by Thomas Digges his sonne*. EEBO. STC 435.47. Thomas Marshe, 1576.

Dives & pauper. EEBO. STC 19213. Westminster: Wynkyn de Worde, 1496.

Elyot, Sir Thomas. *The castell of helthe, gathered, and made by Syr Thomas Elyot knyghte*. EEBO. STC 7642.5. Thomas Berthelet, 1537. Reissued by Berthelet in 1539 as STC 7642.7 and 7643, in 1541 as 7644 and 7645, in 1544 as 7646, in 1547 as 7646.5, and in 1550 as 7647.

The Examination and confession of certaine wytches at Chensforde in the countie of Essex. EEBO. STC 19869.5. William Powell for Wyllyam Pickeringe, 1566.

Fine, Oronce. *Les canons & documens tresamples, touchant lusaige & practique des communs almanachz, que l'on nomme ephemerides*. Bibliothèque nationale de France, département Arsenal 8-S-14288. Paris: Simon de Colines, 1543.

———. *The rules and righte ample documentes, touchinge the vse and practise of the common almanackes, which are named ephemerides*. Translated by Humfrey Baker. EEBO. STC 10878.9. Thomas Marshe, 1558.

Fitzherbert, John. *Here begynneth a newe tracte or treatyse moost profytable for all husbandmen: and very frutefull for all other persons to rede*. EEBO. STC 10994. Richard Pynson, 1523.

Foxe, John. *Actes and monuments of these latter and perillous dayes touching matters of the church*. EEBO. STC 11222. John Day, 1563.

Foxe, John, and Timothy Bright. *An abridgement of the booke of acts and monumentes of the Church*. EEBO. STC 11229. J. Windet for Tim Bright, 1589.

Gale, Thomas. *Certaine workes of chirurgerie, nevvly compiled and published by Thomas Gale, maister in chirurgerie*. EEBO. STC 11529. Rouland Hall [for Thomas Gale], 1563.

Gemini, Thomas, and Andreas Vesalius. *Compendiosa totius anatomie delineatio, Ære exarata: Per Thomam Geminum*. Translated by Nicholas Udall. EEBO. STC 11715.5. Nicholas Hill for Thomas Gemini, 1553. Reissued the same year as STC 11716.

Gerard, John. *The herball or Generall historie of plantes*. EEBO. STC 11750. Edm. Bollifant for Bonham Norton and John Norton, 1597.

Gerard, John, and Thomas Johnson. *The herball or Generall historie of plantes*. EEBO. STC 11751. London: Adam Islip, Joice Norton, and Richard Whitakers, 1633.

Gesner, Konrad, and Peter Morwen. *The treasure of Euonymus*. EEBO. STC 11800. John Day, 1559.

The gospelles of dystaues. EEBO. STC 12091. Wynkyn de Worde, 1510.

The grete herbal. EEBO. STC 13176. Southwark: Peter Treueris, 1526. Reissued again by Treveris in 1529 as STC 13177 and in 1539 as STC 13178 and by John King in 1561 as STC 13179.

Gutenberg, Johannes. *Planeten-Tafel, sive Ephemerides 1448*. USTC 741136. Mainz: Johannes Gutenberg, 1458.

Hakluyt, Richard. *The principall navigations, voiages and discoveries of the English nation made by sea or over land*. EEBO. STC 12625. London: George Bishop and Ralph Newberie, 1589.

Henley, Walter de. *Boke of husbandry*. EEBO. STC 25007. Wynkyn de Worde, 1508.

Hereafter foloweth the judgement of all vrynes. EEBO. STC 14834. Robert Wyer, 1555.

Hereafter foloweth the knowledge, properties, and the vertues of herbes. EEBO. STC 13175.6. Robert Wyer, 1540.

Here begynneth a newe boke of medecynes intytulyd or callyd the treasure of pore men. CUL Sel.5.175. STC 24199. Richard Banckes, 1526. Reissued in 1539 by Robert Redman as STC 24200 and in 1540 as STC 24202.5; by Thomas Petyt in 1539 as STC 24201, in 1540 as STC 24202, and in 1546 as STC 24203.3; by William Middleton in 1544 as STC 24203 and in 1551 as STC 24204; by Richard Lant in 1547 as STC 24203.5; by Robert Copland in 1548 as STC 24203.7; by John Wayland in 1555 as STC 24204.5 and in 1556 as STC 24205; and by Thomas Colwell in 1560 as STC 24206a, in 1565 as STC 24206a.5, and in 1575 as STC 24207.

Here begynnyth a newe mater, the whiche sheweth and treateth of ye vertues & proprytes of herbes. EEBO. STC 13175.1. Richard Banckes, 1525. Reissued by Richard Banckes in 1526 as STC 13175.2.

Here begynneth the kalender of shepherdes. EEBO. STC 22408. Richard Pynson, 1506. Reissued by Pynson in 1510 as STC 22409.3 and in 1517 as STC 22409.7.

Here begynneth the proprytees and medycynes for hors. EEBO. STC 20439.5. Wynkyn de Worde, [1497?]. Reissued by Wynkyn de Worde in 1502 as STC 20439.3.

Here begynneth the seynge of urynes. EEBO. STC 22153. [J. Rastell] for Richard Banckes, 1525. Reissued by Richard Banckes in 1526 as 22153a.

Hutten, Ulrich von. *De morbo gallico*. Translated by Thomas Paynell. EEBO. STC 14024. Thomas Berthelet, 1533. Reissued by Berthelet under the title *Of the wood called guiaiacum* in 1536 as STC 14025, in 1539 as STC 14026 and 14026.5, and in 1540 as STC 14027.

Joannes, Jacobi. *A passing gode lityll boke necessarye & behouefull agenst the pestilence*. Translated by Benggt Knuttson. EEBO. STC 4591. William de Machlinia, [1485?].

———. *Here begynneth a litill boke necessarye & behouefull agenst the pestilence*. Translated by Bengt Knuttson. EEBO. STC 4590. William de Machlinia, [1485?].

———. *Here begynneth a litil boke the whiche traytied and reherced many gode thinges necessaries for the infirmite & grete sekenesse called pestilence*. Translated by Bengt Knuttson. EEBO. STC 4589. William de Machlinia, [1485?].

———. *Here begynneth a treatyse agaynst pestelence and of ye infirmities*. Translated by Bengt Knuttson. EEBO. STC 4592. Enprynted by Wynkyn de Worde, 1509. Reissued by De Worde in 1511 as STC 4592.5.

———. *Regime de l'epidimie*. USTC 71138. Lyon: Guillaume le Roy, [1476?].

J.A. *A Perfyte pronostycacion perpetuall*. EEBO. STC 406.3. Robert Wyer, 1556.

John of Burgundy. *Here begynneth a lytell treatyse called the gouernall of helthe with the medecyne of the stomacke*. EEBO. STC 12139. Wynkyn de Worde, 1506.

———. *In this tretyse that is cleped gouernayle of helthe*. EEBO. STC 12138. Westminster: William Caxton, 1490.

The kalender of shepeherdes. EEBO. STC 22409. Wynkyn de Worde, 1516. Reissued by De Worde in 1516 as STC 22409.5 and in 1528 as STC 22411.

Laet, Jasper. *The pnostication of Maister Jasp Laet, practised in the towne of Antuerpe, for the yere of Our Lorde, M.D.XX.* Edited and translated by Richard Pynson. EEBO. STC 470.6. Richard Pynson, 1520.

Lanfranc of Milan. "Chirurgia Magna." In *Ars Chirurgica Guidonis Cauliaci*, fols. 207–61. Venice: Iuntas, 1546.

Le compost et kalendrier des bergiers. USTC 70063. Paris: Guy Marchant, 1491. Reissued another six times by Marchant between 1491 and 1497 as USTC 70063–67 and 95900. Issued in Lyon by Guillaume Balsarin in 1499 as USTC 89882.

Le traittie intitule les evangiles des quenouilles faittes a l'onneur et exaucement des dames. USTC 70278. Brugge: Colard Mansion, 1479.

Lupton, Thomas. *A thousand notable things, of sundry sortes.* EEBO. STC 16955. London: John Charlewood for Hugh Spooner, 1579.

A lytel herball of the properties of herbes newly amended and corrected . . . by A. Askham. EEBO. STC 13175.13. William Powell, 1550. Reissued by John King in 1561 as STC 13175.19.

Macers herbal: Practysyd by Doctor Lynacro. EEBO. STC 13175.13c. Robert Wyer, 1552.

Medicines for horses. EEBO. STC 20439.7. William Copeland, 1565.

Monardes, Nicolás. *Ioyfull nevves out of the newe founde worlde.* Translated by John Frampton. EEBO. STC 18005a. William Norton, 1577.

———. *Ioyfull newes out of the newfound world, wherein are declared the rare and singular vertues of diuerse and sundrie hearbs, trees, oyles, plantes, & stones.* Translated by John Frampton. John Carter Brown Library. STC 18006.5. Thomas Dawson for William Norton, 1580.

More, Sir Thomas. *A dyaloge of syr Thomas More knyghte.* EEBO. STC18085. William Rastell, 1530.

Moulton, Thomas. *This is the myrour or glasse of helth.* EEBO. STC 18214a. Robert Wyer, 1531. Reissued by Robert Wyer in 1536 as STC 18214a.3, 18214a.5, and 18214a.7.

———. *This is the myrour or glasse of helthe.* EEBO. STC 18216. Robert Redman, 1540. Reissued in 1541 by Elizabeth Redman (Pickering) as STC 18219 and in 1545 by William Middleton as STC 18220.

Moulton, Thomas, and Robert Wyer. *This is the glasse of helth, a great treasure for pore men, necessary and nedefull for euery person to loke in.* EEBO. STC 18225.2. Robert Wyer, 1540. Reissued by Robert Wyer in 1545 as STC 18225.4, in 1547 as STC 18225.6, and in 1555 as STC 18225.8.

A newe herball of Macer, translated out of Laten in to Englysshe. EEBO. STC 13175.8c. Robert Wyer, 1543.

Parron, William. *A prognostication.* EEBO. STC 385.7. Richard Pynson, 1498.

———. *A prognostication for 1498.* EEBO. STC 385.3. Wynkyn de Worde, 1498.

Partridge, John. *The treasurie of commodious conceits, & hidden secrets and may be called, the huswiues closet.* EEBO. STC 19425.5. Richard Jones, 1573.

Proctor, John. *The historie of Wyates rebellion with the order and maner of resisting the same.* EEBO. STC 20407. Robert Caly, 1554.

The pronostycacyon for euer of Erra Pater. EEBO. STC 439.3 Robert Wyer, 1540. Reissued by Robert Wyer in 1545 as STC 339.5; in 1550 as STC 439.7; in 1552 as STC 439.9; in 1554 as STC 439.11; in 1555 as STC 439.13; by Thomas Colwll in 1562 as STC 439.15; and by Thomas East in 1581 as STC 439.17 and in 1598 as STC 439.18.

Ps.-Albertus Magnus. *Liber aggregationis seu liber secretorum Alberti magni de virtutibus herbarum lapidum et animalium*. EEBO. STC 258. William Machlinia, [1483?].

———. *Secreta mulierum et virorum*. EEBO. STC 273. William de Machlinia, [1483?].

Rede, William. *Almanach ephemerides in anno domini M. d. vii. in latitudo Oxonia*. EEBO. STC 504. R. Pynson, 1507.

Regimen sanitatis Salerni. EEBO. STC 21596. Thomas Berthelet, 1528. Reissued again by Berthelet in 1530 as STC 21597, in 1535 as STC 21598, and in 1541 as STC 21599.

Regiomontanus, Joannes. *Ephemerides, 1475–1506*. Bayerische Staatsbibliothek, München. USTC 748421. Nürnberg: Johannes Regiomontanus, 1474.

———. *Kalendarium*. Bayerische Staatsbibliothek, München. USTC 748412. Nürnberg: Johannes Müller of Königsberg, 1474.

Regiomontanus, Joannes, and Hans Biefftruck. *Calendarium*. Library of Congress Incun. 1474. M82. Nuremberg: Sold by Hans Biefftruck?, 1474.

Rösslin, Eucharius, and Richard Jonas. *The byrth of mankynde, newly translated out of Laten into Englysshe*. Edited and translated by Richard Jonas. EEBO. STC 21153. Thomas Raynald, 1540.

Rösslin, Eucharius, Richard Jonas, and Thomas Raynald. *The byrth of mankynde, otherwyse named the womans booke*. EEBO. STC 21154. By Thomas Raynalde, 1545.

Ruscelli, Girolamo. *The secretes of the reverende Maister Alexis of Piemount*. Translated by William Ward. EEBO. STC 293. John Kingstone for Nicolas Inglande, 1558.

———. *Secreti del reverendo donno Alessio piemontese*. USTC 853883. Venice: Per Sigismondo Bordogna, 1555.

To them that before this image of pity devoutly say v. pater noster, v. aves & a credo piteously. EEBO. STC 14077c.6. Westminster: William Caxton, 1487.

A treatyse: contaynynge the orygynall causes, and occasions of the diseases, growynge on horses. EEBO. STC 24237.5. John King, 1560.

Turner, William. *A new herbal*. EEBO. STC 14365. Steven Mierdman, 1551.

Tyndale, William. *An answere vnto Sir Thomas Mores dialoge made by Vvillyam Tindale*. EEBO. STC 24437 Antwerp: S. Cock, 1531.

Vesalius, Andreas. *Radicis chynae Usus*. USTC 149871. Lyon: Jean Frellon, 1547.

Vigo, Giovanni da. *This lytell practyce of Iohannes de Vigo in medycyne*. EEBO. STC 24725. Robert Wyer, 1550.

Voragine, Jacobus. *Thus endeth the legende named in latyn legenda aurea, that is to saye in englysshe the golden legende*. EEBO. STC 24873. Westminster: William Caxton, 1483. Reissued in 1484 by Caxton as STC 24874; and by Wynkyn de Worde in 1493 as STC 24875 and in 1498 as STC 24876.

Xylographic calendar. Bodleian Library Douce A 632. s.n., 1522. Other exemplars are British Library C.36.aa.5 (1537); British Library C.41.a.28 (1538); and British Library C.29.c.6 (1542).

Modern Editions of Primary Sources

Almeida, Fransisco Alonso, ed. *A Middle English Medical Remedy Book*. Middle English Texts 50. Heidelberg: Winter, 2014.

Aquinas, Thomas. *Scriptura Super Sententiis Magistri Petri Lombardi, Tomus III.* Edited by M. F. Moos. Paris: P. Lethielleux, 1933.

Arderne, John. *De Arte Phisicali et de Cirurgia of Master John Arderne, Surgeon of Newark. Dated 1412.* Edited by D'Arcy Power and Eric Miller. New York: W. Wood, 1922.

———. *Treatises of Fistula in Ano, Haemorrhoids, and Clysters by John Arderne.* Edited by D'Arcy Power. EETS, Original Series 139. London: Oxford University Press, 1919.

Arrer, Edward, ed. *A Transcript of the Registers of the Company of Stationers of London. Electronic Reproduction.* Vol. 1: 1554–1640. New York: Columbia University Press, 2007. http://www.columbia.edu/cu/lweb/digital/collections/cul/texts/ldpd_6177070_001/index.html.

Ayoub, Lois. "John Crophill's Books: An Edition of British Library MS Harley 1735." PhD diss., University of Toronto, 1994.

Bacon, Roger. *Opus Majus, Volumes 1 and 2.* Translated by Robert Belle Burke. Philadelphia: University of Pennsylvania Press, 1928.

Bede. *Bede: The Reckoning of Time.* Edited by Faith Wallis. Translated Texts for Historians 29. Liverpool: Liverpool University Press, 1999.

The Bible: Authorized King James Version. Edited by Robert Carroll and Stephen Prickett. Oxford: Oxford University Press, 2008.

Brewer, J. S., ed. "6652: Prophecies." In *Letters and Papers, Foreign and Domestic, of the Reign of Henry VIII*, vol. 4. Part III: 1529–1530: 2997. London: Longman, 1876.

Brodin, Gösta, ed. *Agnus castus: A Middle English Herbal Reconstructed from Various Manuscripts.* Essays and Studies on English Language and Literature 6. Upsala: Lundequistska bokhandeln, 1950.

Calendar of the Patent Rolls Preserved in the Public Record Office: Henry IV, Vol. III, 1405–1408. London: Mackie and Co, 1907.

Calendar of the Patent Rolls Preserved in the Public Record Office. Philip and Mary, Vol. I: 1553–1554. London: His Majesty's Stationery Office, 1937.

Calendar of the Patent Rolls Preserved in the Public Record Office. Philip and Mary, Vol. II: 1554–1555. London: His Majesty's Stationery Office, 1936.

Chaucer, Geoffrey. "The Nun's Priest's Prologue, Tale, and Epilogue." In *The Riverside Chaucer*, edited by Larry Benson, 252–261. 3rd ed. Oxford: Oxford University Press, 2008.

———. *A Treatise on the Astrolabe.* Edited by W. W. Skeat. London: N. Trübner & Co., for the Chaucer Society, 1872.

Chauliac, Guy de. *The Cyrurgerie of Guy de Chauliac, Volume I: Text.* Edited by Margaret S. Ogden. Early English Text Society 265. London: Oxford University Press, 1971.

———. *An Interpolated Middle English Version of* The Anatomy of Guy de Chauliac. Edited by Björn Wallner. Lund: Lund University Press, 1995.

———. *Inventarium Sive Chirurgia Magna, Volume One: Text.* Edited by Michael McVaugh. Studies in Ancient Medicine 14. Leiden: Brill, 1997.

———. *The Middle English Translation of Guy de Chauliac's* Treatise on Wounds, *Book III of* The Great Surgery. Edited by Björn Wallner. Lund: CWK Gleerup, 1976.

Cobb, Cyril, ed. *The Rationale of Ceremonial, 1540–1543, with Notes and Appendices and an Essay on the Regulation of Ceremonial during the Reign of King Henry VIII*. London: Longmans, Green, and Co., 1910.

Cobb, H. S., ed. *The Overseas Trade of London: Exchequer Customs Accounts, 1480–1*. British History Online. London: London Record Society, 1990. https://www.british-history.ac.uk/london-record-soc/vol27.

Collins, Francis, ed. *Register of the Freemen of the City of York, from the City Records. Vol. 1: 1272–1558*. Durham: Andrews and Co., 1896.

Copenhaver, Brian P., ed. *Hermetica: The Greek Corpus Hermeticum and the Latin Asclepius in a New English Translation, with Notes and Introduction*. Cambridge: Cambridge University Press, 1992.

Crisp, Frederick Arthur, ed. *The Parish Register of Cropthorne, Worcestershire: 1557–1717*. London: Printed at the Private Press of F. A. Crisp, 1896.

Cylkowski, David G. "A Middle English Treatise on Horticulture: Godfridus Super Palladium." In *Popular and Practical Science of Medieval England*, edited by Lister M. Matheson, 301–29. Medieval Texts and Studies; No. 11. East Lansing, MI: Colleagues Press, 1994.

Davis, Norman, ed. *The Paston Letters: A Selection in Modern Spelling*. Oxford World's Classics. Oxford: Oxford University Press, 1999.

Dee, John. *The Private Diary of Dr. John Dee, and the Catalog of His Library of Manuscripts*. Edited by J. O. (James Orchard) Halliwell-Phillipps. London: The Camden Society, 1892.

Esteban-Segura, Laura. "The Middle English Circa Instans: A Pharmacopoeia from Glasgow, University Library, MS Hunter 307." *Manuscripta* 59, no. 1 (January 1, 2015): 29–60.

Frere, W. H., and W. P. M. Kennedy, eds. *Visitation Articles and Injunctions of the Period of the Reformation, Vol. II: 1536–1558*. London: Longmans, Green & Company, 1910.

———, eds. *Visitation Articles and Injunctions of the Period of the Reformation, Vol. III: 1559–1575*. London: Longmans, Green & Company, 1910.

Getz, Faye. *Healing and Society in Medieval England: A Middle English Translation of the Pharmaceutical Writings of Gilbertus Anglicus*. Madison: University of Wisconsin Press, 2010.

Green, Monica H. *The Trotula: A Medieval Compendium of Women's Medicine*. The Middle Ages Series. Philadelphia: University of Pennsylvania Press, 2001.

Gregory I, *Registrum Epistolarum Tomus II: Libri VIII–XIV cum indicibus et praefatione*. Edited by Paul Edwald and Ludovic Hartmann. Monumenta Germaniae Historica. Berlin: Weidmann, 1899.

Griffin, Carrie, ed. *The Middle English "Wise Book of Philosophy and Astronomy": A Parallel-Text Edition. Edited from London, British Library, MS Sloane 2453 with a Parallel Text from New York, Columbia University, MA Plimpton 260*. Middle English Texts 47. Heidelberg: Winter, 2013.

Griffiths, Jeremy, A. S. G. Edwards, and Nicolas Barker, eds. *The Tollemache Book of Secrets: a descriptive index and complete facsimile with an introduction and transcriptions together with Catherine Tollemache's Receipts of pastery, confectionary & c*. London: Roxburghe Club, 2001.

Grymonprez, Pol, ed. *Here Men May Se the Vertues of Herbs: A Middle English Herbal (MS. Bodley 483, ff. 57r–67v)*. Scripta: Medieval and Renaissance Texts and Studies 3. Brussels: UFSAL, 1981.

Hale, William, ed. *A Series of Precedents and Proceedings in Criminal Causes Extending from the Year 1475 to 1640: Extracted from Act-books of Ecclesiastical Courts in the Diocese of London, Illustrative of the Discipline of the Church of England*. London: Francis & John Rivington, 1847.

Heinrich, Fritz. *Ein Mittelenglisches Medizinbuch*. Halle: Niemeyer, 1896.

Hippocrates of Cos. "Airs, Waters, Places." In *Hippocrates, Volume I*, translated by Paul Potter, 64–143. Loeb Classical Library 147. Cambridge, MA: Harvard University Press, 2022.

Hunt, Tony. *Anglo-Norman Medicine I: Roger Frugard's Chirurgia, The Practica brevis of Platearius*. Woodbridge, Suffolk: Boydell & Brewer, 1994.

Jeay, Madeleine, and Kathleen Garay. *The Distaff Gospels: A First Modern English Edition of Les Évangiles Des Quenouilles*. Peterborough, ON: Broadview Press, 2006.

Lanfranc of Milan. *Lanfranck's "Science of Cirurgie."* Edited by Robert von Fleischhacker. EETS, Original Series 102. London: Oxford University Press, 1894.

Langland, William. *William Langland's The Vision of Piers Plowman*. London and New York: J. M. Dent and E. P. Dutton, 1978.

Liuzza, R. M. *Anglo-Saxon Prognostics: An Edition and Translation of Texts from London, British Library, MS Cotton Tiberius A.iii*. Anglo-Saxon Texts 8. Cambridge: D. S. Brewer, 2011.

Lynn, Nicholas. *The Kalendarium of Nicholas of Lynn*. Edited by Sigmund Eisner. The Chaucer Library. Athens: University of Georgia Press, 1980.

Macer. *A Middle English Translation of* Macer Floridus De Viribus Herbarum. Edited by Gösta Frisk. Uppsala: Almqvist & Wiksells, 1949.

Madan, F., ed. "Day-Book of John Dorne, Bookseller in Oxford, A.D. 1520." *Collecteana*, edited by Charles Robert Lesley Fletcher, 72–177. First series 5. Oxford: Oxford Historical Society at Clarendon, 1885.

Making and Knowing Project, Pamela H. Smith, Naomi Rosencranz, Tianna Helen Uchacz, Tillmann Taape, Clément Godbarge, Sophie Pitman, et al., eds. *Secrets of Craft and Nature in Renaissance France: A Digital Critical Edition of BnF Ms. Fr. 640*. New York: Making and Knowing Project, 2020. https://edition640.makingandknowing.org.

Ogden, M.S., ed. *The "Liber de Diversis Medicinis" in the Thornton Manuscript (MS Lincoln Cathedral A.5.2)*. Early English Text Society, Original Series 207. London: Oxford University Press, 1938.

Phillimore, W. P. W., ed. *The Visitation of the County of Worcester Made in the Year 1569: With Other Pedigrees Relating to That County from Richard Mundy's Collection*. London: Harleian Society, 1888.

Pickett, Joseph P. "A Translation of the 'Canutus' Plague Treatise." In *Popular and Practical Science of Medieval England*, edited by Lister M. Matheson, 263–82. Medieval Texts and Studies; No. 11. East Lansing, MI: Colleagues Press, 1994.

Pliny the Elder. *Naturalis Historia*, edited by Karl Friedrich Theodor Mayhoff. Leipzig: Teubner, 1906.

Ps.-Albertus Magnus. *Women's Secrets: A Translation of Pseudo-Albertus Magnus's* De Secretis Mulierum *with Commentaries*. SUNY Series in Medieval Studies. Albany: State University of New York Press, 1992.

Reynes, Robert. *The Commonplace Book of Robert Reynes of Acle: An Edition of Tanner MS 407*. Edited by Cameron Louis. New York: Garland, 1980.

Scott-Macnab, David. *The Middle English Text of "The Art of Hunting" by William Twiti*. Middle English Texts 40. Heidelberg: Winter, 2009.

Somer, John. *The Kalendarium of John Somer*. Edited by Linne R. Mooney. Athens: University of Georgia Press, 1998.

Statutes of the Realm. Vols. 1–4: 1235–1624. Westminster: House of Commons, 1819. www.heinonline.org.

Thomas of Cobham. *Summa Confessorum*. Edited by F. Broomfield. Analecta Mediaevalia Namurcensia 25. Louvain: Editions Nauwelaerts, 1968.

Vicary, Thomas. *Anatomie of the Bodie of Man*. Edited by Frederick J. Furnivall and Percy Furnivall. Early English Text Society, Extra Series 53. Berlin and New York: C. Scriber & Co, 1888.

Wilkins, David, ed. *Concilia Magnae Britanniae et Hiberniae, ab Anno MCCCL ad Annum MDXLV. Volumen Tertium*. London: R. Gosling, F. Gyles, T. Woodward, & C. Davis, 1737.

Wyclif, John. "De Apostasia Cleri." In *Select English Works of John Wyclif, Vol. 3*, edited by T. Arnold, 430–40. Oxford: Clarendon Press, 1871.

———. "De Officio Pastorali." In *The English Works of Wyclif Hitherto Unprinted*, edited by F. D. Matthew, 405–57. Early English Text Society, Original Series 74. London: Trübner & Co., 1880.

Secondary Sources

Angelo, Gretchen V. "Author and Authority in the *Evanglies des Quenouilles*." *Fifteenth-Century Studies* 26 (2001): 21–41.

Aston, Margaret. *England's Iconoclasts: Laws against Images*. Oxford: Clarendon Press, 1988.

———. "English Ruins and English History: The Dissolution and the Sense of the Past." *Journal of the Warburg and Courtauld Institutes* 36 (1973): 231–55.

———. *Lollards and Reformers: Images and Literacy in Late Medieval Religion*. History Series 22. London: Hambledon Press, 1984.

Atkin, Tamara, and A. S. G. Edwards. "Printers, Publishers and Promoters to 1558." In *A Companion to the Early Printed Book in Britain, 1476–1558*, edited by Vincent Gillespie and Susan Powell, 27–44. Cambridge: D. S. Brewer, 2014.

Azzolini, Monica. *The Duke and the Stars: Astrology and Politics in Renaissance Milan*. I Tatti Studies in Italian Renaissance History. Boston: Harvard University Press, 2013.

Baggs, A. P., Ann J. Kettle, S. J. Lander, A. T. Thacker, and David Wardle. "Houses of Cistercian Monks: The Abbey of Combermere." In *A History of the County of Chester: Volume 3*, edited by C. R. Elrington and B. E. Harris, 150–56. London: Victoria County History, 1980.

Bahr, Arthur. "Miscellaneity and Variance in Medieval Books." In *The Medieval Manuscript Book: Cultural Approaches*, edited by Michael Johnston and Michael Van Dussen, 181–98. Cambridge: Cambridge University Press, 2015.

Bailey, Michael David. "The Feminization of Magic and the Emerging Idea of the Female Witch in the Late Middle Ages." *Essays in Medieval Studies* 19, no. 1 (2002): 120–34.

Barber, M. J. "The Books and Patronage of Learning of a Fifteenth-Century Prince." *The Book Collector* 12, no. 3 (1963): 308–15.

Barker, Sheila, and Sharon Strocchia. "Household Medicine for a Renaissance Court: Caterina Sforza's Ricettario Reconsidered." In *Gender, Health, and Healing, 1250–1550*, edited by Sara Ritchey and Sharon Strocchia, 139–65. Amsterdam: Amsterdam University Press, 2020.

Barnes, Robin B. *Astrology and Reformation*. New York: Oxford University Press, 2015.

Barron, Caroline M. "What Did Medieval London Merchants Read?" In *Medieval Merchants and Money: Essays in Honour of James L. Bolton*, edited by Martin Allen and Matthew Davies, 43–70. London: Institute for Historical Research, 2016.

Baxandall, Michael. *Painting and Experience in Fifteenth-Century Italy: A Primer in the Social History of Pictorial Style*. 2nd ed. Oxford: Oxford University Press, 1988.

Beaven, Alfred P., ed. "Alderman of the City of London: Cheap Ward." In *The Alderman of the City of London Temp. Henry III–1912*, British History Online, 99–106. London: Corporation of the City of London, 1908. http://www.british-history.ac.uk/no-series/london-aldermen/hen3-1912/pp99-106.

Beer, Barrett L. "Stow, John (1524/5–1605)." In *Oxford Dictionary of National Biography*. Oxford: Oxford University Press, 2004. https://doi.org/10.1093/ref:odnb/26611.

Bennett, H. S. "Science and Information in English Writings of the Fifteenth Century." *The Modern Language Review* 39, no. 1 (1944): 1–8.

Benskin, M., et al., eds. *eLALME: An Electronic Version of A Linguistic Atlas of Late Mediaeval Middle English*. Edinburgh: University of Edinburgh, 2013. http://www.lel.ed.ac.uk/ihd/elalme/elalme.html.

Black, William Henry. *A Descriptive, Analytical, and Critical Catalogue of the Manuscripts Bequeathed unto the University of Oxford by Elias Ashmole*. Oxford: Oxford University Press, 1845.

Blair, Ann M. "Authorship in the Popular 'Problemata Aristotelis.'" *ESM: Early Science and Medicine* 4, no. 3 (1999): 189–227.

———. *Too Much to Know: Managing Scholarly Information before the Modern Age*. New Haven, CT: Yale University Press, 2010.

Blake, N. F. "Worde, Wynkyn de (d. 1534/5)." In *Oxford Dictionary of National Biography*. Oxford: Oxford University Press, 2008. https://doi.org/10.1093/ref:odnb/29968.

Blayney, Peter W. M. *The Bookshops in Paul's Cross Churchyard*. Occasional Papers of the Bibliographical Society 5. London: The Bibliographical Society, 1990.

———. *The Stationers' Company and the Printers of London, 1501–1537*. 2 vols. Cambridge: Cambridge University Press, 2013.

Blomefield, Francis, ed. "Clavering Hundred: Toft." In *An Essay Towards a Topographical History of the County of Norfolk: Volume 8*. British History Online, 61–64. London: W. Miller, 1808. https://british-history.ac.uk/topographical-hist-norfolk/vol8/pp61-64.

Boffey, Julia. *Manuscript and Print in London, c. 1475–1530*. London: The British Library, 2012.

Borland, Jennifer. *Visualizing Household Health: Medieval Women, Art, and Knowledge in the Régime Du Corps*. State College: Pennsylvania State University Press, 2022.

Bourgain, Pascale. "The Circulation of Texts in Manuscript Culture." In *The Me-*

dieval Manuscript Book: Cultural Approaches, 140–59. Cambridge: Cambridge University Press, 2015.

Bower, Hannah. *Middle English Medical Recipes and Literary Play, 1375–1500*. Oxford: Oxford University Press, 2022.

Broedel, Hans Peter. *The Malleus Maleficarum and the Construction of Witchcraft: Theology and Popular Belief*. Manchester: Manchester University Press, 2003.

Bühler, Curt F. *The Fifteenth-Century Book: The Scribes, the Printers, the Decorators*. The A. S. W. Rosenbach Fellowship in Bibliography. Philadelphia: University of Pennsylvania Press, 1960.

Burnett, Charles. "Astrology and Medicine in the Middle Ages." *The Bulletin of the Society for the Social History of Medicine* 37 (1985): 16–18.

———. *The Introduction of Arabic Learning into England*. The Panizzi Lectures. London: The British Library, 1997.

Burnett, Charles, and Danielle Jacquart, eds. *Constantine the African and ʻAlī Ibn Al-ʻAbbās al-Magūsī: The Pantegni and Related Texts*. Studies in Ancient Medicine, 925–1421, no. 10. New York: E. J. Brill, 1994.

Cabré, Montserrat, and Fernando Salmón. "Blood, Milk, and Breastbleeding: The Humoral Economy of Women's Bodies in Medieval Medicine." In *Gender, Health, and Healing, 1250–1550*, edited by Sara Ritchey and Sharon Strocchia, 93–117. Amsterdam: Amsterdam University Press, 2020.

Cameron, M. L. *Anglo-Saxon Medicine*. Cambridge Studies in Anglo-Saxon England, No. 7. New York: Cambridge University Press, 1993.

Capp, Bernard. *English Almanacs, 1500–1800: Astrology and the Popular Press*. Ithaca, NY: Cornell University Press, 1979.

Carey, Hilary M. "Astrological Medicine and the Medieval English Folded Almanac." *Social History of Medicine* 17, no. 3 (December 1, 2004): 345–63.

———. *Courting Disaster: Astrology at the English Court and University in the Later Middle Ages*. Hampshire, UK: Macmillan, 1992.

———. "Henry VII's Book of Astrology and the Tudor Renaissance." *Renaissance Quarterly* 65, no. 3 (2012): 661–710.

———. "What Is the Folded Almanac? The Form and Function of a Key Manuscript Source for Astro-Medical Practice in Later Medieval England." *Social History of Medicine* 16, no. 3 (December 1, 2003): 481–509.

Carlin, Martha. "Parron, William (b. before 1461, d. 1518)." In *Oxford Dictionary of National Biography*. Oxford: Oxford University Press, 2004. https://doi.org/10.1093/ref:odnb/52677.

Carruthers, Mary. *The Book of Memory: A Study of Memory in Medieval Culture*. 2nd ed. Cambridge: Cambridge University Press, 2008.

Cavallo, Sandra, and Tessa Storey. "Regimens, Authors and Readers: Italy and England Compared." In *Conserving Health in Early Modern Culture*, edited by Sandra Cavallo and Tessa Storey, 23–52. Bodies and Environments in Italy and England. Manchester: Manchester University Press, 2017.

Cavanaugh, Susan Hagen. "A Study of Books Privately Owned in England: 1300–1450." PhD diss., University of Pennsylvania, 1980.

Chapman, Allison A. "Marking Time: Astrology, Almanacs, and English Protestantism." *Renaissance Quarterly* 60, no. 4: 1257–90.

Chardonnens, Sándor. *Anglo-Saxon Prognostics, 900–1100: Study and Texts*. Leiden: Brill, 2007.

Chartier, Roger. *The Order of Books: Readers, Authors, and Libraries in Europe between the Fourteenth and Eighteenth Centuries.* Translated by Lydia G. Cochrane. Stanford, CA: Stanford University Press, 1994.

Chow, Kai-Wing. "Reinventing Gutenberg: Woodblock and Moveable-Type Printing in Europe and China." In *Agent of Change: Print Culture Studies after Elizabeth L. Eisenstein,* edited by Eric N. Lindquist, Eleanor F. Shevlin, and Sabrina Alcorn Baron, 169–92. Amherst: University of Massachusetts Press, 2007.

Christianson, C. Paul. *A Directory of London Stationers and Book Artisans, 1300–1500.* New York: The Bibliographical Society of America, 1990.

———. "The Rise of London's Book Trade." In *The Cambridge History of the Book in Britain, Vol. III: 1400–1557,* edited by Lotte Hellinga and J. B. Trapp, 128–47. Cambridge: Cambridge University Press, 1999.

Clanchy, M. T. *From Memory to Written Record: England 1066–1307.* Oxford: Blackwell Press, 1993.

Clark, Stuart. "The 'Gendering' of Witchcraft in French Demonology: Misogyny or Polarity?" *French History* 5, no. 4 (December 1, 1991): 426–37.

Cole, Garold. "The Historical Development of the Title Page." *Journal of Library History (1966–1972)* 6, no. 4 (1971): 303–16.

Coleman, Joyce. "Interactive Parchment: The Theory and Practice of Medieval English Aurality." *The Yearbook of English Studies* 25 (1995): 63–79.

Collinson, Patrick. *From Iconoclasm to Iconophobia: The Cultural Impact of the Second English Reformation.* Reading: University of Reading Press, 1986.

———. *This England: Essays on the English Nation and Commonwealth in the Sixteenth Century.* Politics, Culture and Society in Early Modern Britain. Manchester: Manchester University Press, 2011.

Colson, Justin, and Robert Ralley. "Medical Practice, Urban Politics and Patronage: The London 'Commonalty' of Physicians and Surgeons of the 1420s." *English Historical Review* (October 29, 2015): 1102–31.

Connelly, Erin. "'My Written Books of Surgery in the Englishe Tonge': The London Company of Barber-Surgeons and the Lylye of Medicynes." *Manuscript Studies: A Journal of the Schoenberg Institute for Manuscript Studies* 2, no. 2 (2017): 369–91.

Connolly, Margaret. *Sixteenth-Century Readers, Fifteenth-Century Books.* Cambridge Studies in Paleography and Codicology 16. Cambridge: Cambridge University Press, 2019.

Connolly, Margaret, and Raluca Radulescu, eds. *Insular Books: Vernacular Manuscript Miscellanies in Late Medieval Britain.* Oxford: Oxford University Press, 2015.

Cooper, Alix. *Inventing the Indigenous: Local Knowledge and Natural History in Early Modern Europe.* Cambridge: Cambridge University Press, 2007.

Cressy, David. *Bonfires and Bells: National Memory and the Protestant Calendar in Elizabethan and Stuart England.* London: Weidenfeld and Nicolson, 1989.

———. *Literacy and the Social Order: Reading & Writing in Tudor & Stuart England.* Cambridge: Cambridge University Press, 1980.

Crosby, Ruth. "Oral Delivery in the Middle Ages." *Speculum* 11, no. 1 (1936): 88–110.

Da Rold, Orietta. *Paper in Medieval England: From Pulp to Fictions.* Cambridge Studies in Medieval Literature 112. Cambridge: Cambridge University Press, 2020.

Dale, Marian K. "The London Silkwomen of the Fifteenth Century." *Economic History Review* 4, no. 3 (October 1933): 324–35.

Daston, Lorraine J., and Katharine Park. *Wonders and the Order of Nature, 1150–1750*. New York: Zone Books, 2001.

Davis, Natalie Zemon. *Society and Culture in Early Modern France*. Stanford, CA: Stanford University Press, 1975.

Dean, Ruth J., and Maureen B. M. Boulton. *Anglo-Norman Literature: A Guide to Texts and Manuscripts*. Anglo-Norman Text Society, Occasional Series 3. London: Anglo-Norman Text Society, 1999.

Dear, Peter. "Miracles, Experiments, and the Ordinary Course of Nature." *Isis* 81, no. 4 (December 1990): 663–83.

DeVun, Leah. *Prophecy, Alchemy, and the End of Time: John of Rupescissa in the Late Middle Ages*. New York: Columbia University Press, 2009.

Dodwell, B. "The Foundation of Norwich Cathedral." *Transactions of the Royal Historical Society* 7 (1957): 1–18.

Drimmer, Sonja. *The Art of Allusion: Illuminators and the Making of English Literature, 1403–1476*. Philadelphia: University of Pennsylvania Press, 2019.

———. "The Shapes of History: Houghton Library, MS Richardson 35 and Chronicles of England in Codex and Roll." In *Beyond Words: New Research on Manuscripts in Boston Collections*, edited by Jeffrey F. Hamburger, Lisa Fagin Davis, Anne-Marie Eze, Nancy Netzer, and William P. Stoneman, 253–68. Studies and Texts 221; Text Image Context: Studies in Medieval Manuscript Illumination 8. Toronto: PIMS, 2021.

Driver, Martha W. "Ideas of Order: Wynkyn de Worde and the Title Page." In *Texts and Their Contexts: Papers from the Early Book Society*, edited by V. J. Scattergood and Julia Boffey, 87–149. Dublin: Four Courts Press, 1997.

———. *The Image in Print: Book Illustration in Late Medieval England and Its Sources*. London: The British Library, 2004.

———. "Woodcuts and Decorative Techniques." In *A Companion to the Early Printed Book in Britain, 1476–1558*, edited by Vincent Gillespie and Susan Powell, 95–123. Woodbridge, Suffolk: D. S. Brewer, 2014.

Duff, Edward Gordon. *The Printers, Stationers, and Bookbinders of Westminster and London from 1476 to 1535*. Cambridge: Cambridge University Press, 1906.

Duffy, Eamon. *Marking the Hours*. New Haven, CT: Yale University Press, 2006.

———. *The Stripping of the Altars: Traditional Religion in England, 1400–1580*. New Haven, CT: Yale University Press, 1992.

Duggan, Lawrence G. "Reflections on 'Was Art Really the "Book of the Illiterate"?'" In *Reading Images and Texts*, edited by Mariëlle Hageman and Marco Mostert, 109–19. Utrecht Studies in Medieval Literacy 8. Turnhout: Brepols Publishers, 2005.

———. "Was Art Really the 'Book of the Illiterate'?" *Word and Image* 5, no. 3 (January 1989): 227–51.

Dyer, Christopher. *An Age of Transition? Economy and Society in England in the Later Middle Ages*. Oxford: Oxford University Press, 2005.

———. *Standards of Living in the Later Middle Ages: Social Change in England c. 1200–1500*. Rev. ed. Cambridge: Cambridge University Press, 1998.

Eamon, William. *Science and the Secrets of Nature: Books of Secrets in Medieval and Early Modern Culture*. Princeton, NJ: Princeton University Press, 1994.

Eatough, Geoffrey. "Paynell, Thomas (d. 1564?)." In *The Oxford Dictionary of National Biography*. Oxford: Oxford University Press, 2004. http://doi.org/10.1093/ref:odnb/21661.

Eisenstein, Elizabeth L. *The Printing Revolution in Early Modern Europe*. 2nd ed. Cambridge: Cambridge University Press, 2012.

Elliott, Brent. "The World of the Renaissance Herbal." *Renaissance Studies* 25, no. 1 (2011): 24–41.

Evenden, Elizabeth, and Thomas S. Freeman. *Religion and the Book in Early Modern England: The Making of Foxe's "Book of Martyrs."* Cambridge Studies in Early Modern British History. Cambridge: Cambridge University Press, 2011.

Ezell, Margaret J. M. "Cooking the Books, or, the Three Faces of Hannah Woolley." In *Reading and Writing Recipe Books, 1550–1800*, edited by Sara Pennell and Michelle DiMeo, 159–78. Manchester: Manchester University Press, 2013.

Faracovi, Ornella. "The Return to Ptolemy." In *A Companion to Astrology in the Renaissance*, edited by Brendan Dooley, 87–98. Brill's Companions to the Christian Tradition 49. Leiden: Brill, 2014.

Fein, Susanna, and Michael Johnston, eds. *Robert Thornton and His Books: Essays on the Lincoln and London Manuscripts*. New York: Boydell & Brewer, 2022.

Fissell, Mary E. *Vernacular Bodies: The Politics of Reproduction in Early Modern England*. Oxford: Oxford University Press, 2004.

Flint, Valerie Irene Jane. *The Rise of Magic in Early Medieval Europe*. Princeton, NJ: Princeton University Press, 2020.

Flood, John L. "'Safer on the Battlefield Than in the City': England, the 'Sweating Sickness,' and the Continent." *Renaissance Studies* 17, no. 2 (2003): 147–76.

Foster, Joseph. *Alumni Oxonienses: The Members of the University of Oxford, 1500–1714*. 4 vols. Oxford, 1891.

Fox, Adam. *Oral and Literate Culture in England, 1500–1700*. Oxford: Clarendon Press, 2000.

French, Roger. "Foretelling the Future: Arabic Astrology and English Medicine in the Late Twelfth Century." *Isis* 87, no. 3 (1996): 453–80.

Friedman, John B. "Harry the Haywarde and Talbat His Dog: An Illustrated Girdle Book from Worcestershire." In *Art into Life: Collected Papers from the Kresge Art Museum Medieval Symposia*, edited by Carol Fisher and Kathleen L. Scott, 115–53. East Lansing: Michigan State University Press, 1995.

Gerrish, B. A. "Priesthood and Ministry in the Theology of Luther." *Church History* 34, no. 4 (December 1965): 404–22.

Getz, Faye. "Gilbertus Anglicus Anglicized." *Medical History* 26, no. 4 (October 1982): 436–42.

———. "Medical Practitioners in Medieval England." *Social History of Medicine* 3, no. 2 (August 1, 1990): 245–83.

———. *Medicine in the English Middle Ages*. Princeton, NJ: Princeton University Press, 1998.

Grafton, Anthony. *Defenders of the Text: The Traditions of Scholarship in an Age of Science, 1450–1800*. Cambridge, MA: Harvard University Press, 1991.

———. *Forgers and Critics: Creativity and Duplicity in Western Scholarship*. New ed. Princeton, NJ: Princeton University Press, 2019.

———. *Inky Fingers: The Making of Books in Early Modern Europe*. Cambridge, MA: Harvard University Press, 2020.

Green, Ian. *Print and Protestantism in Early Modern England*. Oxford: Oxford University Press, 2000.

Green, Monica H. "From 'Diseases of Women' to 'Secrets of Women': The Transformation of Gynecological Literature in the Later Middle Ages." *Journal of Medieval and Early Modern Studies* 30, no. 1 (January 1, 2000): 5–40.

———. *Making Women's Medicine Masculine: The Rise of Male Authority in Pre-Modern Gynaecology*. Oxford: Oxford University Press, 2008.

———. "Medical Books." In *The European Book in the Twelfth Century*, edited by Erik Kwakkel and Rodney M. Thomson, 277–92. Cambridge Studies in Medieval Literature 101. Cambridge: Cambridge University Press, 2018.

———. "Obstetrical and Gynecological Texts in Middle English." *Studies in the Age of Chaucer* 14, no. 1 (1992): 53–88.

———. "The Sources of Eucharius Rösslin's 'Rosegarden for Pregnant Women and Midwives' (1513)." *Medical History* 53, no. 2 (April 2009): 167–92.

Green, Monica H., and Linne R. Mooney. "The Sickness of Women." In *Sex, Aging, and Death in a Medieval Medical Compendium: Trinity College Cambridge MS R.14.52, Its Texts, Language, and Scribe, Vol. II*, edited by M. Teresa Tavormina. Medieval and Renaissance Texts and Studies 292. Tempe, AZ: ACMRS, 2006.

Green, Monica H., and Daniel Lord Smail. "The Trial of Floreta d'Ays (1403): Jews, Christians, and Obstetrics in Later Medieval Marseille." *Journal of Medieval History* 34, no. 2 (June 2008): 185–211.

Gumbert, J. P. *Bat Books: A Catalogue of Folded Manuscripts Containing Almanacs or Other Texts*. Bibliologia 41. Turnhout, Belgium: Brepols Publishers, 2016.

Haigh, Christopher. *English Reformations: Religion, Politics, and Society Under the Tudors*. Oxford: Clarendon Press, 1993.

Haller, William. *The Elect Nation: The Meaning and Relevance of Foxe's Book of Martyrs*. New York: Harper & Row, 1963.

Hammond, Frederick. "Odington, Walter (Fl. c. 1280–1301), Benedictine Monk and Scholar." In *Oxford Dictionary of National Biography*. Oxford: Oxford University Press, 2004. https://doi.org/10.1093/ref:odnb/28631.

Hanawalt, Barbara. *Growing Up in Medieval London: The Experience of Childhood in History*. Oxford: Oxford University Press, 1993.

Harkness, Deborah E. *The Jewel House: Elizabethan London and the Scientific Revolution*. New Haven, CT: Yale University Press, 2008.

Harley, David. "Historians as Demonologists: The Myth of the Midwife-Witch." *Social History of Medicine* 3, no. 1 (1990): 1–26.

Herendeen, Wyman H. "Camden, William (1551–1623)." In *Oxford Dictionary of National Biography*. Oxford: Oxford University Press, 2004. https://doi.org/10.1093/ref:odnb/4431.

Hodnett, Edward. *English Woodcuts, 1480–1535*. Oxford: Oxford University Press, 1973.

———. *English Woodcuts, 1480–1535: Additions & Corrections*. London: The Bibliographical Society, 1973.

Horden, Peregrine. "What's Wrong with Early Medieval Medicine?" *Social History of Medicine* 24, no. 1 (April 1, 2011): 5–25.

Hourihane, Colum. *Time in the Medieval World: Occupations of the Months and*

Signs of the Zodiac in the Index of Christian Art. State College: Pennsylvania State University Press, 2007.

Hull, Suzanne W. *Chaste, Silent, & Obedient: English Books for Women, 1475–1640*. San Marino, CA: Huntington Library Press, 1982.

Hunt, Tony. "Early Anglo-Norman Receipts for Colours." *Journal of the Warburg and Courtauld Institutes* 58 (1995): 203.

———. *Popular Medicine in Thirteenth-Century England: Introduction and Texts*. Woodbridge, Suffolk: Boydell & Brewer Ltd., 1990.

Jacquart, Danielle. "Theory, Everyday Practice, and Three Fifteenth-Century Physicians." *Osiris* 6 (1990): 140–60.

Johns, Adrian. *The Nature of the Book: Print and Knowledge in the Making*. Chicago: University of Chicago Press, 2000.

———. *Piracy: The Intellectual Property Wars from Gutenberg to Gates*. Chicago: University of Chicago Press, 2009.

Johnson, Francis R. "'A Newe Herball of Macer' and Banckes's 'Herball': Notes on Robert Wyer and the Printing of Cheap Handbooks of Science in the Sixteenth Century." *Bulletin of the History of Medicine* 15, no. 3 (March 1944): 246–60.

Jones, Claire. "Discourse Communities and Medical Texts." In *Medical and Scientific Writing in Late Medieval English*, edited by Irma Taavitsainen and Päivi Pahta, 23–36. Cambridge: Cambridge University Press, 2004.

Jones, Karen, and Michael Zell. "'The Divels Speciall Instruments': Women and Witchcraft before the 'Great Witch-Hunt.'" *Social History* 30, no. 1 (2005): 45–63.

Jones, Lori. "Itineraries and Transformations: John of Burgundy's Plague Treatise." *Bulletin of the History of Medicine* 95, no. 3 (2021): 277–314.

Jones, Peter Murray. "Crophill, John (d. in or after 1485)." In *Oxford Dictionary of National Biography*. Oxford: Oxford University Press, 2004. http://doi.org/10.1093/ref:odnb/6780.

———. "Four Middle English Translations of John of Arderne." In *Latin and Vernacular: Studies in Late Medieval Manuscripts*, edited by Alastair Minnis, 61–89. Woodbridge, Suffolk: Boydell & Brewer, 1989.

———. "Gemini [Geminus, Lambrit], Thomas (Fl. 1540–1562), Engraver, Printer, and Instrument Maker." In *The Oxford Dictionary of National Biography*. Vol. 1. Oxford: Oxford University Press, 2004. https://doi.org/10.1093/ref:odnb/10513.

———. "Medicine and Science." In *The Cambridge History of the Book in Britain, Vol. III: 1400–1557*, edited by Lotte Hellinga and J. B. Trapp, 433–48. Cambridge: Cambridge University Press, 1999.

———. *Medieval Medicine in Illuminated Manuscripts*. Rev. ed.. London: The British Library, 1998.

———. "University Books and the Sciences, c. 1250–1400." In *The Cambridge History of the Book in Britain, Vol. II: 1100–1400*, edited by Nigel Morgan and Rodney M. Thomson, 453–62. Cambridge: Cambridge University Press, 2008.

Jones, Peter Murray, and Lea T. Olsan. "Performative Rituals for Conception and Childbirth in England, 900–1500." *Bulletin of the History of Medicine* 89, no. 3 (October 27, 2015): 406–33.

Jordan, Mark D. "Medicine as Science in the Early Commentaries on 'Johannitius.'" *Traditio* 43 (1987): 121–45.

Kassell, Lauren. *Medicine and Magic in Elizabethan London: Simon Forman: Astrologer, Alchemist, and Physician*. Oxford: Oxford University Press, 2005.

Kavey, Allison. *Books of Secrets: Natural Philosophy in England, 1550–1600*. Urbana and Chicago: University of Illinois Press, 2007.

Keiser, George R. *A Manual of the Writings in Middle English, 1050–1500. Volume X: Works of Science and Information*. New Haven, CT: Connecticut Academy of Arts and Sciences, 1998.

———. "Medicines for Horses: The Continuity from Script to Print." *Yale University Library Gazette* 69, no. 3–4 (1995): 111–28.

———. "Robert Thornton's *Liber de Diversis Medicinis*: Text, Vocabulary, and Scribal Confusion." In *Rethinking Middle English: Linguistic and Literary Approaches*, edited by Nikolaus Ritt and Herbert Schendl, 30–41. Frankfurt: Peter Lang, 2005.

———. "Two Medieval Plague Treatises and Their Afterlife in Early Modern England." *Journal of the History of Medicine* 58 (July 2003): 292–324.

Kerby-Fulton, Kathryn. *Books Under Suspicion: Censorship and Tolerance of Revelatory Writing in Late Medieval England*. Notre Dame, IN: University of Notre Dame Press, 2006.

Kessler, Herbert L. "Gregory the Great and Image Theory in Northern Europe during the Twelfth and Thirteenth Centuries." In *A Companion to Medieval Art: Romanesque and Gothic in Northern Europe*, edited by Conrad Rudolph, 151–72. New York: John Wiley & Sons, Ltd., 2006.

Kieckhefer, Richard. *European Witch Trials: Their Foundations in Popular and Learned Culture, 1300–1500*. Berkeley: University of California Press, 1976.

King, Helen. *Hippocrates' Woman: Reading the Female Body in Ancient Greece*. London: Routledge, 2002.

Kowaleski, Maryanne, and Judith M. Bennett. "Crafts, Guilds, and Women in the Middle Ages: Fifty Years after Marian K. Dale." *Signs* 14, no. 2 (Winter 1989): 474–501.

Kristeller, Paul Oskar. "The School of Salerno." *Bulletin of the History of Medicine* 17 (January 1, 1945): 138–92.

Kumler, Aden. *Translating Truth: Ambitious Images and Religious Knowledge in Late Medieval France and England*. New Haven, CT: Yale University Press, 2011.

Kusukawa, Sachiko. *Picturing the Book of Nature: Image, Text, and Argument in Sixteenth-Century Human Anatomy and Medical Botany*. Chicago: University of Chicago Press, 2012.

Lacey, Kay. "The Production of 'Narrow Ware' by Silkwomen in Fourteenth and Fifteenth Century England." *Textile History* 18, no. 2 (November 1987): 187–204.

Larkey, Sanford V. "The Vesalian Compendium of Geminus and Nicholas Udall's Translation: Their Relation to Vesalius, Caius, Vicary, and De Mondeville." *The Library* s4-XIII, no. 4 (1933): 367–95.

Lawrence-Mathers, Anne. "Domesticating the Calendar: The Hours and the Almanac in Tudor England." In *Women and Writing, c. 1340–1650: The Domestication of Print Culture*, edited by Anne Lawrence-Mathers and Phillipa Hardman, 34–61. Woodbridge, Suffolk: Boydell & Brewer, 2010.

———. *Medieval Meteorology: Forecasting the Weather from Aristotle to the Almanac*. 1st ed. Cambridge University Press, 2019.

Leong, Elaine. "Collecting Knowledge for the Family: Recipes, Gender and Practical Knowledge in the Early Modern English Household." *Centaurus; Interna-*

tional Magazine of the History of Science and Medicine 55, no. 2 (May 2013): 81–103.

———. "'Herbals She Peruseth': Reading Medicine in Early Modern England." *Renaissance Studies* 28, no. 4 (September 2014): 556–78.

———. "Making Medicines in the Early Modern Household." *Bulletin of the History of Medicine* 82, no. 1 (Spring 2008): 145–68.

———. *Recipes and Everyday Knowledge*. Chicago: University of Chicago Press, 2018.

Life, Page. "Bright, Timothy (1549/50–1615)." In *Oxford Dictionary of National Biography*. Oxford: Oxford University Press, 2008. https://doi.org/10.1093/ref:odnb/3424.

Long, Pamela O. *Artisan/Practitioners and the Rise of the New Sciences, 1400–1600*. Corvallis: Oregon State University Press, 2011.

———. *Openness, Secrecy, Authorship: Technical Arts and the Culture of Knowledge from Antiquity to the Renaissance*. Baltimore: Johns Hopkins University Press, 2001.

Love, Harold. *Scribal Publication in Seventeenth-Century England*. Oxford: Clarendon Press, 1993.

Lowden, John. *The Making of the Bibles Moralisées*. Vol. 1: The Manuscripts. University Park: Pennsylvania State University Press, 2000.

Loysen, Kathleen. *Conversation and Storytelling in Fifteenth- and Sixteenth-Century French Nouvelles*. New York: Peter Lang, 2004.

Lyall, R. J. "Materials: The Paper Revolution." In *Book Production and Publishing in Britain, 1375–1475*, edited by Jeremy Griffiths and Derek Pearsall, 11–29. Cambridge: Cambridge University Press, 1989.

Lyon, Gregory B. "Baudouin, Flacius, and the Plan for the Magdeburg Centuries." *Journal of the History of Ideas* 64, no. 2 (2003): 253–72.

Macfarlane, Alan. *Witchcraft in Tudor and Stuart England*. New York: Harper & Row, 1970.

Mackman, Jonathan, and Matthew Stevens, eds. *Court of Common Pleas: The National Archives, Cp40 1399–1500*. British History Online. London: Centre for Metropolitan History, 2010. https://www.british-history.ac.uk/no-series/common-pleas/1399-1500.

Marcus, Hannah. *Forbidden Knowledge: Medicine, Science, and Censorship in Early Modern Italy*. Chicago: University of Chicago Press, 2020.

Martin, Charles Trice, and Rachel Davies. "Sir William Butts (c. 1485–1545)." In *Oxford Dictionary of National Biography*. Oxford: Oxford University Press, 2004. https://doi.org/10.1093/ref:odnb/4241.

Masschaele, James. "The Public Life of the Private Charter in Thirteenth-Century England." In *Commercial Activity, Markets and Entrepreneurs in the Middle Ages: Essays in Honour of Richard Britnell*, edited by Ben Dodds and Christian Drummond Liddy, 199–216. Woodbridge, Suffolk: Boydell & Brewer Ltd., 2011.

———. "The Renaissance Depression Debate: The View from England." *The History Teacher* 27, no. 4 (August 1994): 405–16.

Matheson, Lister M. *The Prose Brut: The Development of a Middle English Chronicle*. Medieval & Renaissance Texts & Studies 180. Tempe, AZ: Medieval & Renaissance Texts & Studies, 1998.

McFarlane, K. B. *John Wycliffe and the Beginnings of English Nonconformity*. New York: Macmillan, 1953.

McIntosh, Marjorie K. *Working Women in English Society 1300–1620*. Cambridge: Cambridge University Press, 2005.

McMahon, Madeline. "Matthew Parker and the Practice of Church History." In *Confessionalisation and Erudition in Early Modern Europe: An Episode in the History of the Humanities*, edited by Nicholas Hardy and Dmitri Levitin, 116–53. Proceedings of the British Academy. Oxford and New York: Oxford University Press, 2020.

McSparran, Frances, et al., eds. *Middle English Dictionary*. Online ed. Ann Arbor: University of Michigan Press, 2000–2018. http://quod.lib.umich.edu/m/middle-english-dictionary/.

McVaugh, Michael. *The Rational Surgery of the Middle Ages*. Micrologus' Library 15. Firenze: SISMEL, 2006.

Meale, Carol M. "Patrons, Buyers and Owners: Book Production and Social Status." In *Book Production and Publishing in Britain, 1375–1475*, 201–38. Cambridge: Cambridge University Press, 1989.

Michael, M. A. "Urban Production of Manuscript Books and the Role of the University Towns." In *The Cambridge History of the Book in Britain, Vol. II: 1100–1400*, edited by Nigel Morgan and Rodney M. Thomson, 168–94. Cambridge: Cambridge University Press, 2008.

Millstone, Noah. *Manuscript Circulation and the Invention of Politics in Early Stuart England*. Cambridge Studies in Early Modern British History. Cambridge: Cambridge University Press, 2016.

Minnis, Alastair. *Medieval Theory of Authorship: Scholastic Literary Attitudes in the Later Middle Ages*. Philadelphia: University of Pennsylvania Press, 2012.

Mooney, Linne R. "English Almanacs from Script to Print." In *Texts and Their Contexts: Papers from the Early Book Society*, edited by V. J. Scattergood and Julia Boffey, 11–25. Dublin: Four Courts Press, 1997.

———. "Vernacular Literary Manuscripts and Their Scribes." In *The Production of Books in England, 1350–1500*, edited by Alexandra Gillespie and Daniel Wakelin, 192–211. Cambridge: Cambridge University Press, 2014.

Mooney, Linne R., et al., eds. *The DIMEV: An Open-Access, Digital Edition of the Index of Middle English Verse*. www.dimev.net.

Moran, Bruce T. "The 'Herbarius' of Paracelsus." *Pharmacy in History* 35, no. 3 (1993): 99–127.

Neville, Sarah. *Early Modern Herbals and the Book Trade: English Stationers and the Commodification of Botany*. 1st ed. Cambridge: Cambridge University Press, 2022.

Neville-Sington, Pamela. "Pynson, Richard (c. 1449–1529/30)." In *Oxford Dictionary of National Biography*. Oxford: Oxford University Press, 2008. https://doi.org/10.1093/ref:odnb/22935.

Noel, Will. "The Needham Calculator." The Schoenberg Institute for Manuscript Studies, University of Pennsylvania Libraries. Accessed 23 June 2023. http://www.needhamcalculator.net.

Noel, Will, and George Gordon. "The Needham Calculator (1.0) and the Flavors of Fifteenth-Century Paper." The Schoenberg Institute for Manuscript Studies, University of Pennsylvania Libraries. 30 January 2017. https://schoenberginstitute.org/2017/01/30/the-needham-calculator-1-0-and-the-flavors-of-fifteenth-century-paper/.

Norri, Juhani. *Dictionary of Medical Vocabulary in English, 1375–1550: Body Parts, Sicknesses, Instruments, and Medicinal Preparations*. New York: Routledge, 2016.

Nutton, Vivian. "'A Diet for Barbarians': Introducing Renaissance Medicine to Tudor England." In *Natural Particulars: Nature and the Disciplines in Renaissance Europe*, edited by Anthony Grafton and Nancy G. Siraisi, 275–94. Cambridge, MA: MIT Press, 1999.

———. "Linacre, Thomas (c. 1460–1524), Humanist Scholar and Physician." In *Oxford Dictionary of National Biography*. Oxford: Oxford University Press, 2004. https://doi.org/10.1093/ref:odnb/16667.

O'Boyle, Cornelius. "Astrology and Medicine in Later Medieval England: The Calendars of John Somer and Nicholas of Lynn." *Sudhoffs Archiv* 89, no. 1 (2005): 1–22.

Olsan, Lea T. "Charms and Prayers in Medieval Medical Theory and Practice." *Social History of Medicine* 16, no. 3 (December 1, 2003): 343–66.

———. "The Corpus of Charms in the Middle English Leechcraft Remedy Books." In *Charms, Charmers and Charming: International Research on Verbal Magic*, edited by Jonathan Roper, 214–37. Palgrave Historical Studies in Witchcraft and Magic. New York: Palgrave Macmillan, 2009.

———. "The Marginality of Charms in Medieval England." In *The Power of Words: Studies on Charms and Charming in Europe, edited by* James Kapalo, Éva Pócs, and William Francis Ryan, 122–46. Budapest and New York: Central European University Press, 2013.

Orme, Nicholas. *Medieval Schools: From Roman Britain to Renaissance England*. New Haven, CT: Yale University Press, 2006.

Oschinsky, Dorothea. "Medieval Treatises on Estate Management." *Economic History Review* 8, no. 3 (1956): 296–309.

Overty, Joanne Filippone. "The Cost of Doing Scribal Business: Prices of Manuscript Books in England, 1300–1483." *Book History* 11, no. 1 (September 12, 2008): 1–32.

Ovitt, George. "The Status of the Mechanical Arts in Medieval Classifications of Learning." *Viator* 14 (January 1, 1983): 89–105.

Page, William, ed. "Alien Houses: The Priory of Toft Monks." In *A History of the County of Norfolk: Volume 2*. British History Online, 464–65. London: Victoria County History, 1906. http://www.british-history.ac.uk/vch/norf/vol2/pp464-465.

———, ed. "Houses of Benedictine Monks: The Abbey of Winchcombe." In *A History of the County of Gloucester: Volume 2*. British History Online, 66–72. London: Victoria County History, 1907. https://www.british-history.ac.uk/vch/glos/vol2/pp66-72.

———, ed. "Houses of Benedictine Monks: The Priory of Luffield." In *A History of the County of Buckingham: Volume 1*. British History Online, 347–50. London: Victoria County History, 1905. https://www.british-history.ac.uk/vch/bucks/vol1/pp347-350.

Pantzer, Katherine, Alfred W. Pollard, and G. R. Regrave, eds. *A Short-Title Catalogue of Books Printed in England, Scotland, and Ireland, and of English Books Printed Abroad, 1475–1640*. Revised and enlarged 2nd ed. 3 vols. London: Bibliographical Society, 1976.

"Parishes: Cropthorne." In *A History of the County of Worcester: Volume 3*, British

History Online, 322–29. London: Victoria County History, 1913. https://www.british-history.ac.uk/vch/worcs/vol3/pp322-329.

Park, Katharine. *Secrets of Women: Gender, Generation, and the Origins of Human Dissection*. New York: Zone Books, 2006.

Parkes, M. B. *Scribes, Scripts, and Reader: Studies in the Communication, Presentation, and Dissemination of Medieval Texts*. London: Hambledon Press, 1991.

Pennell, Sara, and Michelle DiMeo, eds. *Reading and Writing Recipe Books, 1550–1800*. Manchester: Manchester University Press, 2013.

Phillpott, Matthew. *The Reformation of England's Past: John Foxe and the Revision of History in the Late Sixteenth Century*. Routledge Research in Early Modern History. New York: Routledge, 2018.

Plomer, Henry Robert. *A Short History of English Printing, 1476–1898*. Edited by Alfred W Pollard. The English Bookman's Library. London: Kegan Paul, Trench, Trübner, and Company, 1900.

Prus, Caleb. "John Tyryngham: An Ordinary Medical Practitioner in Ashmole 1481." Master's Thesis, University of Oxford, 2022.

Rampling, Jennifer M. *The Experimental Fire: Inventing English Alchemy, 1400–1700*. Synthesis. Chicago: University of Chicago Press, 2020.

Rankin, Alisha. "New World Drugs and the Archive of Practice: Translating Nicolás Monardes in Early Modern Europe." *Osiris* 37 (June 1, 2022): 67–88.

———. *Panaceia's Daughters: Noblewomen as Healers in Early Modern Germany*. Chicago: University of Chicago Press, 2013.

Rawcliffe, Carole. *Medicine and Society in Later Medieval England*. Stroud, Gloucestershire: Alan Sutton, 1995.

Reeves, Marjorie. *The Influence of Prophecy in the Later Middle Ages: A Study in Joachimism*. Oxford: Clarendon Press, 1969.

Reynolds, Melissa. "'Here Is a Good Boke to Lerne': Practical Books, the Coming of the Press, and the Search for Knowledge, ca. 1400–1560." *Journal of British Studies* 58, no. 2 (April 2019): 259–88.

———. "How to Cure a Horse, or, the Experience of Knowledge and the Knowledge of Experience." *Historical Studies in the Natural Sciences* 52, no. 4 (August 2022): 546–52.

———. "The Sururgia of Nicholas Neesbett: Writing Medical Authority in Later Medieval England." *Social History of Medicine* 35, no. 1 (February 2022): 144–69.

Rheinberger, Hans-Jörg. *Toward a History of Epistemic Things: Synthesizing Proteins in the Test Tube*. Stanford, CA: Stanford University Press, 1997.

Rider, Catherine. "Common Magic." In *The Cambridge History of Magic and Witchcraft in the West*, edited by S. J. Collins, 303–31. Cambridge: Cambridge University Press, 2015.

———. "Medical Magic and the Church in Thirteenth-Century England." *Social History of Medicine: The Journal of the Society for the Social History of Medicine / SSHM* 24, no. 1 (April 1, 2011): 92–107.

Robbins, Rossell Hope. "English Almanacks of the Fifteenth Century." *Philological Quarterly* 18, no. 4 (October 1939): 321–31.

———. "Medical Manuscripts in Middle English." *Speculum* 45, no. 3 (July 1, 1970): 393–415.

Roberts, R. Julian. "Dee, John (1527–1609)." In *Oxford Dictionary of National*

Biography. Oxford: Oxford University Press, 2004. https://doi.org/10.1093/ref:odnb/7418.

Robinson, Benedict Scott. "'Darke Speech': Matthew Parker and the Reforming of History." *The Sixteenth Century Journal* 29, no. 4 (1998): 1061–83.

Robinson, P. R. "The Format of Books—Books, Booklets and Rolls." In *The Cambridge History of the Book in Britain, Vol. II: 1100–1400*, edited by Nigel Morgan and Rodney M. Thomson, 39–54. Cambridge: Cambridge University Press, 2008.

———. "'Lewdecalendars' from Lynn." In *Tributes to Kathleen L. Scott, English Medieval Manuscripts: Readers, Makers, and Illuminators*, edited by Marlene Villalobos Hennessy. London: Harvey Miller Publishers, 2009.

Roper, Lyndal. "Witchcraft and Fantasy in Early Modern Germany." In *Oedipus and the Devil: Witchcraft, Religion, and Sexuality in Early Modern Europe*, 200–227. New York: Taylor & Francis, 1994.

Rudy, Kathryn M. "Dirty Books: Quantifying Patters of Use in Medieval Manuscripts Using a Densitometer." *Journal of Historians of Netherlandish Art* 2, no. 1–2 (Summer 2010), https://doi.org/10.5092/jhna.2010.2.1.1.

———. "Touching the Book Again: The Passional of Abbess Kunigunde of Bohemia," in *Codex und Material*, Wolfenbütteler Mittelalter-Studien 34 (Wiesbaden: Harrassowitz Verlag in Kommission, 2018), 247–58.

Saenger, Paul. "Books of Hours and the Reading Habits of the Later Middle Ages." In *The Culture of Print: Power and the Uses of Print in Early Modern Europe*, edited by Roger Chartier, translated by Lydia G. Cochrane, 141–73. Princeton, NJ: Princeton University Press, 1987.

———. "Colard Mansion and the Evolution of the Printed Book." *The Library Quarterly* 45, no. 4 (October 1975): 405–18.

Scott, Kathleen L. "Past Ownership: Evidence of Book Ownership by English Merchants in the Later Middle Ages." In *Makers and Users of Medieval Books: Essays in Honour of A. S. G. Edwards*, edited by Carol M. Meale and Derek Pearsall, 150–75. Woodbridge, Suffolk: Boydell & Brewer Ltd., 2014.

Shagan, Ethan H. *Popular Politics and the English Reformation*. Cambridge: Cambridge University Press, 2003.

Shapin, Steven. "The House of Experiment in Seventeenth-Century England." *Isis* 79, no. 3 (September 1988): 373–404.

———. *Never Pure: Historical Studies of Science as If It Was Produced by People with Bodies, Situated in Time, Space, Culture, and Society, and Struggling for Credibility and Authority*. Baltimore: Johns Hopkins University Press, 2010.

———. *The Social History of Truth: Civility and Science in Seventeenth-Century England*. Chicago and London: University of Chicago Press, 1994.

Sherman, William. *Used Books: Marking Readers in Renaissance England*. Philadelphia: University of Pennsylvania Press, 2009.

Siraisi, Nancy. "The Faculty of Medicine." In *A History of the University in Europe*, edited by Hilde de Ridder-Symoens, 360–87. Cambridge: Cambridge University Press, 1991.

Skemer, Don C. *Binding Words: Textual Amulets in the Middle Ages*. Magic in History. State College: Pennsylvania State University Press, 2006.

Slack, Paul. "Mirrors of Health and Treasures of Poor Men: The Uses of the Vernacular Medical Literature of Tudor England." In *Health, Medicine and Mortality in the Sixteenth Century*, edited by Charles Webster, 237–73. Cambridge: Cambridge University Press, 1979.

Smith, Pamela H. *The Body of the Artisan: Art and Experience in the Scientific Revolution*. Chicago: University of Chicago Press, 2006.

———. *From Lived Experience to the Written Word: Reconstructing Practical Knowledge in the Early Modern World*. Chicago: University of Chicago Press, 2022.

———. "In the Workshop of History: Making, Writing, and Meaning." *West 86th* 19, no. 1 (2012): 4–31.

Smoller, Laura Ackerman. *History, Prophecy, and the Stars: The Christian Astrology of Pierre d'Ailly, 1350–1420*. Princeton, NJ: Princeton University Press, 1994.

Spufford, Margaret. "First Steps in Literacy: The Reading and Writing Experiences of the Humblest Seventeenth-Century Spiritual Autobiographers." *Social History* 4, no. 3 (1979): 407–35.

Stanley, E. G. "Directions for Making Many Sorts of Laces." In *Chaucer and Middle English Studies in Honour of Rossell Hope Robbins*, edited by Beryl Rowland, 89–103. Kent, OH: Kent State University Press, 1974.

Star, Sarah, ed. *Henry Daniel and the Rise of Middle English Medical Writing*. Toronto: University of Toronto Press, 2022.

Sterry, Sir Wasey, ed. *The Eton College Register, 1441–1698, Alphabetically Arranged and Edited with Biographical Notes*. Eton: Spottiswoode, Ballantyne, & Co., Ltd., 1943.

Stock, Brian. "Antiqui and Moderni as 'Giants' and 'Dwarfs': A Reflection of Popular Culture?" *Modern Philology* 76, no. 4 (1979): 370–74.

Storey, Tessa. *Italian Books of Secrets Database*. Leicester: University of Leicester, 2008. https://hdl.handle.net/2381/4335.

Stratford, Jenny. "The Manuscripts of John, Duke of Bedford: Library and Chapel." In *England in the Fifteenth Century: Proceedings of the 1986 Harlaxton Symprium*, edited by Daniel Williams, 329–51. Woodbridge, Suffolk: Boydell & Brewer, 1987.

Strocchia, Sharon T. *Forgotten Healers: Women and the Pursuit of Health in Late Renaissance Italy*. I Tatti Studies in Italian Renaissance History. Cambridge, MA: Harvard University Press, 2019.

Summit, Jennifer. *Memory's Library: Medieval Books in Early Modern England*. Chicago: University of Chicago Press, 2008.

Taavitsainen, Irma. "The Identification of Middle English Lunary MSS." *Neuphilologische Mitteilungen* 88, no. 1 (1987): 18–26.

———. "Scriptorial 'House-Styles' and Discourse Communities." In *Medical and Scientific Writing in Late Medieval English*, edited by Irma Taavitsainen and Päivi Pahta, 209–40. Cambridge: Cambridge University Press, 2004.

———. "Storia Lune and Its Paraphrase in Prose: Two Versions of a Middle English Lunary." In *Neophilologica Fennica*, edited by Leena Kahlas-Tarkka, 521–55. Mémoires de La Société Néophilologique de Helsinki 45. Helsinki: Société Philologique, 1987.

Talbot, Charles H., and Eugene Ashby Hammond. *The Medical Practitioners in Medieval England: A Biographical Register*. Wellcome Historical Medical Library, 1965.

Tavormina, M. Teresa. "The Middle English Letter of Ipocras." *English Studies* 88, no. 6 (December 2007): 632–52.

———. "Uroscopy in Middle English: A Guide to the Texts and Manuscripts." *Studies in Medieval and Renaissance History* 11 (2014): 1–154.

Taylor, Andrew. "Into His Secret Chamber: Reading and Privacy in Late Medieval England." In *The Practice and Representation of Reading in England*, edited by James Raven, Helen Small, and Naomi Tadmor, 41–61. Cambridge: Cambridge University Press, 2007.

———. "Manual to Miscellany: Stages in the Commercial Copying of Vernacular Literature in England." *The Yearbook of English Studies* 33 (2003): 1–17.

Thomas, Keith. "The Meaning of Literacy in Early Modern England." In *The Written Word: Literacy in Transition*, edited by Gerd Baumann, 97–131. Oxford: Clarendon Press, 1986.

———. *Religion and the Decline of Magic*. Reprint. London: Penguin Books, 1991.

Thorndike, Lynn. *A History of Magic and Experimental Science: During the First Thirteen Centuries of Our Era, Vol. II*. New York: Columbia University Press, 1923.

Voet, Leon, and Jenny Voet-Grisolle. *The Plantin Press (1555–1589): A Bibliography of the Works Printed and Published by Christopher Plantin at Antwerp and Leiden, Vols. I–V*. Amsterdam: Van Hoeve, 1980.

Voigts, Linda Ehrsam. "The Master of the King's Stillatories." In *The Lancastrian Court: Proceedings of the 2001 Harlaxton Symposium*, edited by Jenny Stratford, 232–55. Harlaxton Medieval Studies 13. Donington, Lincolnshire: Shaun Tyas, 2003.

———. "Scientific and Medical Books." In *Book Production and Publishing in Britain, 1375–1475*, edited by Derek Pearsall and Jeremy Griffiths, 345–402. Cambridge: Cambridge University Press, 1989.

———. "The 'Sloane Group': Related Scientific and Medical Manuscripts from the Fifteenth Century in the Sloane Collection." *British Library Journal* 16, no. 1 (1990): 26–57.

———. "What's the Word? Bilingualism in Late-Medieval England." *Speculum* 71, no. 4 (1996): 813–26.

Voigts, Linda Ehrsam, and Patricia Deery Kurtz. *Scientific and Medical Writings in Old and Middle English: An Electronic Reference*. Ann Arbor: University of Michigan Press, 2000. http://cctr1.umkc.edu/search.

Wallis, Faith. "The Experience of the Book: Manuscripts, Texts, and the Role of Epistemology in Early Medieval Medicine." In *Knowledge and the Scholarly Medical Traditions*, edited by Don Bates, 101–26. Cambridge: Cambridge University Press, 1995.

———. "The Ghost in the Articella: A Twelfth-Century Commentary on the Constantinian Liber Graduum." In *Herbs and Healers from the Ancient Mediterranean through the Medieval West: Essays in Honor of John M. Riddle*, edited by Ann Van Arsdall and Timothy Graham, 107–52. Aldershot: Ashgate, 2012.

———. "Medicine in Medieval Calendar Manuscripts." In *Manuscript Sources of Medieval Medicine. A Book of Essays*, edited by M. R. Schleissner, 105–43. New York: Garland, 1995.

Walsham, Alexandra. *Providence in Early Modern England*. Oxford: Oxford University Press, 1999.

———. *The Reformation of the Landscape: Religion, Identity, and Memory in Early Modern Britain and Ireland*. Oxford: Oxford University Press, 2011.

Warburg, Aby. *The Renewal of Pagan Antiquity: Contributions to the Cultural History of the European Renaissance*. Translated by David Britt. Getty Research Publications. Los Angeles: Getty Research Institute, 1999.

Watt, Diane. "Reconstructing the Word: The Political Prophecies of Elizabeth Barton (1506–1534)." *Renaissance Quarterly* 50, no. 1 (1997): 136–63.

Watt, Tessa. *Cheap Print and Popular Piety, 1550–1640*. Cambridge: Cambridge University Press, 1991.

Wear, Andrew. *Knowledge and Practice in Early Modern English Medicine, 1550–1680*. Cambridge: Cambridge University Press, 2000.

Westman, Robert. *The Copernican Question: Prognostication, Skepticism, and Celestial Order*. Berkeley: University of California Press, 2011.

White, Eric Marshall. "A Census of Print Runs for Fifteenth-Century Books." *Consortium for European Research Libraries* (blog), 2012. https://www.cerl.org/_media/resources/links_to_other_resources/printruns_intro.pdf.

White, Lynn. "Theophilus Redivivus." *Technology and Culture* 5, no. 2 (1964): 224–33.

Wilson, Adrian, and Joyce Lancaster Wilson. *A Medieval Mirror: Speculum Humanae Salvationis 1324–1500*. Berkeley: University of California Press, 1985.

Wilson, Janet. "A Catalogue of the 'Unlawful' Books Found in John Stow's Study on 21 February 1568/9." *British Catholic History* 20, no. 1 (May 1990): 1–30.

Wrightson, Keith. "Estates, Degrees, and Sorts: Changing Perceptions of Society in Tudor and Stuart England." In *Language, History, and Class*, edited by P. J. Corfield, 30–52. Oxford: Blackwell Press, 1991.

Wrightson, Keith, and David Levine. *Poverty and Piety in an English Village: Terling, 1525–1700*. 2nd ed. Oxford: Clarendon Press, 1995.

Yates, Frances Amelia. *Giordano Bruno and the Hermetic Tradition*. London: Routledge and Kegan Paul, 1964.

Zanden, Jan Luiten van. "Wages and the Cost of Living in Southern England (London), 1450–1700." Institute of the Royal Netherlands Academy of Arts and Sciences. International Institute of Social History, n.d. http://www.iisg.nl/hpw/dover.php.

Index

Page numbers in italics refer to figures and tables.